D1539153

ANNALS OF
THE NEW YORK ACADEMY
OF SCIENCES

Volume 415

EDITORIAL STAFF

Executive Editor
BILL BOLAND

Managing Editor
JOYCE HITCHCOCK

Associate Editor
SHEILA KATES TREITLER

The New York Academy of Sciences
2 East 63rd Street
New York, New York 10021

CATALYTIC TRANSITION METAL HYDRIDES

ANNALS OF THE NEW YORK ACADEMY OF SCIENCES

Volume 415

CATALYTIC TRANSITION METAL HYDRIDES

Edited by D. W. Slocum and William R. Moser

The New York Academy of Sciences
New York, New York
1983

Library of Congress Cataloging in Publication Data

Main entry under title:

Catalytic transition metal hydrides.

    (Annals of the New York Academy of Sciences; v. 415)
    Papers from a conference held Nov. 15–17, 1982 by the
New York Academy of Sciences.
    Bibliography: p.
    Includes index.
    1. Transition metal hydrides—Congresses. 2. Tran-
sition metal catalysts—Congresses. I. Slocum, D. W.
(Donald Warren), 1933–   II. Moser, William R.
III. New York Academy of Sciences. IV. Series.
Q11.N5 vol. 415 [QD172.T6] 500s [546'.6] 83–26909

ISBN 0–89766–224–5
ISBN 0–89766–225–3 (pbk.)

S/PCP
Printed in the United States of America
ISBN 0-89766-224-5 (Cloth)
ISBN 0-89766-225-3 (Paper)
ISSN 0077-8923

ANNALS OF THE NEW YORK ACADEMY OF SCIENCES

VOLUME 415

December 30, 1983

CATALYTIC TRANSITION METAL HYDRIDES*

Editors and Conference Organizers
D. W. SLOCUM and WILLIAM R. MOSER

International Advisory Committee
D. ASTRUC, R. AUGUSTINE, J. EISCH, D. FAHEY, J. FALLER, D. FORSTER,
M. L. H. GREEN, W. HERRMANN, O. R. HUGHES, T. KOBYLINSKI, P. KRUSIC,
K. MOEDRITZER, C. U. PITTMAN, M. RAUSCH, D. SINGLETON, and I. WENDER

CONTENTS

* This volume is the result of a conference entitled Catalytic Transition Metal Hydrides, held on November 15–17, 1982 by The New York Academy of Sciences.

Part II. Hydrides in Carbon Monoxide Reactions

Part III. Hydrides in Alkene and Alkyne Reactions

Part IV. Hydrides in Hydrogenation of Organic Substrates

Financial assistance was received from:

- ALLIED CORPORATION
- CELANESE RESEARCH COMPANY
- E. I. DU PONT DE NEMOURS & COMPANY
- W. R. GRACE & CO.
- HOFFMANN-LA ROCHE, INC.
- PERGAMON PRESS INC.
- SHELL DEVELOPMENT COMPANY

Acknowledgments:

Special thanks are in order to those who served as session chairmen and discussion leaders for the conference. These are Irving Wender, Marv Rausch, John Bradley, Ken Barnett, Jim Lyons, John Armour, Bob Crabtree, Dick Hughes, Wayne Pretzer, Bob Augustine, and Robert Wade.

TWO TRIBUTES

Usually bland or hype offerings on the contents of a volume constitute the substance of editorial remarks. We would like to depart from this tradition and pay tribute to two noted chemists, Professor Rowland Pettit and Professor Minoru Tsutsui, both of whom died during the year preceding our November 1982 conference.

Rollie Pettit obtained a Ph.D. from the University of Adelaide in 1953 and a second Ph.D. from the University of London in 1957. That same year he became associated with the University of Texas where he rose to the rank of full professor in 1963. Rollie had a long and continuing interest in The Academy, having afforded distinguished contributions to the meetings held in 1973, 1976, and 1979 and to the Annals resulting therefrom. He surely would have liked to have been with us to describe his latest endeavors in the modeling of intermediates for the Fisher-Tropsch reaction, work which has, as have most of the earlier contributions from his group, attracted worldwide attention. He will be missed not only for his extraordinary contributions to the field, but also for the insight and wit he brought to his presentations and to the general air of bonhomie he presented to the world.

In 1941 Minoru Tsutsui received an M.S. degree from the University of Tokyo. After coming to the United States, he earned M.S. and Ph.D. degrees in organic chemistry at Yale University where he studied with H. H. Zeiss. His dissertation on "polyphenylchromium" described in detail the syntheses of these compounds (first described by F. Hein in 1918) and correctly formulated them as π-bonded dibenzenechromium complexes. Knowledge of these complexes together with other events such as the discovery of ferrocene fueled the genesis of the modern field of transition metal organometallic chemistry.

Minoru joined Texas A & M University as Professor after eight years at New York University. During his tenure in New York he founded the organometallic subdivision of The New York Academy of Sciences under whose auspices the meetings are presented and the Annals are published. Minoru had always evinced an interest in science policy and the international exchange of scientific information and personalities. He was convinced that basic research can only survive through the broad concern and participation of individual scientists. In this regard he organized many conferences and symposia, notably the recent Trilateral Seminar in Organometallic Chemistry held in the People's Republic of China in 1980.

Both of these men shall be missed for their original contributions to organometallic chemistry, for their dynamic devotion, each in his own way, to improving relationships with colleagues here and abroad, and lastly and most importantly for the friendship they offered to all who knew them.

D. W. Slocum
Chemistry and Biochemistry
Southern Illinois University

William R. Moser
Chemical Engineering
Worcester Polytechnic Institute

STEREOCHEMICAL ANALYSIS OF VARIOUS TRANSITION METAL HYDRIDE CLUSTERS AND RESULTING BONDING IMPLICATIONS*

Lawrence F. Dahl

*Department of Chemistry
University of Wisconsin—Madison
Madison, Wisconsin 53706*

INTRODUCTION

In 1965–66, x-ray crystallographic studies[1–4] provided definite structural evidence for the existence of symmetrical three-center, two-electron (3c–2e) metal-hydrogen-metal bonds in transition metal clusters.[1–4] Since that time the structures of a large number of transition metal hydride clusters have been determined from x-ray diffraction measurements. A relatively small number of these species have also been analyzed by neutron diffraction (particularly by Koetzle and Bau at Brookhaven National Laboratory and by Williams, Schultz, and Petersen at Argonne National Laboratory). Because hydrogen positions can be obtained much more accurately by neutron diffraction, this technique has played an especially important role in uncovering structural-bonding interrelationships in transition metal hydride complexes.

It is the purpose of this report to provide an overview of the detailed structural information that has been obtained about the nature of hydrogen bonding in a variety of transition metal hydride complexes. In a considerable number of the x-ray diffraction studies (especially the earlier ones), the positions of the hydrogen atoms were not directly located but were readily inferred by their geometrical effects on M–M bond lengths and/or on the disposition of the other adjacent ligands. Hydride sites have also been indirectly located in metal cluster complexes via a potential-energy procedure (in which optimal positions are found for each postulated hydride site by minimization of the potential energy of the intramolecular nonbonded interactions involving the hydride ligand).[5,6]

Because there are a number of excellent structural reviews of transition metal hydrides,[7–15] particularly a recent comprehensive one by Teller and Bau,[15] this presentation focuses on an historical development of the underlying stereochemical and bonding principles gleaned from crystallographic diffraction analyses of certain metal hydride complexes. Only a limited number of examples are described; the choices reflect the special interests and prejudices of the author. The following groups of compounds are discussed: (1) metal complexes with terminal hydride ligands; (2) metal complexes with doubly bridging hydride ligands; (3) metal clusters with triply bridging hydride ligands; and (4) metal clusters with interstitial hydrogen atoms.

* Supported by the National Science Foundation.

METAL COMPLEXES WITH TERMINAL HYDRIDE LIGANDS

The stereochemical nature of the terminal hydride ligand in transition metal hydride complexes was the subject of considerable confusion until diffraction investigations in the 1960s, especially those of K_2ReH_9,[16] $Rh(CO)(PPh_3)_3H$,[17,18] and $Mn(CO)_5H$,[19] established unequivocally that terminal hydride ligands occupy regular metal coordination sites with normal "covalent" M—H bonds. An illuminating historical account of this controversy is given by Ibers,[7] who included the initial interpretations of spectroscopic studies which supported the notion that the hydride ligand exerted little steric influence and was imbedded as a proton within the metal orbitals. In 1971 Frenz and Ibers also pointed out that the observed geometrical deviations from ideal configurations found from diffraction determinations of mononuclear metal hydrides could be readily rationalized from the reasonable premise that a terminal hydrogen ligand is stereochemically active but that as a rule the other nonhydrogen ligands adjacent to the hydride ligand are displaced toward the hydrogen due to its relatively small steric requirements.[8]

COMPLEXES WITH DOUBLY BRIDGING HYDRIDE LIGANDS

The crystal structures of a considerable variety of di-, tri-, and polynuclear metal complexes containing one or more hydrogen atoms bridging two transition metals have been investigated by diffraction methods. Brief descriptions of the salient structural features and concomitant bonding interpretations are presented below for a select group of these complexes, especially those that in the author's eyes had an important impact in the chronological development of the stereochemical-bonding principles for doubly bridging ligands. These examples include clusters possessing localized $M(\mu_2\text{-}H)_nM$ arrays ($n = 1, 2, 3, 4$). A qualitative MO bonding description of $M(\mu_2\text{-}H)_nM$ systems ($n = 1, 2$) was put forth in 1973 by Mason and Mingos;[20,21] an MO examination of the bonding and rotational barriers of hydrido-bridged dimers ($n = 2$–4) and their unsupported analogues was presented in 1979 by Dedieu, Albright, and Hoffmann.[22]

$Mo_2(\eta^5\text{-}C_5H_5)_2(CO)_4(\mu_2\text{-}H)(\mu_2\text{-}PMe_2)$

A photographic x-ray diffraction investigation by Doedens and Dahl of this compound in 1965 provided the first crystallographic evidence for a symmetrical, bent, 3c-2e M—H—M bond.[1] Previous proton NMR solution studies,[23–25] which indicated the presence of metal-bonded hydrogen atoms in this and related dimetal-hydride complexes, had led to the proposal that there was either a rapid intramolecular exchange of the hydrogen atom between two metal atoms or that the hydrogen was equally associated with both metal atoms. The probable location of the bridging hydrogen atom on the molecular pseudo-twofold axis at a regular coordination site about each Mo atom was readily deduced from the close resemblance of the overall molecular geometry (FIGURE 1) with those of related mononuclear molybdenum complexes. At this time it was emphasized that a "bent 3c-2e Mo—H—Mo bond involving one electron from the two

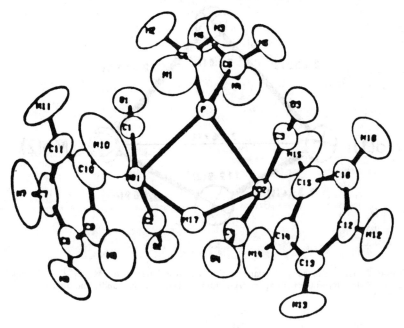

FIGURE 1. "Neutron-determined" configuration of $Mo_2(\eta^5\text{-}C_5H_5)_2(CO)_4(\mu_2\text{-}H)(\mu_2\text{-}PMe_2)$, which possesses a pseudo-twofold axis passing through the bridging phosphorus and hydrogen atoms.[26]

molybdenum atoms and one from the hydrogen accounts for the compound's diamagnetism without the invoking of a separate Mo—Mo bond."[1]

A subsequent neutron diffraction study of the 34-electron $Mo_2(\eta^5\text{-}C_5H_5)_2(CO)_4$-$(\mu_2\text{-}H)(\mu_2\text{-}PMe_2)$ by Petersen, Dahl, and Williams in 1974 corroborated the proposed location of the bridging hydrogen atom and provided the first direct crystallographic evidence that the hydrogen nucleus in a 3c–2e M—H—M bond of a transition metal complex can be symmetrically located between the two metal atoms in what was presumed to be a minimum potential well.[26] The difficulty in distinguishing between a single-minimum and double-minimum potential well for a crystallographically symmetric bridging hydrogen atom was initially indicated by Ibers in 3c–4e O—H—O and F—H—F systems[27-29] and later amplified by Doedens, Robinson, and Ibers in connection with their x-ray structural determination of the 34-electron $Mn_2(CO)_8(\mu_2\text{-}H)(\mu_2\text{-}PPh_2)$,[30] in which the hydrogen atom was directly located from a difference Fourier map on a crystallographic twofold axis.

The conclusion that the hydrogen nucleus in $Mo_2(\eta^5\text{-}C_5H_5)_2(CO)_4(\mu_2\text{-}H)$-$(\mu_2\text{-}PMe_2)$ is in a single-minimum potential well rather than in a double-minimum potential well was based on a detailed analysis of the sizes, shapes, and orientations of the determined atomic thermal ellipsoids of the planar Mo_2HP ring.[26] The fact that the isotropic thermal cross sections of the H and P atoms within the planar Mo_2HP core (FIGURE 2) are similar to those of the Mo atoms (coupled with

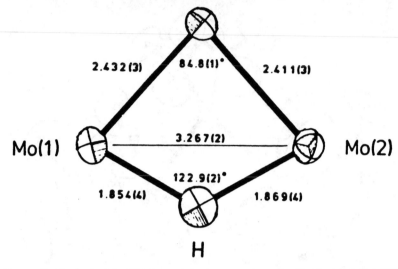

FIGURE 2. The mean Mo_2HP plane of the "neutron-determined" $Mo_2(\eta^5\text{-}C_5H_5)_2(CO)_4$-$(\mu_2\text{-}H)(\mu_2\text{-}PMe_2)$ molecule with appropriate internuclear distances and bond angles.[26]

the much greater thermal anisotropy of the bridging hydrogen and phosphorus atoms perpendicular to the Mo_2HP plane) provided support for a single equilibrium hydrogen site. The fact that the Mo—Mo distance of 3.267(2) Å is only slightly longer than the electron-pair Mo—Mo bond of 3.235(1) Å in the structurally related $Mo_2(\eta^5\text{-}C_5H_5)_2(CO)_6$ molecule[31] led to the proposal of extensive metal-metal bonding character in the bent 3c–2e Mo—H—Mo bond.[1,26]

X-Ray diffractometry data subsequently obtained for $Mo_2(\eta^5\text{-}C_5H_5)_2(CO)_4$-$(\mu_2\text{-}H)(\mu_2\text{-}PMe_2)$ by Petersen and Williams[32] in 1978 allowed a comparative analysis, which showed that the average "x-ray-determined" Mo—H distance of 1.79(15) Å is ca. 0.1 Å shorter than the average "neutron-determined"Mo—H distance of 1.86(1) Å due to the displacement of the "x-ray-determined" electron-density hydrogen peak by ca. 0.2 Å from the "neutron-determined" hydrogen nucleus toward the Mo—Mo internuclear vector. A concomitant effect of this 0.1 Å shortening of the Mo—H bond lengths is a 10° increase in the Mo—H—Mo bond angle from 123° to an estimated 133°. It was concluded that for other closed-type, bent M—H—M bonds one should similarly expect to find shorter average M—H bond distances and larger M—H—M bond angles by x-ray diffraction than by neutron diffraction.[32]

The stereochemical consequences of the deprotonation of the bent Mo—H—Mo bond in $Mo_2(\eta^5\text{-}C_5H_5)_2(CO)_4(\mu_2\text{-}H)(\mu_2\text{-}PMe_2)$ were presented by Petersen and Stewart in 1980 from a combined x-ray diffraction spectroscopic study of $[Ph_4As]^+[Mo_2(\eta^5\text{-}C_5H_5)_2(CO)_4(\mu_2\text{-}PMe_2)]^-$. The deprotonated monoanion retains the basic overall configuration of the neutral hydride parent with a pseudo-twofold axis still passing through the bridging phosphido group, but expectedly the Mo—Mo bond length in the monoanion is 0.10 Å shorter than that in the neutral parent.

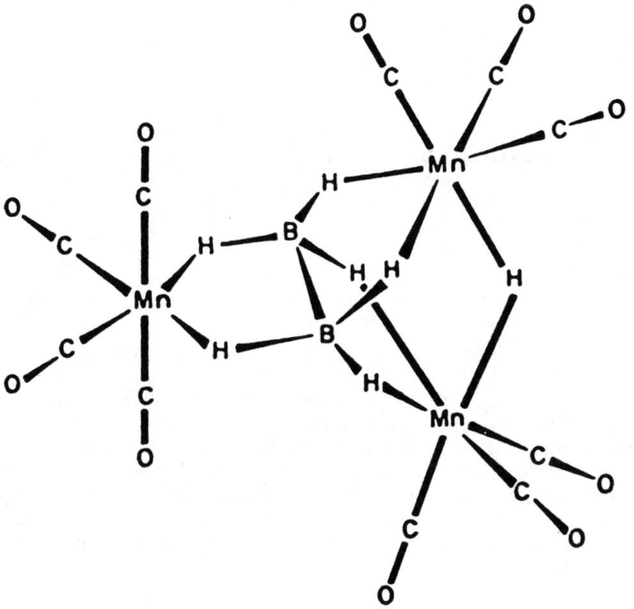

FIGURE 3. "X-Ray determined" configuration of the $Mn_3(CO)_{10}(B_2H_6)(\mu_2$-H) molecule, which possesses a crystallographic mirror plane passing through both B atoms, one Mn atom, two CO ligands, and three H atoms. This molecule contains one 3c–2e Mn—H—Mn bond and six 3c–2e B—H—Mn bonds.[2]

$Mn_3(CO)_{10}(B_2H_6)(\mu_2$-H)

The preparation and x-ray structural determination (via photographic data) of this first example of a polyborane transition metal carbonyl complex were reported by Kaesz, Dahl, and co-workers in 1965.[2] All four independent bridging hydrogen atoms were located from difference Fourier maps. The molecular configuration (FIGURE 3) of crystallographic C_S-m site symmetry and of idealized C_{2v} geometry consists of a (B—B)-bonded $(BH_3)_2$ fragment linked to an $Mn(CO)_4$ and an $Mn_2(CO)_6(\mu_2$-H) fragment by six 3c–2e B—H—Mn bonds. With the inclusion of the 3c–2e Mn—H—Mn bond, the electronic configuration of each of the two kinds of Mn atoms conforms to a closed-shell configuration. Besides representing the first example in which all three hydrogens of a BH_3 group are coordinated to other atoms via 3c–2e bonds, this structure provided further support for the presence of symmetrical hydrogen bridging between two transition metal atoms.

The $[M_2(CO)_{10}(\mu_2$-H)]$^-$ Monoanions (M = Cr, W)

The variable solid-state geometries exhibited by these particular hydrido-bridged monoanions in different salts have provided considerable insight into the nature

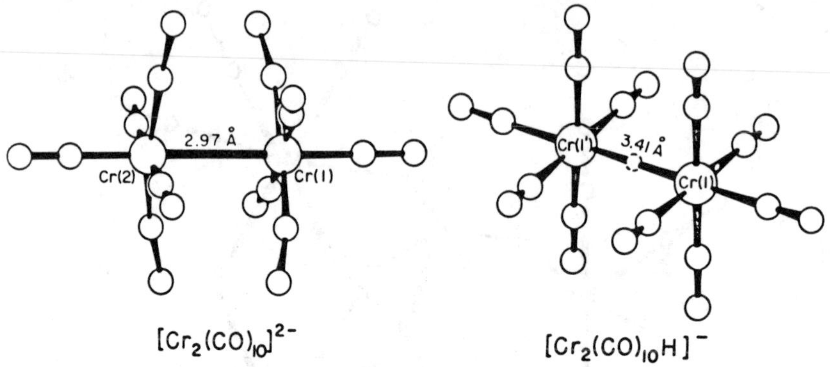

$$[Cr_2(CO)_{10}]^{2-}$$ $$[Cr_2(CO)_{10}H]^-$$

FIGURE 4. "X-Ray-determined" configurations showing the observed geometrical change from the D_{4d} carbonyl-staggered configuration of the $[Cr_2(CO)_{10}]^{2-}$ dianion to the D_{4h} carbonyl-eclipsed configuration in the $[Cr_2(CO)_{10}H]^-$ monoanion upon *protonation* of the electron-pair Cr—Cr bond to give a 3c–2e M—H—M bond. The bridging hydrogen atom was not located from the x-ray study, but its presumed position on the crystallographic center of symmetry to give a linear Cr—H—Cr bond was based on the assumption (later proven to be incorrect) that it should occupy a regular octahedral coordination site about each Cr atom.[4,34]

of an unsupported M—H—M bond and its geometrical flexibility in the crystalline state.

In 1966 a photographic x-ray study of $[NEt_4]^+[Cr_2(CO)_{10}(\mu_2\text{-}H)]^-$ by Handy, Dahl, and co-workers showed the monoanion to exhibit a D_{4h} carbonyl-eclipsed geometry (FIGURE 4);[4] although not directly established from the x-ray work, it was proposed that the bridging hydrogen atom lay on a crystallographically required center of symmetry in a regular octahedral coordination site about each Cr atom and was collinear with and equidistant from the two symmetry-equivalent chromium atoms.

This prototype structure represented the first known *linear* M—H—M molecular system stabilized by a 3c–2e bond. In 1970 a comparative x-ray analysis of the $[Cr_2(CO)_{10}(\mu_2\text{-}H)]^-$ monoanion in the $[NEt_4]^+$ salt and its corresponding nonprotonated $[Cr_2(CO)_{10}]^{2-}$ dianion in the $[(Ph_3P)_2N]^+$ salt was presented by Handy, Ruff, and Dahl.[34] Deprotonation was found to produce a marked 0.44 Å shortening of the Cr—Cr bond distance as well as a conformational change (due to carbonyl repulsions) in the two sets of equatorial carbonyl ligands from an eclipsed to a staggered D_{4d} $Mn_2(CO)_{10}$-type structure (FIGURE 4). It was stated that "these metal-metal bonding distances in the M—H—M systems emphasize the large direct metal-metal bonding character present in the three-center electron-pair bond."[34] This conclusion was supported by theoretical calculations,[34] which showed considerable direct overlap of metal orbitals in the presumed linear Cr—H—Cr core of the $[Cr_2(CO)_{10}(\mu_2\text{-}H)]^-$ monoanion.

In 1977 a room-temperature neutron-diffraction investigation of $[NEt_4]^+$-$[Cr_2(CO)_{10}(\mu_2\text{-}H)]^-$ by Petersen, Williams, Dahl, and co-workers surprisingly revealed a *nonlinear* Cr—H—Cr fragment in the monoanion (FIGURE 5) with the bridging hydrogen randomly distributed at two centrosymmetrically related

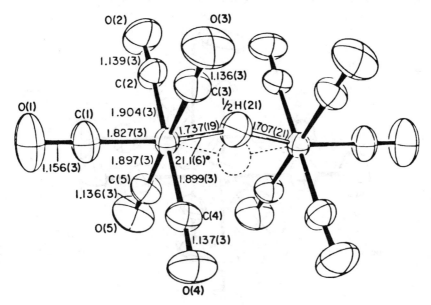

FIGURE 5. "Neutron-determined" architecture of the $[Cr_2(CO)_{10}(\mu_2\text{-H})]^-$ monoanion in the $[NEt_4]^+$ salt of crystallographic C_i-$\bar{1}$ symmetry showing the approximate D_{4h} geometry of the metal carbonyl framework and the two centrosymmetrically related (half-weighted) sites of the bridging hydrogen atom in the bent Cr—H—Cr molecular fragment.[35]

disordered sites displaced by ca. 0.3 Å from the crystallographic center of symmetry.[35] The resulting slightly *bent* Cr—H—Cr bond was found to possess a bond angle of 158.9(6)° with experimentally equivalent Cr—H internuclear distances of 1.707(21) and 1.737(19) Å. Another view (FIGURE 6) of the $[Cr_2(CO)_{10}(\mu_2\text{-H})]^-$ monoanion shows that the two half-occupied hydrogen sites are *staggered* with respect to the eclipsed configuration of the equatorial carbonyl ligands. It was then apparent that the carbonyl-eclipsed geometry is sterically induced by the bridging hydrogen. A further examination of the sizes, shapes, and orientations of the nuclear thermal ellipsoids of the axial carbonyl atoms (FIGURE 6) provided convincing evidence that the thermal anisotropy observed for the carbonyl structure most likely is a consequence of the near superposition of two slightly bent, identical $[Cr_2(CO)_{10}(\mu_2\text{-H})]^-$ structures of idealized C_{2v} geometry. This composite model is consistent with the bridging hydrogen atom not being positioned exactly at an octahedral metal coordination site. Instead, it is offset by approximately 0.2 Å from this site in a direction that is in accordance with the 3c–2e Cr—H—Cr

interaction being a closed-type $\overset{\displaystyle H}{\underset{\displaystyle M \quad M}{\diagdown}}$ bond.

The ability of the $[M_2(CO)_{10}(\mu_2\text{-H})]^-$ monoanion to also adopt a considerably bent, carbonyl-staggered geometry was uncovered by Bau, Koetzle, and co-workers from neutron diffraction studies in 1974 of two crystal forms of the electronically equivalent, neutral (crystal-disordered) $W_2(CO)_9(NO)(\mu_2\text{-H})$ molecule[36] and in 1976 of a corresponding phosphine-substituted (crystal-ordered) $W_2(CO)_8(NO)$-

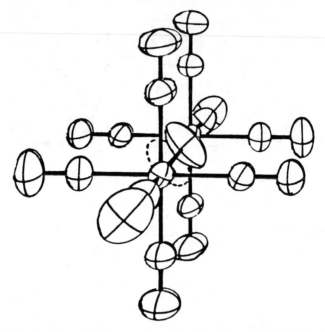

FIGURE 6. Another view of the "neutron-determined" geometry of the $[Cr_2(CO)_{10}(\mu_2\text{-H})]^-$ monoanion in the $[NEt_4]^+$ salt showing the two centrosymmetrically disordered, half-weighted hydrogen sites, which are staggered relative to the eclipsed arrangement of the two sets of equatorial carbonyl ligands.[35]

$(P(OMe)_3)$ molecule[37] (FIGURE 7). They observed that both crystal forms of the former molecule show essentially the same structure with a single bent $W-H-W$ linkage of bond angle 125.5° and with equivalent $W-H$ distances of 1.873 Å (average). For the $W_2(CO)_8(NO)(P(OMe)_3)(\mu_2\text{-H})$ structure, which possesses an analogous backbone with staggered equatorial carbonyl ligands, the $W-H-W$ framework was found to be similarly bent [129.4(3)°] with slightly different $W-H$ distances of 1.859(6) and 1.894(6) Å [the hydrogen being closer to the more electron-deficient $W(CO)_5$ group]. These neutron diffraction studies showed that the axial ligand–metal vectors in these similarly bent $M-H-M$ systems approximately point at the centroid of the WHW triangle and *not* at the bridging hydrogen atom. This consistency of molecular parameters led to their formulation of a closed-type

$$\overset{\displaystyle H}{\underset{M\diagup\diagdown M}{|}}$$

bond with a significant amount of metal-metal interaction for these highly bent $M-H-M$ bonds.

At the same time Bau and co-workers found from x-ray diffraction studies that the $[W_2(CO)_{10}(\mu_2\text{-H})]^-$ monoanion exists in both the carbonyl-eclipsed and carbonyl-staggered geometries.[38] In $[NEt_4]^+[W_2(CO)_{10}(\mu_2\text{-H})]^-$, whose crystals are isomorphous with those of $[NEt_4]^+[Cr_2(CO)_{10}(\mu_2\text{-H})]^-$, the tungsten monoanion has an overall D_{4h} carbonyl-eclipsed configuration at room temperature, whereas in $[(Ph_3N)_2N]^+[W_2(CO)_{10}(\mu_2\text{-H})]^-$ it has a carbonyl-staggered configuration.

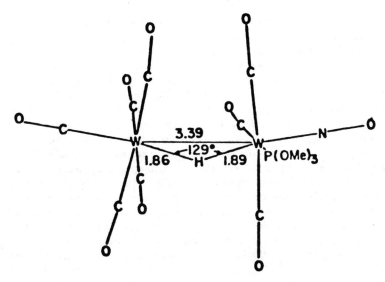

FIGURE 7. "Neutron-determined" architecture of the crystal-ordered $W_2(CO)_8(NO)$-$(P(OMe)_3)(\mu_2$-H) molecule, which possesses a carbonyl-staggered geometry with a highly bent W—H—W fragment.[37]

However, a subsequent neutron diffraction study of $[NEt_4]^+[W_2(CO)_{10}(\mu_2$-H)]$^-$ at low temperature (14K) revealed a disorder superposition of two slightly bent carbonyl-eclipsed monoanions together with a considerably bent and highly asymmetrical W—H—W bond.[39] The bridging hydrogen was found to be situated 0.71 Å from the *average* center of symmetry at two highly asymmetric disordered sites such that the resulting W—H distances are 1.718(12) and 2.070(12) Å. This marked bond-length variation is anomalous for M—H—M arrays, but nevertheless a 3c–2e instantaneous description is still applicable for such asymmetric bonding. In contrast, a neutron diffraction study of $[PPh_4]^+[W_2(CO)_{10}(\mu_2$-H)]$^-$ showed the monoanion to possess an ordered carbonyl-staggered structure which closely resembles that of $W_2(CO)_9(NO)(\mu_2$-H).[39] It is not clear to this author whether the drastic structural variation found in $[NEt_4]^+[W_2(CO)_{10}(\mu_2$-H)]$^-$ at low temperature can be attributed to the bridging hydrogen in the unsupported, nonrigid W—H—W framework undergoing discrete phase change(s) from symmetrical to highly unsymmetrical sites. Nevertheless, these results are consistent with the observed large temperature-dependent variations of the M—H—M vibrational modes in low-temperature IR[40–42] and Raman[41,42] spectra of $[NEt_4]^+[W_2(CO)_{10}(\mu_2$-H)]$^-$ being due primarily to structural asymmetry of the hydrogen atom at low temperature in multiple-well potential minima.[41–43]

In order to determine whether the solid-state structure of the $[Cr_2(CO)_{10}(\mu_2$-H)]$^-$ monoanion is also influenced by a change of counterion and concomitant change in packing forces, a combined x-ray and neutron diffraction study of $[(Ph_3P)_2N]^+[Cr_2(CO)_{10}(\mu_2$-H)]$^-$ was carried out at room temperature by Petersen, Dahl, Williams, and co-workers.[44] In complete contrast to the bent, carbonyl-staggered geometry of the tungsten monoanion in the $[(Ph_3P)_2N]^+$ salt, the pseudo-D_{4h} carbonyl-eclipsed geometry is maintained in the $[Cr_2(CO)_{10}(\mu_2\text{-}H)]^-$ monoanion which lies on a crystallographic center of symmetry. The observed large, abnormal disk-shaped thermal ellipsoid of the hydrogen nucleus is completely consistent with there being a fourfold crystallographic disorder involving a slightly bent $[Cr_2(CO)_{10}(\mu_2\text{-}H)]^-$ monoanion of the "instantaneous" C_{2v} geometry (previously found in the $[NEt_4]^+$ salt) being distributed among four equivalent sites. The disk-shaped hydrogen ellipsoid then is the result of the superposition of the four disordered, off-axis hydrogen sites (of the bent Cr—H—Cr units), which are staggered with respect to the essentially composite carbonyl-eclipsed framework. A subsequent low-temperature (17K) neutron diffraction study of the isomorphous $[(Ph_3P)_2N]^+[Cr_2(CO)_{10}(\mu_2\text{-}D)]^-$ (ca. 85% deuterated) indicated no disorder of the metal carbonyl framework.[45] Although the relatively large thermal ellipsoid for the bridging deuterium nucleus was not resolved into the four disordered off-axis sites (see above), least-squares refinement demonstrated that the disordered four-site hydrogen model was accommodated by the diffraction data. These results are in agreement with an extensive variable-temperature IR-Raman spectral analysis by Shriver and co-workers,[40] who concluded that the Cr—H—Cr part of the $[Cr_2(CO)_{10}(\mu_2\text{-}H)]^-$ monoanion is similarly bent in both the $[NEt_4]^+$ and $[(Ph_3P)_2N]^+$ salts.

A neutron diffraction analysis in 1981 by Petersen, Brown, and Williams of $[K(2,2,2\text{-crypt})]^+[Cr_2(CO)_{10}(\mu_2\text{-}H)]^-$ at 20K revealed that the two C_{4v} Cr(CO)$_5$ halves of the monoanion are rotated by ca. 19° about the Cr—Cr line from an eclipsed carbonyl conformation toward a bent structure.[46] This twisting-type distortion of the carbonyl framework, which was ascribed to an appreciable cation-anion interaction, is accompanied by a smaller Cr—H—Cr bond angle of 145.2(3)° and a concomitant decrease of the Cr—Cr distance to 3.300(4) Å in accordance with a stronger metal-metal bonding component in the "closed" 3c–2e Cr—H—Cr bond. An ordered bridging H atom was found to reside in a symmetrical electronic site; the equivalent Cr—H distances of 1.723(5) and 1.735(5) Å are comparable to those in the $[NEt_4]^+$ salt. These structural data indicate that the solid-state structure of the $[Cr_2(CO)_{10}(\mu_2\text{-}H)]^-$ monoanion is affected by its crystalline environment to a much lesser degree than is its tungsten analogue. This may be ascribed to the stronger metal-metal bonding interaction in the M—H—M fragment of the tungsten analogue, which favors the considerably bent, carbonyl-staggered geometry with a smaller M—M distance.

Several different symbolisms (FIGURE 8) have been utilized to represent the so-called closed 3c–2e M—H—M bonds; each formulation incorporates the widely accepted notion of appreciable M—M bonding character (as illustrated by I). Designation II was originally formulated by Dahl and co-workers[1–4,26,34] based on the premise that in general an M—H—M bond can be considered as a protonated M—M bond that still retains considerable M—M bond character. The equivalent dashed-line designation IV was originally suggested by Churchill and co-workers to emphasize M—M as well as M—H interactions in a 3c–2e bond.[14,47,48] One obvious problem with this convention is that others have invariably redrawn these dashed lines as solid lines, which thereby implies (at least to this author) the incorrect notion (when solid lines between atoms are presumed to denote bonds) of a separate M—M electron-pair bond in addition to the 3c–2e bond. The closed-type designation III adapted by Bau, Koetzle, and co-

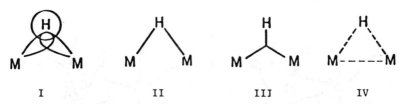

$$\text{I} \qquad\qquad \text{II} \qquad\qquad \text{III} \qquad\qquad \text{IV}$$

FIGURE 8. Equivalent designations (II, III, and IV) of the three-center two-electron "bent" M—H—M bond, which as a protonated M—M bond in general contains appreciable metal-metal bonding orbital character (denoted by I). As emphasized by the closed-type designation III, the off-axial position (shown in I) of the bridging H atom from the regular coordination site of each M (determined by the intersection of two axial ligand-metal vectors) was ascertained from several neutron diffraction studies including those of $W_2(CO)_9(NO)(\mu_2\text{-H})$, $W_2(CO)_8(NO)(P(OMe)_3)(\mu_2\text{-H})$, and $[PPh_4]^+[W_2(CO)_{10}\text{-}(\mu_2\text{-H})]^-$ (in which the monoanion possesses a carbonyl-staggered geometry) and of $[NEt_4]^+[Cr_2(CO)_{10}(\mu_2\text{-H})]^-$ (in which the monoanion possesses a carbonyl-eclipsed geometry).

workers[11,12,36] from boron hydride chemistry is a good representation of these bent M—H—M bonds. In any event, it must be emphasized that designations II, III, and IV are all equivalent in their representation of a "closed" 3c–2e M—H—M array irrespective of the M—H—M bond angle.

To date all evidence suggests that a truly linear M—H—M geometry is inherently less stable than a bent geometry since the latter permits a smaller M—M distance with a stronger 3c–2e bonding interaction. The M—H—M linkages in the "neutron-determined" structures of $Ru_3(CO)_9(\mu_3\text{-C}\equiv CBu^t)(\mu_2\text{-H})$[49] and $Os_3(CO)_9(\mu_2\text{-H})_2(\mu_3\text{-S})$[50] were later described as "open" (instead of "closed") 3c–2e bent bonds. This description, which implies little or no direct M—M interaction, was based upon the fact that the extension of the two M—C vectors of the carbonyl ligands *trans* to each hydrogen atom intersect very close to the hydrogen atom. Nevertheless, the fact that the hydrogen-bridged M—M distances in these molecules are within the range of single-bond M—M distances leads one to presume from orbital overlap considerations that there is still extensive direct M—M bonding character in each M—H—M unit. Hence, it is the author's prejudice that each of the M—H—M units in these molecules can likewise be described as protonated M—M bonds in accordance with designations II or IV (FIGURE 8).

The $[Fe_3(CO)_{10}(\mu_2\text{-CO})(\mu_2\text{-H})]^-$ Monoanion

An x-ray photographic analysis of $[NEt_3H]^+[Fe_3(CO)_{11}H]^-$ in 1965 was of particular importance in that the idealized mirror-plane configuration (FIGURE 9) of the monoanion enabled Dahl and Blount[3] to propose the correct molecular configuration for the structurally controversial triiron dodecacarbonyl,[51–54] $Fe_3(CO)_{10}(\mu_2\text{-CO})_2$, by the formal substitution of a bridging carbonyl ligand in place of the bridging H^- ligand. The presence of a symmetrically equivalent bridging hydrogen atom at its indicated coordination site (which completes an octahedral environment about each attached Fe atom) was not directly established by the x-ray work but was inferred from stereochemical and bonding

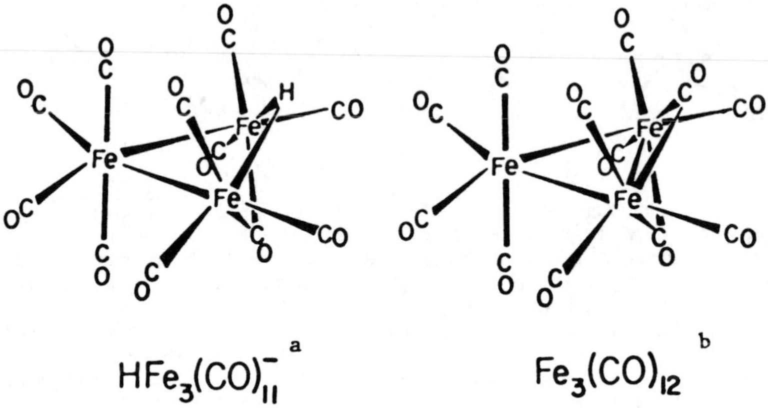

$$HFe_3(CO)_{11}^{-} \qquad Fe_3(CO)_{12}$$

FIGURE 9. "X-Ray-determined" configurations of the electronically equivalent and structurally analogous $[Fe_3(CO)_{11}H]^-$ monoanion and $Fe_3(CO)_{12}$ molecule, which possess a basic $M_3(CO)_{10}(\mu_2\text{-}X)(\mu_2\text{-}Y)$-type geometry.[3]

considerations. The similarity of the Fe—Fe distances in the $[Fe_3(CO)_{10}(\mu_2\text{-}CO)\text{-}(\mu_2\text{-}H)]^-$ monoanion to those in $Fe_3(CO)_{10}(\mu_2\text{-}CO)_2$ led to the proposal that the bent Fe—H—Fe system in the monoanion may be considered to arise from protonation of an Fe—Fe single bond in the nonprotonated $[Fe_3(CO)_{11}]^{2-}$ dianion.[3] The major geometrical change resulting from deprotonation of the mono-anion to give the $[Fe_3(CO)_{11}]^{2-}$ dianion involves a marked tipping of the cis-$Fe(CO)_4$ group in order to transform a terminal axial carbonyl into a triply bridging carbonyl, which thereby occupies the missing coordination site about each of the two equivalent iron atoms.[55]

The $[Fe_3(CO)_{10}(\mu_2\text{-}H)(\mu_2\text{-}CO)]^-$-type geometry has subsequently been found in a considerable number of electronically equivalent (48-electron) ruthenium[48] and osmium[56−64] clusters of general formula $M_3(CO)_{10}(\mu_2\text{-}H)(\mu_2\text{-}L)$, where the two-electron–donating bridging carbonyl and negative charge have been formally replaced by a three-electron–donating L ligand (see below).

$Os_3(CO)_{10}(PPh_3)H(\mu_2\text{-}H)$

The crystal-ordered structure (FIGURE 10) of this 48-electron equatorial phosphine–substituted complex was obtained from an x-ray analysis by Churchill and DeBoer in 1977.[65] The corresponding isostructural $Os_3(CO)_{11}H(\mu_2\text{-}H)$ cluster[66] contains a disorder of the terminal axial hydride ligand and the axial carbonyl ligand on the same Os atom. Unlike the corresponding $H_2Fe_3(CO)_{11}$ cluster, which was formulated from an extensive NMR analysis as $Fe_3(CO)_{10}(\mu_2\text{-}H)\text{-}(\mu_2\text{-}COH)$ with the second hydrogen being oxygen bound,[67] the overall geometry of $Os_3(CO)_{11}H(\mu_2\text{-}H)$ is closely related to that of $Os_3(CO)_{12}$.[66,68] The replacement of a two-electron–donating axial carbonyl ligand by a one-electron–donating hydrogen atom necessitates the addition of the second one-electron–donating hydrogen atom as part of a 3c–2e Os—H—Os bond in the plane of the three osmium atoms.

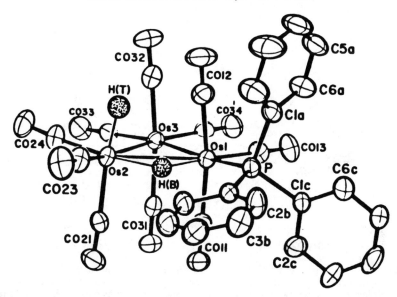

FIGURE 10. "X-Ray-determined" configuration of the triphenylphosphine-substituted derivative of $Os_3(CO)_{11}H(\mu_2\text{-}H)$, which possesses an $Os_3(CO)_{12}$-type structure with the bridging hydrogen occupying an equatorial, edge-bridged site. Both hydride ligands were located directly.[65,66]

The $[Re_3(CO)_{12}(\mu_2\text{-}H)_x]^{n-}$ Anions (x = 1, n = 2; x = 2, n = 1)

The polyhedral carbonyl arrangement of each of these two 48-electron anions (FIGURE 11), reported in 1968 and 1972 by Churchill and co-workers from x-ray analyses, also resembles that in the isoelectronic $Os_3(CO)_{12}$ molecule with the hydrogen atoms occupying equatorial, edge-bridged sites of the triangular metal core. The hydrogen positions were indirectly determined via displacement of the equatorial carbonyl ligands from their ideal orientations. The proposed hydrogen positions are completely consistent with the observed edge-bridged sites of the three hydrogen atoms in the corresponding isostructural $Mn_3(CO)_{12}(\mu_2\text{-}H)_3$ molecule.[70] These early analyses (coupled with those of the $[Cr_2(CO)_{10}(\mu_2\text{-}H)]^-$, $[Cr_2(CO)_{10}]^{2-}$ pair) were important in establishing the empirical generalization (emphasized primarily by Churchill from his elegant x-ray studies on a number of important metal hydride clusters)[14,48,56,63-66,69,71] that "a single, *unsupported* bridging hydride ligand normally causes a lengthening of a metal-metal bond."[48]

$Os_3(CO)_{10}(\mu_2\text{-}H)_{2-n}(\mu_2\text{-}L)_n$ Series (n = 0,1,2; L = OMe)

X-Ray[62,71] and neutron[72,73] diffraction analyses of the $Os_3(CO)_{10}(\mu_2\text{-}H)_2$ molecule (FIGURE 12), an important precursor to a number of other osmium clusters, revealed an $Fe_3(CO)_{12}$-type geometry with two equivalent $Os(CO)_3$ fragments linked by two hydride ligands. This electron-deficient 46-electron system possesses two 3c-2e Os—H—Os bonds such that the resulting $Os(\mu_2\text{-}H)_2Os$ unit

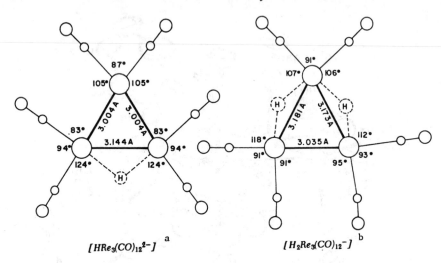

FIGURE 11. "X-Ray-determined" geometries of the equatorial planes of the $[HRe_3(CO)_{12}]^{2-}$ and $[H_2Re_3(CO)_{12}]^-$ anions showing the approximate positions of the edge-bridged hydride ligands. The geometrical effect characteristic of an *unsupported* μ_2-hydride ligand is illustrated here with the location of each edge-bridged hydrogen deduced from the marked relative increase in the Re—Re distance of the Re—H—Re bond together with the observed angular displacement of the adjacent equatorial carbonyl ligands.[14, 69]

may be conceptually considered as a diprotonated M—M double bond (of M—M bond order 2.0). The fact that the dibridged Os—Os bond of 2.682 Å is substantially shorter than the two equivalent nonbridged Os—Os single bonds of 2.815 Å is in accordance with there being Os—Os double-bond character in the

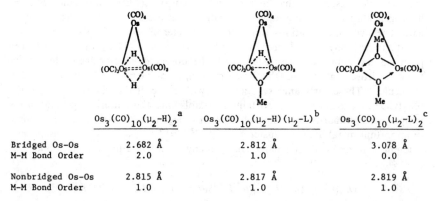

	$Os_3(CO)_{10}(\mu_2\text{-H})_2$ [a]	$Os_3(CO)_{10}(\mu_2\text{-H})(\mu_2\text{-L})$ [b]	$Os_3(CO)_{10}(\mu_2\text{-L})_2$ [c]
Bridged Os–Os	2.682 Å	2.812 Å	3.078 Å
M–M Bond Order	2.0	1.0	0.0
Nonbridged Os–Os	2.815 Å	2.817 Å	2.819 Å
M–M Bond Order	1.0	1.0	1.0

FIGURE 12. This $Os_3(CO)_{10}(\mu_2\text{-H})_{2-n}(\mu_2\text{-L})_n$ series ($n = 0, 1, 2$; L = OMe) illustrates the changes in M—M distances as a three-electron–donating, bridging L ligand (L = OMe) is formally substituted in place of a one-electron–donating bridging H atom, which forms part of a localized 3c-2e M—H—M bond. The di(μ-hydrido)-bridged system may be considered as a diprotonated M—M double bond and the mono(μ-hydrido)-bridged system as a monoprotonated M—M single bond.[62, 74]

$Os(\mu_2\text{-}H)_2Os$ system. Alternatively, the entire $Os(\mu_2\text{-}H)_2Os$ interaction may be viewed as a delocalized 4c–4e bond.[73]

$Os_3(CO)_{10}(\mu_2\text{-}H)(\mu_2\text{-}OMe)$ (FIGURE 12) was expectedly shown from an x-ray structural determination by Churchill and Wasserman in 1980 to exhibit an $[Fe_3(CO)_{10}(\mu_2\text{-}H)(\mu_2\text{-}CO)]^-$-type geometry.[74] It differs from $Os_3(CO)_{10}(\mu_2\text{-}H)_2$ by the formal substitution of a three-electron–donating, bridging L ligand (L = OMe) in place of the one-electron–donating, bridging hydrogen atom to give a normal 48-electron species. The methoxy oxygen atom forms a localized electron-pair σ-bond with each of the two Os atoms, which are also connected by the remaining 3c–2e M—H—M bond. The observed increase in the dibridged Os—Os bond from 2.682 Å to 2.817 Å is consistent with the formal reduction of the M—M bond order from 2.0 to 1.0 with the bent Os—H—Os system corresponding to a protonated M—M single bond.

The 50-electron $Os_3(CO)_{10}(\mu_2\text{-}OMe)_2$ complex[62] (FIGURE 12) is produced by the formal replacement of the other bridging hydrogen with a second three-electron–donating L ligand (again with L = OMe). Electronic considerations predict a nonbonding Os—Os distance (with an M—M bond order of 0.0) for the dibridged $Os(\mu_2\text{-}L)_2Os$ system in accordance with the observed value of 3.078 Å for the dialkoxy-bridged system.

A further examination in FIGURE 12 of the pattern of M—M distances vs. M—M bond orders for the three members of the $Os_3(CO)_{10}(\mu_2\text{-}H)_{2-n}(\mu_2\text{-}OMe)_n$ series ($n = 0, 1, 2$) reveals that the nonbridged Os—Os single bonds for the three complexes are essentially identical; this M—M bond-length invariance emphasizes that steric interactions in this series of closely related clusters are relatively unimportant in affecting the direct M—M electron-pair bonds. The fact that the dibridged Os—Os bond of 2.812 Å in the $Os_3(CO)_{10}(\mu_2\text{-}H)(\mu_2\text{-}L)$ complex (with L = OMe) is virtually identical with the nonbridged Os—Os single bonds is more fortuitous in that its protonated M—M bond length expectedly depends upon the nature of the bridgehead atom(s) of the bridging L ligand.[56–64] From their extensive systematic studies, Churchill and co-workers have pointed out that when two metal atoms in an M—H—M bond are additionally linked by another bridging ligand, the resulting M—M bond length may be shorter than, longer than, or equivalent to a normal nonbridged M—M bond length.[14,48,56]

The $[W_2(CO)_8(\mu_2\text{-}H)_2]^{2-}$ Dianion

This dianion in the $[NEt_4]^+$ salt was found by Churchill and Chang in 1974 from an x-ray diffraction investigation (which directly located and refined the two bridging hydrogen atoms) to conform closely to an edge-bridged bioctahedral D_{2h} geometry[75] (FIGURE 13). In 1982 Bau, Kirtley, Koetzle, and co-workers reported a low-temperature (28K) neutron diffraction study of $[(Ph_3P)_2N]_2^+$-$[W_2(CO)_8(\mu_2\text{-}H)_2]^{2-}$, which not only substantiated the x-ray results but also expectedly provided much more precise parameters for the planar $W(\mu_2\text{-}H)_2W$ core.[76] Electronic considerations suggest that the dibridged $M(\mu_2\text{-}H)_2M$ system in this 32-electron dianion and in the isoelectronic and structurally analogous $Re_2(CO)_8(\mu_2\text{-}H)_2$ molecules[77] may be regarded as possessing a diprotonated M—M double bond. Evidence for multiple M—M bond character in both the

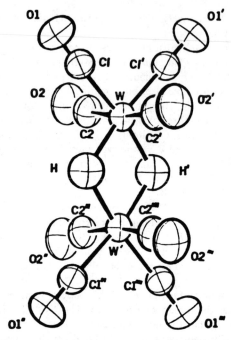

FIGURE 13. "X-Ray-determined" configuration of the 32-electron $[W_2(CO)_8(\mu_2\text{-}H)_2]^{2-}$ dianion in the $[NEt_4]^+$ salt. Its edge-bridged bioctahedral geometry of crystallographic C_{2h}-2/m site symmetry conforms closely to D_{2h} symmetry. The planar $W(\mu_2\text{-}H)_2W$ fragment formally contains a diprotonated M—M double bond.[75] A relatively precise "neutron-determined" geometry of this dianion in the $[(Ph_3P)_2N]^+$ salt was recently reported.[76]

$W(\mu_2\text{-}H)_2W$ and $Re(\mu_2\text{-}H)_2Re$ systems is given by each of the M—M distances being considerably shorter than a corresponding normal (nonbridged) M—M single-bond distance—viz., 3.016(1) Å[75] and 3.010(2) Å[76] in the ditungsten dianion vs. 3.222(1) Å in $W_2(\eta^5\text{-}C_5H_5)_2(CO)_6$[78] and 2.896(3) Å in $Re_2(CO)_8(\mu_2\text{-}H)_2$[77] vs. 3.02(1) Å in $Re_2(CO)_{10}$.[79] The corresponding "x-ray-determined" $Re_2(CO)_6$-$(Ph_2PCH_2PPh_2)(\mu_2\text{-}H)_2$,[80] a derivative of $Re_2(CO)_8(\mu_2\text{-}H)_2$ with two of the axial carbonyls replaced by the diphosphine ligand, has a virtually identical diprotonated Re—Re double-bond length of 2.893(2) Å. An even shorter diprotonated Re—Re double-bonded distance of 2.797(4) Å was found from an x-ray crystallographic examination of the 46-electron $[H_3Re_3(CO)_{10}]^{2-}$ dianion,[81] an isoelectronic and presumed structural analogue of $Os_3(CO)_{10}(\mu_2\text{-}H)_2$ (FIGURE 12) with the third hydride ligand randomly disordered between the two other edge-bridged equatorial sites.

Since one μ_2-hydride ligand causes a *net increase* of ca. 0.15 Å in an Re—Re distance (see FIGURE 11), these results are consistent with the premise that two μ_2-hydrido ligands between two metals cause a *net increase* of an M—M distance to a value considerably *greater* than a normal (nonconstrained) M—M double-bond value but yet *less* than a normal M—M single-bond value. Such an M—M

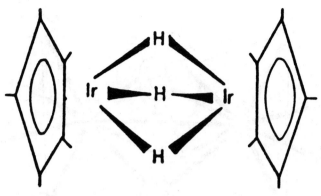

FIGURE 14. "Neutron-determined" configuration of the 30-electron $[Ir_2(\eta^5\text{-}C_5Me_5)_2$-$(\mu_2\text{-}H)_3]^+$ cation in the $[BF_4]^-$ salt. The $M(\mu_2\text{-}H)_3M$ unit of this confacial bioctahedral cation may be viewed as a triprotonated $M-M$ triple bond.[82]

bond-length trend directly correlates with the vibrational assignment by Shriver and co-workers of the relative $M-M$ stretching frequencies of $M{=}M > M$-$(\mu_2\text{-}H)_2M > M-M > M(\mu_2\text{-}H)M$ in (otherwise unsupported) dinuclear metal hydride complexes.[40]

The $[Ir_2(\eta^5\text{-}C_5Me_5)_2(\mu_2\text{-}H)_3]^+$ Monocation

A neutron diffraction examination by Bau, Koetzle, and co-workers showed that this cation has a confacial bioctahedral geometry (FIGURE 14) with the pentamethylcyclopentadienyl ligands occupying three coordination sites and the Ir atoms symmetrically linked by the three hydrogen atoms.[82] The formulation of the $Ir(\mu_2\text{-}H)_3Ir$ unit in this 30-electron monocation as a triprotonated $M-M$ triple bond is compatible with the observed short Ir—Ir distance of 2.458(6) Å. Geometrically analogous, 30-electron iridium dimers also containing triprotonated $M-M$ triple bonds (with the hydrogen positions indirectly determined from combined x-ray and NMR analysis) are the $[(IrH(PPh_3)_2)_2(\mu_2\text{-}H)_3]^+$ mono-cation[83,84] and $[(IrH(dppp))_2(\mu_2\text{-}H)_3]^+$ monocation[85] [where dppp denotes $Ph_2P(CH_2)_3PPh_2$], which possess virtually identical Ir—Ir bond lengths of 2.518 and 2.514(1) Å, respectively.

Other known species with an $M(\mu_2\text{-}H)_3M$ array corresponding to a triprotonated $M-M$ triple bond include the 30-electron confacial bioctahedral $[Re_2(CO)_6$-$(\mu_2\text{-}H)_3]^-$ monoanion[86] (characterized by ^1H NMR and IR) and the x-ray-determined, confacial bioctahedral, 30-electron $[Fe_2(p_3)_2(\mu_2\text{-}H)_3]^+$ monocation[87] [where (p_3) denotes $MeC(CH_2PPh_2)_3$]. Electron-counting procedures indicate that the corresponding 32-electron $[Co_2(as_3)_2(\mu_2\text{-}H)_3]^+$ monocation[87] [where (as_3) denotes $MeC(CH_2AsPh_2)_3$] is *not* a triprotonated $M-M$ triple-bonded species but instead a *net* $M-M$ double-bonded species. Although the three bridging hydrogen atoms may be considered as forming three localized 3c-2e $M-H-M$ bonds of $M-M$ bond order 3.0 (corresponding under assumed D_{3h} symmetry

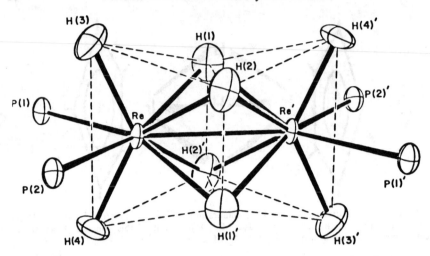

FIGURE 15. "Neutron-determined" configuration of the central $Re_2P_4H_4(\mu_2\text{-}H)_4$ fragment of the 30-electron $Re_2(PEt_2Ph)_4H_4(\mu_2\text{-}H)_4$ molecule. The coordination about each Re atom may be described as approximating a tricapped trigonal prism with the six hydrogens forming a distorted trigonal prism whose rectangular faces are capped by two P atoms and the other Re atom. Electronic considerations are consistent with its $M(\mu_2\text{-}H)_4M$ unit possessing a *net* M—M triple bond due to the extra electron pair occupying a dimetal-antibonding HOMO.[88]

to three delocalized dimetal-bonding MO's), the doubly degenerated HOMO's containing the extra two unpaired electrons (arising when the two Fe atoms are replaced by Co atoms) are both dimetal antibonding, which reduces the *net* M—M bond order from 3.0 to 2.0. The short Fe—Fe and significantly longer Co—Co distances of 2.332(3) and 2.377(8) Å, respectively, are in accordance with this bonding formulation.

$Re_2(PEt_2Ph)_4H_4(\mu_2\text{-}H)_4$

Both the composition and atomic arrangement of this first example of an $M(\mu_2\text{-}H)_4M$ system were revealed from a neutron diffraction analysis by Bau, Koetzle, and co-workers in 1977.[88] Its molecular geometry (FIGURE 15) of crystallographic C_i-$\bar{1}$ site symmetry may be conceptually considered as composed of two square-antiprismatic $H_2P_2ReH_4$ units sharing a common square H_4 face. Bonding considerations point to this 30-electron, tetrahydrogen-bridged dimer having a *net* Re—Re bond order of 3.0. This *net* M—M triple bond is a consequence of the extra electron pair occupying a nondegenerate HOMO which is dimetal antibonding. The M—M bond order of 4.0 resulting from the four relatively low energy 3c–2e M—H—M bonds [which under D_{2h} symmetry for the $Re_2P_4H_4(\mu_2\text{-}H)_4$ core are primary components of four delocalized dimetal-bonding MO's] is thereby decreased to a net value of 3.0. The observed Re—Re distance of 2.538(4) Å is intermediate between that of 2.241(7) Å in the Re—Re quadruple bond of the classical $[Re_2Cl_8]^{2-}$ dianion[89] and that of 2.896(3) Å in the diprotonated double bond of the $Re_2(CO)_8(\mu_2\text{-}H)_2$ molecule.[77]

COMPLEXES WITH TRIPLY BRIDGING HYDRIDE LIGANDS

The relatively few metal clusters having one or more hydrogen atoms coordinated to three transition metal atoms included the 60-electron $FeCo_3(CO)_9(P(OMe)_3)_3(\mu_2\text{-}H)$,[82,90,91] the 56-electron $Re_4(CO)_{12}(\mu_3\text{-}H)_4$[92] and 60-electron $Co_4(\eta^5\text{-}C_5H_5)_4(\mu_3\text{-}H)_4$[93] cubanelike clusters, and the 63-electron $Ni_4(\eta^5\text{-}C_5H_5)_4(\mu_3\text{-}H)_3$.[94–96] The electronic structures of $M_4(CO)_{12}H_n$ and $M_4(\eta^5\text{-}C_5H_5)_4H_n$ clusters ($n = 3, 4, 6$), which may possess edge- or face-bridging hydrogens, were theoretically analyzed by Hoffmann et al. who, in deriving the metal orbitals, considered these tetrahedral metal clusters as deprotonated species.[97] The similarity of the M—M distances in these cobalt and nickel clusters with those in the metals themselves led to the suggestion by Bau, Koetzle, and co-workers that the geometrical disposition of the hydrogen atoms above the tetrahedral metal cluster faces may parallel that of hydrogen atoms chemisorbed on metal surfaces containing triangular metal arrays.[11,12,82,90]

X-Ray and NMR data are also consistent with face-bridging hydride ligands for $HRh_3(\eta^5\text{-}C_5H_5)_3(\mu_3\text{-}C_5H_5)$,[98] for the $[(IrH_2(PCy_3)(py))_3(\mu_3\text{-}H)]^{2+}$ dication[99] and $[(IrH(dppp))_3(\mu_2\text{-}H)_3(\mu_3\text{-}H)]^{2+}$ dication,[85] and for two octahedral metal clusters, the $H_2Ru_6(CO)_{18}$ molecule[100,101] and the $[HOs_6(CO)_{18}]^-$ monoanion.[102]

$FeCo_3(CO)_9(P(OMe)_3)_3(\mu_3\text{-}H)$

Neutron diffraction measurements at 90K by Koetzle, Bau, and colleagues[82,90] substantiated the previous x-ray diffraction study at 134K by Kaesz and co-workers[91] who observed an idealized C_{3v} molecular configuration (FIGURE 16) with the hydrogen atom located outside the tetrahedral $FeCo_3$ core in a tricobalt face-capped site on the pseudo-threefold axis.

$Ni_4(\eta^5\text{-}C_5H_5)_4(\mu_3\text{-}H)_3$

This particular cluster (FIGURE 17) is of special theoretical interest because it is paramagnetic. As predicted from the earlier x-ray diffraction study,[94,95] the hydrogen atoms were found from low-temperature (81K) neutron diffraction measurements[96] to bridge three faces of the nickel tetrahedron. The resulting Ni_4H_3 core, which may be envisioned as a cubanelike molecule with one corner vacant, closely conforms to C_{3v} symmetry with both types of Ni—Ni bonds nearly indistinguishable [2.477(8) Å vs. 2.461(5) Å] but with the hydrogen atoms slightly displaced away from the unique nickel atom [1.716(3) Å vs. 1.678(6) Å]. The fact that the Ni—Ni bonds in this 63-electron molecule are longer than electron-pair Ni—Ni bonds in other organometallic species is not inconsistent with the bonding description put forth for this molecule by Hoffmann et al., who showed that the three unpaired electrons occupy a triply degenerate MO of trinickel antibonding character.[97]

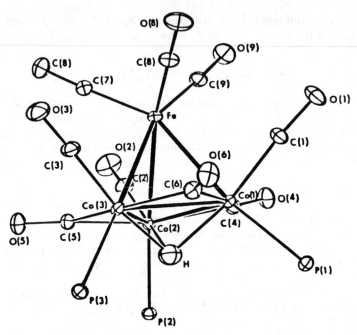

FIGURE 16. "Neutron-determined" configuration of the 60-electron $FeCo_3(CO)_9$-$(P(OMe)_3)_3(\mu_3$-H) molecule which ideally conforms to C_{3v} symmetry.[82,90]

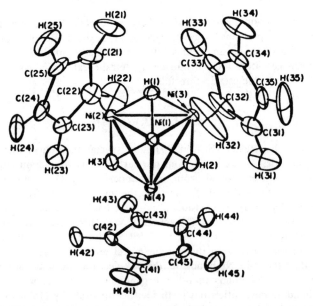

FIGURE 17. "Neutron-determined" configuration of the 63-electron $Ni_4(\eta^5\text{-}C_5H_5)_4(\mu_3\text{-H})_3$ molecule which approximates C_{3v} symmetry.[96]

METAL CLUSTERS CONTAINING INTERSTITIAL HYDROGEN LIGANDS

Although a variety of discrete metal clusters containing interstitial carbon or nitrogen atoms are known, there are still relatively few transition metal clusters that possess interstitial hydrogen atoms. The existence of such an interstitial hydrogen atom was first shown by Simon in 1967 from powder neutron diffraction measurements,[103] which indicated that HNb_6I_{11} and DNb_6I_{11} possess the nonprotonated $[Nb_6I_{11}]^-$-type geometry[104,105] with hydrogen (or deuterium) atoms occupying the octahedral Nb_6 sites. Since then, several other metal clusters have been found to possess hydrogen atoms in octahedral sites (but as yet not tetrahedral sites). These include the "neutron-determined" $[Ru_6(CO)_{18}H]^-$ mono-anion,[106-108] the "neutron-determined" $[Co_6(CO)_{15}H]^-$ monoanion,[109] and the "neutron-determined" $[Ni_{12}(CO)_{21}H_{4-n}]^{n-}$ anions $(n = 2, 3)$.[110] The hydrogen atoms in each of the "x-ray-determined" $[Rh_{13}(CO)_{24}H_{5-n}]^{n-}$ anions $(n = 2, 3)$[111,112] were not directly located but were deduced from an x-ray analysis to occupy semioctahedral (i.e., square pyramidal) cavities on the cluster surface.[112] Parallel structural-bonding relationships between the rhodium and nickel carbonyl hydride anions and simple metallic interstitial hydrides were presented by Chini et al.[113]

The $[Co_6(CO)_{15}H]^-$ Monoanion

A neutron diffraction investigation in 1979 of $[(PPh_3)_2N]^+[Co_6(CO)_{15}H]^-$ at 80K revealed unexpectedly that the hydrogen in the 86-electron monoanion is not carbonyl bound (as suggested from the unusual low-field ^1H NMR resonance of τ -13.2 ppm) but instead is located in the center of the octahedral Co_6 hole[109] (FIGURE 18). The reversible, interstitial protonation of the corresponding $[Co_6(CO)_{15}]^{2-}$ dianion[114] gives rise to a small but distinct expansion (by 2–3%) of the occupied hexacobalt octahedron. Of interest is that the "neutron-determined" 86-electron $[Ru_6(CO)_{18}H]^-$ monoanion,[108] in which the interstitial hydrogen position was initially deduced indirectly from the combined evidence of x-ray and ^1H-^{13}C NMR analyses,[106] also displays an anomalously low-field ^1H NMR signal (τ -6.4), unlike the normal high-field ^1H NMR resonances observed for the interstitial hydrogens in the $[Rh_{13}(CO)_{24}H_{5-n}]^{n-}$ anions $(n = 2, 3)$[111,115] and $[Ni_{12}(CO)_{21}H_{4-n}]^{n-}$ anions $(n = 2, 3)$.[110] The large downfield shifts of the ^1H NMR resonances in the $[Co_6(CO)_{15}H]^-$ and $[Ru_6(CO)_{18}H]^-$ monoanions have been attributed[108] to the symmetrical octahedral metal environments about the interstitial hydrogen atoms in contrast to the highly unsymmetrical metal environments about the interstitial hydrogen atoms in the rhodium and nickel anions.

The $[Ni_{12}(CO)_{21}H_{4-n}]^{n-}$ Anions (n = 2, 3)

X-Ray diffraction studies of the $[AsPh_4]^+$, $[(Ph_3P)_2N]^+$, and $[PPh_4]^+$ salts of the dihydrido dianion and of the $[AsPh_4]^+$ salt of the monohydrido trianion and subsequent neutron diffraction measurements of $[PPh_4]_2^+[Ni_{12}(CO)_{21}H_2]^{2-}$ and $[AsPh_4]_3^+[Ni_{12}(CO)_{21}H]^{3-}$ provided the first crystallographic description of the interaction of interstitial hydrogens with a close-packed fragment of metal atoms.[110] The x-ray results revealed that the anions possess the same overall D_{3h}

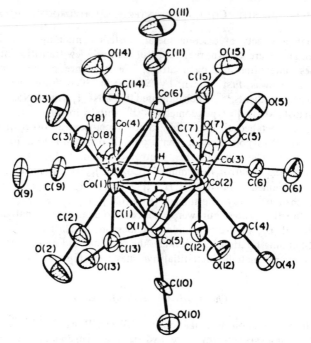

FIGURE 18. "Neutron-determined" configuration of the 86-electron $[Co_6(CO)_{15}H]^-$ mono-anion in the $[(Ph_3P)_2N]^+$ salt. The interstitial hydrogen atom is located in the center of the octahedral Co_6 core.[109]

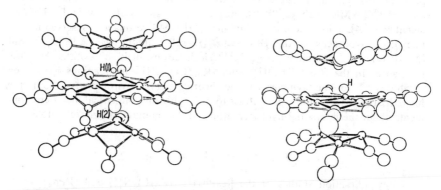

FIGURE 19. "Neutron-determined" architectures of the $[Ni_{12}(CO)_{21}H_2]^{2-}$ dianion (left) in the $[PPh_4]^+$ salt and the $[Ni_{12}(CO)_{21}H]^{3-}$ trianion (right) in the $[AsPh_4]^+$ salt. The two hydrogen atoms in the dihydrido dianion were found to occupy the two octahedral interstices, while the single hydrogen atom in the monohydrido trianion was determined to be localized in one of the two octahedral interstices instead of being crystallographically disordered in both octahedral holes.[110]

geometry (FIGURE 19) consisting of a dodecanickel framework encapsulated by 9 terminal and 12 bridging carbonyls. This 12-atom metal fragment of an hcp metal lattice contains two octahedral and six tetrahedral interstices. Neutron diffraction results not only ascertained the conclusions (based on Ni—Ni bond-length variations) of exclusive hydrogen occupation of one octahedral hole in the monohydridic trianion and both octahedral holes in the dihydridic dianion but also furnished considerable stereochemical insight concerning interstitial proton migration in reversible protonation-deprotonation reactions of these multi-hole metal clusters.

ACKNOWLEDGMENT

It is a pleasure to thank the past and present co-workers who made possible our research contributions to this important area of chemistry.

REFERENCES

1. DOEDENS, R. J. & L. F. DAHL. 1965. J. Am. Chem. Soc. **37:** 2576–2581.
2. KAESZ, H. D., W. FELLMANN, G. R. WILKES & L. F. DAHL. 1965. J. Am. Chem. Soc. **87:** 2753–2755.
3. DAHL, L. F. & J. F. BLOUNT. 1965. Inorg. Chem. **4:** 1373–1375.
4. HANDY, L. B., P. M. TREICHEL, L. F. DAHL & R. G. HAYTER. 1966. J. Am. Chem. Soc. **88:** 366–367.
5. ORPEN, A. G. 1980. J. Chem. Soc. Dalton Trans.: 2509–2516.
6. BERTOLUCCI, A., G. CIANI, M. FRENI, P. ROMITI, V. G. ALBANO & A. ALBINATI. 1976. J. Organomet. Chem. **117:** C37–C40; CIANI, G., D. GIUSTO, M. MANASSERO & A. ALBINATI. 1976. J. Chem. Soc. Dalton Trans.: 1943–1947.
7. IBERS, J. A. 1965. Annu. Rev. Phys. Chem. **16:** 375–396.
8. FRENZ, B. A. & J. A. IBERS. 1971. *In* Transition Metal Hydrides. E. L. Muetterties, Ed.: 33–74. Marcel Dekker, Inc. New York, N.Y.
9. SAILLANT, R. B. & H. D. KAESZ. 1972. Chem. Rev. **72:** 231–281.
10. HUMPHRIES, A. P. & H. D. KAESZ. 1979. Prog. Inorg. Chem. **25:** 145–222.
11. BAU, R. & T. F. KOETZLE. 1978. Pure Appl. Chem. **50:** 55–64.
12. BAU, R., R. G. TELLER, S. W. KIRTLEY & T. F. KOETZLE. 1979. Acc. Chem. Res. **12:** 176–183.
13. IBERS, J. A. 1978. Adv. Chem. Ser. **167:** 26–35.
14. CHURCHILL, M. R. 1978. Adv. Chem. Ser. **167:** 36–60.
15. TELLER, R. G. & R. BAU. 1981. Structure and Bonding **44:** 1–82.
16. ABRAHAMS, S. C., A. P. GINSBERG & K. KNOX. 1964. Inorg. Chem. **3:** 558–567.
17. LA PLACA, S. J. & J. A. IBERS. 1963. J. Am. Chem. Soc. **85:** 3501–3502.
18. LA PLACA, S. J. & J. A. IBERS. 1965. Acta Crystallogr. **18:** 511–519.
19. LA PLACA, S. J., W. C. HAMILTON, J. A. IBERS & A. DAVISON. 1969. Inorg. Chem. **8:** 1928–1935.
20. MASON, R. & D. M. P. MINGOS. 1973. J. Organomet. Chem. **50:** 53–61.
21. TEO, B. K., M. B. HALL, R. F. FENSKE & L. F. DAHL. 1974. J. Organomet. Chem. **70:** 413–420.
22. DEDIEU, A., T. A. ALBRIGHT & R. G. HOFFMANN. 1979. J. Am. Chem. Soc. **101:** 3141–3151.
23. DAVISON, A., W. McFARLANE, L. PRATT & G. WILKINSON. 1962. J. Chem. Soc.: 3653–3666.
24. HAYTER, R. G. 1963. J. Am. Chem. Soc. **85:** 3120–3124.
25. HAYTER, R. G. 1963. Inorg. Chem. **2:** 1031–1035.

26. PETERSEN, J. L., L. F. DAHL & J. M. WILLIAMS. 1974. J. Am. Chem. Soc. **96:** 6610–6620.
27. MCGAW, B. L. & J. A. IBERS. 1963. J. Chem. Phys. **39:** 2677–2684.
28. IBERS, J. A. 1964. J. Chem. Phys. **41:** 25–28.
29. HAMILTON, W. C. & J. A. IBERS. 1968. *In* Hydrogen Bonding in Solids: 108–113. W. A. Benjamin. New York, N.Y.
30. DOEDENS, R. J., W. T. ROBINSON & J. A. IBERS. 1967. J. Am. Chem. Soc. **89:** 4323–4329.
31. WILSON, F. C. & D. P. SHOEMAKER. 1957. J. Chem. Phys. **27:** 809–810.
32. PETERSEN, J. L. & J. M. WILLIAMS. 1978. Inorg. Chem. **17:** 1308–1312.
33. PETERSEN, J. L. & R. P. STEWART, JR. 1980. Inorg. Chem. **19:** 186–191.
34. HANDY, L. B., J. K. RUFF & L. F. DAHL. 1970. J. Am. Chem. Soc. **92:** 7312–7326.
35. ROZIERE, J., J. M. WILLIAMS, R. P. STEWART, JR., J. L. PETERSEN & L. F. DAHL. 1977. J. Am. Chem. Soc. **99:** 4497–4499.
36. OLSEN, J. P., T. F. KOETZLE, S. W. KIRTLEY, M. ANDREWS, D. L. TIPTON & R. BAU. 1974. J. Am. Chem. Soc. **96:** 6621–6627.
37. LOVE, R. A., H. B. CHIN, T. F. KOETZLE, S. W. KIRTLEY, B. R. WHITTLESEY & R. BAU. 1976. J. Am. Chem. Soc. **98:** 4491–4498.
38. WILSON, R. D., S. A. GRAHAM & R. BAU. 1975. J. Organomet. Chem. **91:** C49–C52.
39. HART, D. W., R. BAU & T. F. KOETZLE. (As cited in Reference 12.)
40. COOPER, C. B. III, D. F. SHRIVER & S. ONAKA. 1978. Adv. Chem. Ser. **167:** 233–247.
41. COOPER, C. B. III, D. F. SHRIVER, D. J. DARENSBOURG & J. A. FROELICH. 1979. Inorg. Chem. **18:** 1407–1408.
42. HARRIS, D. C. & H. B. GRAY. 1975. J. Am. Chem. Soc. **97:** 3073–3075.
43. ANDREWS, M. A., S. W. KIRTLEY & H. D. KAESZ. 1978. Adv. Chem. Ser. **167:** 215–231.
44. PETERSEN, J. L., P. L. JOHNSON, J. O'CONNOR, L. F. DAHL & J. M. WILLIAMS. 1978. Inorg. Chem. **17:** 3460–3469.
45. PETERSEN, J. L., R. K. BROWN, J. M. WILLIAMS & R. K. MCMULLAN. 1979. Inorg. Chem. **18:** 3494–3498.
46. PETERSEN, J. L., R. K. BROWN & J. M. WILLIAMS. 1981. Inorg. Chem. **20:** 158–165.
47. CHURCHILL, M. R. & S. W.-Y. NI. 1973. J. Am. Chem. Soc. **95:** 2150–2155.
48. CHURCHILL, M. R., B. G. DEBOER & F. J. ROTELLA. 1976. Inorg. Chem. **15:** 1843–1853.
49. CATTI, M., G. GERVASIO & S. A. MASON. 1977. J. Chem. Soc. Dalton Trans.: 2260–2264.
50. JOHNSON, B. F. G., J. LEWIS, D. PIPPARD, P. R. RAITHBY, G. M. SHELDRICK & K. D. ROUSE. 1979. J. Chem. Soc. Dalton Trans.: 616–618.
51. WEI, C. H. & L. F. DAHL. 1966. J. Am. Chem. Soc. **88:** 1821–1822.
52. WEI, C. H. & L. F. DAHL. 1969. J. Am. Chem. Soc. **91:** 1351–1361.
53. COTTON, F. A. & J. M. TROUP. 1974. J. Am. Chem. Soc. **96:** 4155–4159.
54. DESIDERATO, R., JR. & G. R. DOBSON. 1982. J. Chem. Ed. **59:** 752–756.
55. LO, F. Y.-K., G. LONGONI, P. CHINI, L. D. LOWER & L. F. DAHL. 1980. J. Am. Chem. Soc. **102:** 7691–7701.
56. CHURCHILL, M. R. & B. G. DEBOER. 1977. Inorg. Chem. **16:** 1141–1146.
57. GUY, J. J., B. E. REICHERT & G. M. SHELDRICK. 1976. Acta Crystallogr. **B32:** 3319–3320.
58. ORPEN, A. G., D. PIPPARD, G. M. SHELDRICK & K. D. ROUSE. 1978. Acta Crystallogr. **B34:** 2466–2472.
59. CHURCHILL, M. R. & R. A. LASHEWYCZ. 1979. Inorg. Chem. **18:** 848–853.
60. CHURCHILL, M. R., F. J. HOLLANDER, J. R. SHAPLEY & J. B. KEISTER. 1980. Inorg. Chem. **19:** 1272–1277.
61. BURGESS, K., B. F. G. JOHNSON, J. LEWIS & P. R. RAITHBY. 1982. J. Chem. Soc. Dalton Trans.: 263–269.
62. ALLEN, V. F., R. MASON & P. B. HITCHCOCK. 1977. J. Organomet. Chem. **140:** 297–307.
63. CHURCHILL, M. R. & R. A. LASHEWYCZ. 1979. Inorg. Chem. **18:** 1926–1930.
64. CHURCHILL, M. R. & R. A. LASHEWYCZ. 1979. Inorg. Chem. **18:** 3261–3267.
65. CHURCHILL, M. R. & B. G. DEBOER. 1977. Inorg. Chem. **16:** 2397–2403.
66. CHURCHILL, M. R. & B. G. DEBOER. 1977. Inorg. Chem. **16:** 878–884.
67. HODALI, H. A., D. F. SHRIVER & C. A. AMMLUNG. 1978. J. Am. Chem. Soc. **100:** 5239–5240.

68. COREY, E. R. & L. F. DAHL. 1962. Inorg. Chem. 1: 521–526.
69. CHURCHILL, M. R., P. H. BIRD, H. D. KAESZ, R. BAU & B. FONTAL. 1968. J. Am. Chem. Soc. 90: 7135–7136.
70. KIRTLEY, S. W., J. P. OLSEN & R. BAU. 1973. J. Am. Chem. Soc. 95: 4532–4536.
71. CHURCHILL, M. R., F. J. HOLLANDER & J. P. HUTCHINSON. 1977. Inorg. Chem. 16: 2697–2700.
72. ORPEN, A. G., A. V. RIVERS, E. G. BRYAN, D. PIPPARD, G. M. SHELDRICK & K. D. ROUSE. 1978. J. Chem. Soc. Chem. Commun.: 723–724.
73. BROACH, R. W. & J. M. WILLIAMS. 1979. Inorg. Chem. 18: 314–319.
74. CHURCHILL, M. R. & H. J. WASSERMAN. 1980. Inorg. Chem. 19: 2391–2395.
75. CHURCHILL, M. R. & S. W.-Y. CHANG. 1974. Inorg. Chem. 13: 2413–2419.
76. WEI, C.-Y., M. W. MARKS, R. BAU, S. W. KIRTLEY, D. E. BISSON, M. E. HENDERSON & T. F. KOETZLE. 1982. Inorg. Chem. 21: 2556–2565.
77. BENNETT, M. J., W. A. G. GRAHAM, J. K. HOYANO & W. L. HUTCHEON. 1972. J. Am. Chem. Soc. 94: 6232–6233.
78. ADAMS, R. D., D. M. COLLINS & F. A. COTTON. 1974. Inorg. Chem. 13: 1086–1090.
79. DAHL, L. F., E. ISHISHI & R. E. RUNDLE. 1957. J. Chem. Phys. 26: 1750–1751.
80. PREST, D. W., M. J. MAYS, P. R. RAITHBY & A. G. ORPEN. 1982. J. Chem. Soc. Dalton Trans.: 737–745.
81. BERTOLUCCI, A., M. FRENI, P. ROMITI, G. CIANI, A. SIRONI & V. G. ALBANO. 1976. J. Organomet. Chem. 113: C61–C64.
82. BAU, R., W. E. CARROLL, D. W. HART, R. G. TELLER & T. F. KOETZLE. 1978. Adv. Chem. Ser. 167: 73–92.
83. CRABTREE, R. H., H. FELKIN, G. E. MORRIS, T. J. KING & J. A. RICHARDS. 1976. J. Organomet. Chem. 113: C7–C9.
84. CRABTREE, R. 1979. Acc. Chem. Res. 12: 331–338.
85. WANG, H. H. & L. H. PIGNOLET. 1980. Inorg. Chem. 19: 1470–1480.
86. GINSBERG, A. P. & M. J. HAWKES. 1968. J. Am. Chem. Soc. 90: 5930–5932.
87. DAPPORTO, P., S. MIDOLLINI & L. SACCONI. 1975. Inorg. Chem. 14: 1643–1650.
88. BAU, R., W. E. CARROLL, R. G. TELLER & T. F. KOETZLE. 1977. J. Am. Chem. Soc. 99: 3872–3874.
89. COTTON, F. A. & R. A. WALTON. 1982. Multiple Bonds between Metal Atoms: 265–269. John Wiley & Sons. New York, N.Y.
90. TELLER, R. G., R. D. WILSON, R. K. McMULLAN, T. F. KOETZLE & R. D. BAU. 1978. J. Am. Chem. Soc. 100: 3071–3077.
91. HUIE, B. T., C. B. KNOBLER & H. D. KAESZ. 1975. J. Chem. Soc. Chem. Commun.: 684–685.
92. WILSON, R. D. & R. BAU. 1976. J. Am. Chem. Soc. 98: 4687–4689.
93. HUTTNER, F. & H. LORENZ. 1975. Chem. Ber. 108: 973–983.
94. MÜLLER, J., H. DORNER, F. HUTTNER & H. LORENZ. 1973. Angew. Chem. Int. Ed. 12: 1005–1006.
95. HUTTNER, G. & H. LORENZ. 1974. Chem. Ber. 107: 996–1008.
96. KOETZLE, T. F., J. MÜLLER, D. L. TIPTON, D. W. HART & R. BAU. 1979. J. Am. Chem. Soc. 101: 5631–5637.
97. HOFFMANN, R., B. E. R. SCHILLING, R. BAU, H. D. KAESZ & D. M. P. MINGOS. 1978. J. Am. Chem. Soc. 100: 6088–6093.
98. MILLS, O. S. & E. F. PAULUS. 1968. J. Organomet. Chem. 11: 587–594.
99. CHODOSH, D. F., R. H. CRABTREE, H. FELKIN & G. E. MORRIS. 1978. J. Organomet. Chem. 161: C67–C70.
100. CHURCHILL, M. R., J. WORMALD, J. KNIGHT & M. J. MAYS. 1970. Chem. Commun.: 458–459.
101. CHURCHILL, M. R. & J. WORMALD. 1971. J. Am. Chem. Soc. 93: 5670–5677.
102. McPARTLIN, M., C. R. EADY, B. F. G. JOHNSON & J. LEWIS. 1976. J. Chem. Soc. Chem. Commun.: 883–885.
103. SIMON, A. 1967. Z. Anorg. Allg. Chem. 355: 311–322.
104. BATEMAN, L. R., J. F. BLOUNT & L. F. DAHL. 1966. J. Am. Chem. Soc. 88: 1082–1084.
105. SIMON, A., H.-G. V. SCHNERING & H. SCHÄFER. 1967. Z. Anorg. Allg. Chem. 355: 295–310.

106. EADY, C. R., B. F. G. JOHNSON, J. LEWIS, M. C. MALATESTA, P. MACHIN & M. MCPARTLIN. 1976. J. Chem. Soc. Chem. Commun.: 945–946.
107. EADY, C. R., P. F. JACKSON, B. F. G. JOHNSON, J. LEWIS, M. C. MALATESTA, M. MCPARTLIN & W. J. H. NELSON. 1980. J. Chem. Soc. Dalton Trans.: 383–392.
108. JACKSON, P. F., B. F. G. JOHNSON, J. LEWIS, P. R. RAITHBY, M. MCPARTLIN, W. J. H. NELSON, K. D. ROUSE, J. ALLIBON & S. A. MASON. 1980. J. Chem. Soc. Chem. Commun.: 295–297.
109. HART, D. W., R. G. TELLER, C.-Y. WEI, R. BAU, G. LONGONI, S. CAMPANELLA, P. CHINI & T. F. KOETZLE. 1979. Angew. Chem. Int. Ed. 18: 80–81.
110. BROACH, R. W., L. F. DAHL, G. LONGONI, P. CHINI, A. SCHULTZ & J. M. WILLIAMS. 1978. Adv. Chem. Ser. 167: 93–110.
111. ALBANO, V. G., A. CERIOTTI, P. CHINI, G. CIANI, S. MARTINENGO & W. M. ANKER. 1975. J. Chem. Soc. Chem. Commun.: 859–860.
112. ALBANO, V. G., G. CIANI, S. MARTINENGO & A. SIRONI. 1979. J. Chem. Soc. Dalton Trans.: 978–982.
113. CHINI, P., G. LONGONI, S. MARTINENGO & A. CERIOTTI. 1978. Adv. Chem. Ser. 167: 1–10.
114. ALBANO, V. G., P. CHINI & V. SCATTURIN. 1968. J. Organomet. Chem. 15: 423–432.
115. MARTINENGO, S., B. T. HEATON, R. J. GOODFELLOW & P. CHINI. 1977. J. Chem. Soc. Chem. Commun.: 39–40.

REACTIVITY PATTERNS OF TRANSITION METAL POLYHYDRIDES

Kenneth G. Caulton

Department of Chemistry
Indiana University
Bloomington, Indiana 47405

HIGH-VALENT TRANSITION METAL POLYHYDRIDES

The majority of both fundamental and applied studies of transition metal hydride chemistry are based upon compounds with a metal oxidation state of two and below. Work with the Rh^I/Rh^{III} and Ir^I/Ir^{III} couples exceeds this threshold, but these trivalent complexes generally have no more than two hydride ligands. For the purposes of this discussion, we will arbitrarily refer to L_mMH_n complexes with $n \geq 3$ as *high-valent* polyhydrides. As shown in TABLE 1, compounds of this stoichiometry are abundant. Moreover, they dramatically illustrate the broad range of oxidation states tolerated by phosphine and also hydride ligands. The compounds shown have been isolated for a range of tertiary phosphine ligands,[1] but little is known concerning their reactivity. All compounds shown possess an 18 valence electron configuration. Of particular interest are the low d-electron populations, including d^2 and even d^0. We began this study anticipating that high-valent hydride complexes might exhibit new organometallic reaction patterns (and perhaps mechanisms) involving the catalytically relevant hydride ligand. For example, is it possible that activation of aliphatic hydrocarbons can be achieved by *electrophilic* attack? This is suggested by the successful H/D exchange using Pt(II)/Pt(IV) in acidic medium.[2] Since we have the potential for generating *formally* high-valent unsaturated polyhydrides, will these effect C—H and C—C bond scission? Our work has thus been directed toward exploring and hopefully exploiting the potentially unique reactivity of high-valent metal polyhydrides.

$ReH_5(PMe_2Ph)_3$

Our work with high-valent polyhydrides began with ReH_5P_3 ($P \equiv PMe_2Ph$). The structure of this complex is unknown, but we have shown, by ^1H and ^{31}P NMR at $-130°C$, that the molecule possesses mirror symmetry. A coordination geometry based on a dodecahedron is most consistent with these observations:

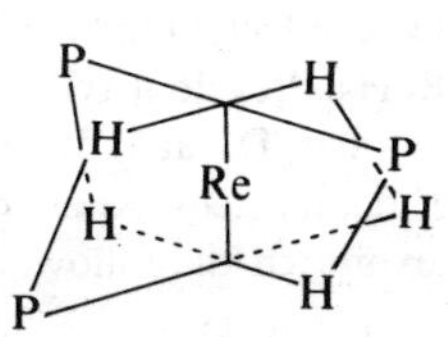

Our reactivity studies employ photochemical activation since ReH_5P_3 is thermally quite stable (e.g., no reaction of $ReH_5(PMe_2Ph)_3$ with pyridine after three hours reflux in acetone!). Our continuous photolysis studies[3] show that $\lambda > 310$ nm radiation efficiently ejects one phosphine, to yield the 16-electron species ReH_5P_2.

27

TABLE 1
SOME KNOWN TRANSITION METAL POLYHYDRIDES*

Configuration	Group			
	V	VI	VII	VIII
d^8				$(HIrP_4)$
d^6			$(HReP_5)$	(H_2OsP_4)	H_3IrP_3
d^4		(H_2MP_5)	H_3ReP_4	H_4OsP_3	H_5IrP_2
d^2		H_4MP_4	H_5ReP_3	H_6OsP_2	
d^0	$H_5Ta(dmpe)_2$	H_6WP_3	H_7ReP_2		

* $P \equiv PMe_2Ph$. M = Mo, W.

The phototransient ReH_5P_2 has been shown[3,4] to:

● Undergo dimerization to $Re_2H_8P_4$. Collaborative conventional and laser flash photolysis studies (with G. Ferraudi)[5] reveal that this proceeds by reaction of ReH_5P_2 with ReH_5P_3 and not by dimerization of two transients.

● Undergo exchange of Re—H bonds with the ring hydrogens of benzene, toluene, naphthalene, and PMe_2Ph; in the last case, only *meta*- and *para*-hydrogens exchange.

● Catalytically hydrogenate 1-hexene at 20°C under 1 atm H_2.

● From stable olefin complexes with cyclopentene and 1,4-cyclooctadiene.

● From $(C_5H_5)ReH_4P$ and $(C_5H_5)ReH_2P_2$ with cyclopentadiene.

● From $(\eta^4\text{-}C_8H_{10})ReH_3P_2$ and then $(\eta^5\text{-}C_8H_{11})ReH_2P_2$ by sequential metal-to-ring hydrogen transfer (*endo* stereochemistry established) upon reaction with cyclooctatetraene. The crystal structure of the final product (FIGURE 1) shows inequivalent hydrides, which are *not* averaged by ring rotation at 70°C.

● Hydrogenate *t*-butyl ethylene to leave a highly unsaturated species (ReH_3P_2??), which can scavenge N_2 [to give $ReN_2(PMe_2C_6H_4)(PMe_2Ph)_3$, FIGURE 2] or benzene [to give $(\eta^6\text{-}C_6H_6)ReP_2(CH_2CH_2{}^tBu)$].

In view of the high reactivity of ReH_5P_2 toward $H—C(sp^2)$ bonds demonstrated above, it was disappointing for us to find that transient $ReH_5(PMe_2Ph)_2$ does not exchange (or react) with cyclohexane-d_{12}. On the other hand, Ferraudi's flash photolysis study revealed unexpected transients,[5] which were proposed to involve π donation to rhenium of aryl ring $C=C$ unsaturation of coordinated PMe_2Ph in a *meta*-stable $ReH_5(PMe_2Ph)_2$. Such internal stabilization by pendant π-electron density is obviously undesirable and suggested that we should attempt to incorporate a trialkylphosphine ligand into a polyhydride. We selected $P(cyclohexyl)_3$ for the added reason that steric effects might allow *thermal* generation of a 16-electron species. In fact, $ReH_7[P(cyclohexyl)_3]_2$ was obtainable and indeed undergoes ReH/CD exchange with C_6D_6 at 80°C. Of interest is the fact that the cyclohexyl rings also become deuterated. After one hour at 80°C in C_6D_6, deuteration was complete and exhibited the following regiospecificity:

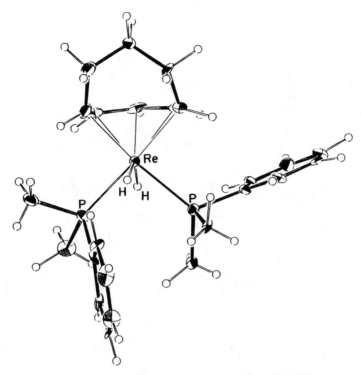

FIGURE 1. Molecular structure of $(\eta^5\text{-}C_8H_{11})ReH_2(PMe_2Ph)_2$. Unlabeled atoms are carbon and hydrogen.

Particularly noteworthy is the fact that only C2 and C3 are deuterated, and each is only monodeuterated. These results are consistent with a wholly intramolecular mechanism, involving only four- and five-membered rings.

$OsH_4(PMe_2Ph)_3$

We have recently extended our studies to include the polyhydride OsH_4P_3. We have recorded the ^1H NMR (two Os—H chemical shifts) and the $^{31}P\{^1H\}$ NMR (a 1:2 intensity pattern) of OsH_4P_3, both at $-120°C$. These data are in agreement with the established pentagonal bipyramidal structure I (by neutron diffraction),[6] although the hydride and phosphine ligands are fluxional above $-60°C$. Photolysis ($\lambda < 300$ nm) of a benzene or tetrahydrofuran (THF) solution of

$$\begin{array}{c} H \quad P \quad H \\ \diagdown \; | \; \diagup \\ P-Os \\ \diagup \; | \; \diagdown \\ H \quad P \quad H \end{array}$$

I

OsH_4P_3 produces an extraordinary array of products. A major product is $(OsH_2P_3)_2$, a dimer of what we feel is the primary photoproduct, the transient OsH_2P_3.

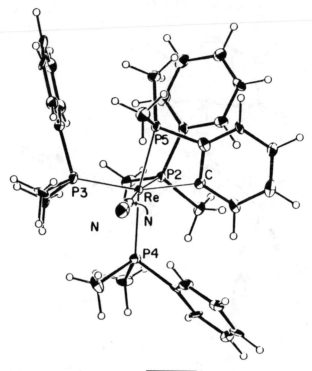

FIGURE 2. Molecular structure of ReN$_2$(PMe$_2$C$_6$H$_4$)(PMe$_2$Ph)$_3$. P5 indicates the phosphorus ligand that has been *ortho*-metallated. Unlabeled atoms are C and H.

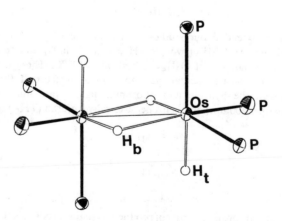

FIGURE 3. Inner coordination sphere of Os$_2$H$_4$(PMe$_2$Ph)$_6$.

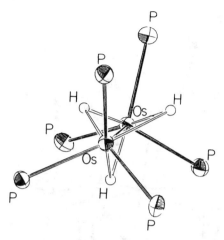

FIGURE 4. Inner coordination sphere of $Os_2H_3(PMe_2Ph)_6{}^+$. The *fac*-H_3OsP_3 octahedral geometry is apparent.

Evidence supporting this claim comes from the fact that photolysis of OsH_4P_3 in the presence of three equivalents of added PMe_2Ph gives *cis*-OsH_2P_4 as the *only* photoproduct. The dimer $(OsH_2P_3)_2$ has both bridging and terminal hydrides (FIGURE 3); the Os/Os distance (2.818 Å) is consistent with the presence of an Os=Os double bond.[7]

Pure solid $(OsH_2P_3)_2$ displays unanticipated behavior in noncoordinating solvents: phosphine dissociation (Equation 1). This is an authentic equilibrium process, which can be shifted to the right by heating or removal of phosphine under vacuum and to the left by adding phosphine. A saturated solution of pure solid $Os_2H_4P_6$ has nearly equal amounts of $Os_2H_4P_6$ and $Os_2H_4P_5$. Since coordinative unsaturation is rarely if ever found in isolable osmium hydrido

$$Os_2H_4P_6 \rightleftharpoons Os_2H_4P_5 + P \qquad (1)$$

phosphine complexes, we conclude that $Os_2H_4P_5$ contains an Os≡Os triple bond and that phosphine dissociation is made possible because of the potential for such internal stabilization. Neutral ligand dissociation/association of the sort in Equation 1 may provide applications in stoichiometric or catalytic substrate modification.

The photochemistry of OsH_4P_3 exhibits another unusual product. Irradiation of a dilute benzene solution of this complex results in precipitation of a yellow crystalline benzene-insoluble product, which has a single line $^{31}P\{^1H\}$ NMR spectrum upon redissolving in THF. This material is a salt containing the cation $Os_2(\mu\text{-}H)_3(PMe_2Ph)_6{}^+$ (FIGURE 4). The Os/Os distance in this cation, 2.558 Å, is consistent with the presence of an Os≡Os bond. This compound is thus a formal analogue of $Os_2H_4P_5$ (replace one P in FIGURE 4 by H^-); such a structure is in agreement with spectral data on $Os_2H_4P_5$ (e.g., an A_2X_3 $^{31}P\{^1H\}$ spectrum). While $Os_2H_3P_6{}^+$ is also formed on treatment of $Os_2H_4P_6$ with HBF_4, we feel that this is not the source of the dimeric monocation in our photolysis studies: intentional addition or removal of sources of H^+ to our photolysis vessels does not alter the

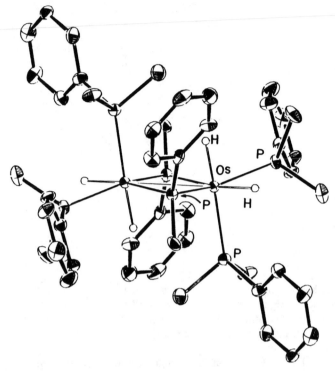

FIGURE 5. Perspective view of $Os_2H_4(\mu\text{-}PMePh)_2(PMe_2Ph)_4$ with "organic" hydrogens deleted.

yield of this product. Formation of an ionic product is unprecedented in transition metal hydride photochemistry, particularly in a nonpolar solvent. The fate of the implied H^- that has been transferred in this reaction currently occupies our attention.

A final surprise in the photolysis of $OsH_4(PMe_2Ph)_3$ is the isolation of a product with (among others) a far downfield ^{31}P chemical shift. This neutral compound also exhibits a very complex set of P—Me proton resonances: five chemical shifts! An x-ray diffraction study of this complex identifies it as $Os_2H_4(\mu\text{-}PMePh)_2(PMe_2Ph)_4$ (FIGURE 5). An 18-electron configuration at osmium requires an Os—Os single bond. It is in just such a situation that large downfield ^{31}P chemical shifts are found.[8] The noteworthy feature of this product is of course the loss of P-methyl groups. This reaction involves, to our knowledge, the mildest conditions yet observed for P—CH_3 bond rupture. All previous examples have employed either high temperature (above 150°C) or alkali metal reductants. This result speaks for the high reactivity of the unsaturated hydride complex produced photochemically.

In summary, we have shown that photolysis is an effective means for transforming thermally inert polyhydride complexes into transients capable of reacting with relatively inert hydrocarbon substrates. Moreover, the method produces progeny, in the sense that, under controlled conditions, photolysis offers a route for condensation of monomeric polyhydrides to hydride-rich dimers.

Cu(I) Hydrides

We consider that there is ample reason to expand transition metal hydride chemistry to the late transition elements, particularly copper. This arises from our earlier activity on the topic of metal-catalyzed synthesis gas $(CO + H_2)$ conversions.[9] In particular, our work with early transition elements suggested that such reductions of CO proceed to make M—O bonds, which are then resistant to rupture by the available reductants, CO and H_2. Our search for a more suitable metallic element upon which to base our continued studies was influenced by several factors. The first was the fact that the "low-pressure" conversion of syngas to methanol is based on a mixed copper/zinc oxide catalyst.[10] The production of an oxygenate product (i.e., retention of the C—O bond) was consistent with the fact that solid CuO is readily reduced to the metal ($\sim 100°C$, 1 atm). Moreover, Cu(I) is known to form carbonyl complexes (e.g., the COSORB process for removal of CO from gas streams),[11] albeit weak ones. Such complexes are appealing since they are labile; ready access of ligands (i.e., substrate) to the coordination sphere is a feature absent among the binary carbonyl clusters of the third transition series [e.g., $Os_3(CO)_{12}$].

We felt at the outset that it was essential to establish that "hydrogenolysis" of a Cu—O bond was indeed possible using H_2 (Equation 2), since such a reaction is precisely the reverse of the behavior expected of early transition elements. If

$$Cu—OR + H_2 \longrightarrow CuH + ROH \qquad (2)$$

successful, this would demonstrate a likely final step in an alcohol synthesis and would simultaneously represent a new synthetic method. The accessibility of copper hydrides for further study was an attractive potential dividend of this study.

We found that $(CuO^tBu)_4$, a compound chosen for its solubility in inert solvents, reacts promptly with H_2 (1 atm) in the presence of PPh_3 at 25°C to give the known compound $(HCuPPh_3)_6$.[12] All hydride ligands are bridging in this complex. The reaction (Equation 3) is quite general for a range of monodentate

$$\tfrac{6}{4}(CuO^tBu)_4 + 6 L + H_2 \longrightarrow (HCuL)_6 + 6\ ^tBuOH] \qquad (3)$$

phosphines, and we were pleased to be able to locate, for the first time, the ^1H NMR resonance of these fluxional polyhydride clusters. The hydride chemical shift lies 2–3.5 ppm *downfield* of TMS, thus suggesting a pattern that both d^0 [e.g., Zr(IV)] and d^{10} complexes tend to have hydride chemical shifts in the "normal" region.

While the synthesis in Equation 3 is quite general, we observed a strong dependence of the reaction rate on the mole ratio R_3P/Cu employed during the reaction. This necessitated an examination of the phosphine binding equilibria of $(CuO^tBu)_4$ in THF. Using ^{31}P NMR and also x-ray diffraction of isolated solids, we find that the 1:1 complex is dimeric: $Cu_2(\mu-O^tBu)_2(PPh_3)_2$. Additional equivalents of PPh_3 give $Cu_2(O^tBu)_2(PPh_3)_3$ and then monomeric $Cu(PPh_3)_2(O^tBu)$. The latter compound shows no tendency to add a third phosphine. Consequently,

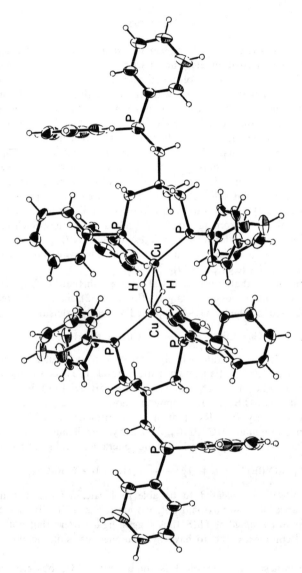

FIGURE 6. Molecular structure of $Cu_2(\mu\text{-}H)_2[\eta^2\text{-}MeC(CH_2PPh_2)_3]_2$.

several coordinatively unsaturated complexes are available for reaction with H_2. In contrast, the smaller ligand $P(OPh)_3$, when employed at a P/Cu ratio of 3:1, shows no reaction with H_2; at a 1:1 ratio, reaction occurs. An attempt to make $(HCuCO)_n$ by hydrogenolysis of $Cu_4(CO)_4(\mu_3\text{-}O^tBu)_4$ yields only unreacted reagent complex; copper is saturated in this complex. Taken together, these results indicate that hydrogenolysis succeeds only when unsaturated Cu(I) and/or an alkoxy lone pair is available. This is consistent with either a mechanism involving oxidative addition to give a Cu(III) dihydride or a transition state involving heterolytic splitting of H_2 (**II**).

$$
\begin{array}{ccc}
 & & \overset{\delta-}{} \quad \overset{\delta+}{} \\
H\!-\!H & & H \cdots H \\
 & & \vdots \qquad \vdots \\
\longrightarrow & & \vdots \qquad \vdots \\
Cu\!-\!\ddot{O}R & & Cu \cdots \ddot{O}R
\end{array}
$$

II

The complex $(HCuPPh_3)_6$ reacts with CO_2 (1 atm, 25°C). Cleanest reactions result when the reaction is carried out in the presence of added PPh_3, yielding $Cu(\eta^2\text{-}O_2CH)(PPh_3)_2$. Formaldehyde is catalytically converted to methyl formate by $(HCuPPh_3)_6$, a reaction that is initiated by the formal addition of the Cu—H bond across the $O\!=\!CH_2$ bond. $(DCuPPh_3)_6$ gives initially HCO_2CH_2D. Such a result aids in understanding the selectivity of the (Cu,Zn)O catalyst for a C_1 product.

Since we find $(HCuPPh_3)_6$ to be unreactive toward CO, we sought a copper complex with a terminally bound hydride as potentially more reactive. To this end, we sought to block aggregation (by hydride bridging) by employing a tridentate ligand in Equation 3. This reaction succeeds, using $MeC(CH_2PPh_2)_3$, to give a complex of empirical formula $HCu(Ph_2PCH_2)_3CMe$. However, the $^{31}P\{^1H\}$ NMR spectrum of this complex exhibits *two* resonances (intensity 2:1), the latter having a chemical shift close to that of the free ligand. The crystal structure of this material shows it to be a dimer in which one "arm" of each ligand is rejected by copper in favor of hydride bridging (FIGURE 6). We can be certain that this result is a consequence of thermodynamic, not steric, factors since the complex reacts with CO_2 to give $Cu(\eta^1\text{-}O_2CH)[\eta^3\text{-}(Ph_2PCH_2)_3CMe]$.

Our results to date in copper hydride chemistry justify optimism that continued work is warranted. Equation 3 is a "double dividend" result. It establishes that the key difficulty of early transition metal/carbon oxide chemistry is eliminated for copper. At the same time, it provides us with a route to copper hydrides with a range of ligand sets. What is currently lacking is a demonstrated attack of such a hydride upon CO. Our current efforts in this direction center on mixed-valence hydrides and mixed Cu/Zn hydrido carbonyl complexes.

ACKNOWLEDGMENTS

This work was supported financially by the National Science Foundation (CHE 80–06331), materially by Cleveland Refractory Metals, and manually and spiritually by the dedicated efforts of David DeWit, Kirsten Folting, Rolf Geerts, Mark Green, Gary Goeden, Scott Horn, John Huffman, Timothy Lemmen, Michaeleen Trimarchi, and Kelle Zeiher.

REFERENCES

1. MUETTERTIES, E. L. 1971. Transition Metal Hydrides. Marcel Dekker. New York, N.Y.
2. WEBSTER, D. E. 1977. Adv. Organomet. Chem. **15**: 147.
3. GREEN, M. A., J. C. HUFFMAN & K. G. CAULTON. 1981. J. Am. Chem. Soc. **103**: 695.
4. GREEN, M. A., J. C. HUFFMAN, K. G. CAULTON, W. K. RYBAK & J. J. ZIOLKOWSKI. 1981. J. Organomet. Chem. **218**: C39.
5. MURALIDHARAN, S., G. FERRAUDI, M. A. GREEN & K. G. CAULTON. 1983. J. Organomet. Chem. **244**: 47.
6. HART, D., R. BAU & T. KOETZLE. 1977. J. Am. Chem. Soc. **99**: 7557.
7. GREEN, M. A., J. C. HUFFMAN & K. G. CAULTON. 1983. J. Organomet. Chem. **243**: C78.
8. KEITER, R. L. & M. J. MADIGAN. 1982. Organometallics **1**: 409.
9. CAULTON, K. G. 1981. J. Mol. Catal. **13**: 71.
10. KUNG, H. L. 1980. Catal. Rev. **22**: 235.
11. WALKER, D. G. 1975. Chemtech. **5**: 308.
12. CHURCHILL, M., S. A. BEZMAN, J. A. OSBORN & J. WORMALD. 1972. Inorg. Chem. **11**: 1818.

HYDROGEN AND CARBON MONOXIDE AS LIGAND COMBINATION FOR TRANSITION ELEMENTS*

Fausto Calderazzo

Institute of General Chemistry
University of Pisa
56100 Pisa, Italy

INTRODUCTION

An important class of reactions catalyzed by transition metals, either heterogeneously or homogeneously, is the reduction of carbon monoxide by molecular hydrogen to give methanol,[1] Fischer-Tropsch products,[2] or ethylene glycol (see Reference 3 and references therein). In the course of these reactions, both hydrogen and carbon monoxide can bind to the metallic catalyst. This is the reason why hydrido metal carbonyls have attracted considerable interest.[4] Moreover, these compounds may be involved in other catalytic reactions, such as the hydroformylation of olefins:[5] in the latter case, $CoH(CO)_4$ or its 16-electron dissociative product $CoH(CO)_3$ has been suggested to be the actual catalyst.[6,7] In view of this interest, it is appropriate to consider somewhat in detail the class of transition metal complexes having hydrogen and CO as ligands. For simplicity we shall consider the monomeric products of general formula $MH(CO)_n$ only, since it is believed that a similar systematic approach can be used in the case of more complex substances of the same class.

In spite of the fact that the first compound of this class was discovered more than 50 years ago, i.e., $FeH_2(CO)_4$,[8] followed by $CoH(CO)_4$ in the next few years,[9-12] and that a huge amount of research work has been devoted to these compounds,[13-16] some fundamental questions concerning them are still unanswered. TABLE 1 reports the compounds that have been isolated and those whose existence is possible on the basis of the inert gas rule. It is interesting to note that the isoelectronic $CoH(CO)_4$ and $[FeH(CO)_4]^-$, once prepared by an independent synthetic procedure, decompose to give the corresponding dimeric 18-electron species:

$$2\ CoH(CO)_4 \rightleftharpoons H_2 + Co_2(CO)_8 \tag{1}$$

$$2\ [FeH(CO)_4]^- \longrightarrow H_2 + [Fe_2(CO)_8]^{2-} \tag{2}$$

The dimeric neutral $Ni_2H_2(CO)_6$, rather than the $[NiH(CO)_3]^-$ anion, has been reported to result from the reaction of $Ni(CO)_4$ with sodium in liquid ammonia.[23]

Some uncertainties still exist about metal carbonyl hydrides because of limited knowledge about some fundamental properties of these compounds. It is the purpose of this paper to report some recent experimental results obtained with

* The author wishes to thank the National Research Council (CNR, Rome) and the Ministry of Education for financial support.

TABLE 1

MONONUCLEAR 18-ELECTRON METAL CARBONYL HYDRIDES OF THE 3d SEQUENCE, OF GENERAL FORMULA MH(CO)$_n$

VH(CO)$_6$	[CrH(CO)$_5$]$^-$	MnH(CO)$_5$	[FeH(CO)$_4$]$^-$
1*	2†	3†	4†
	CoH(CO)$_4$	[NiH(CO)$_3$]$^-$	
	5‡	6*	

* Not isolated.
† References 17–22.
‡ References 9–12.

early transition elements (V, Nb, Ta) and to indicate possible ways to overcome the limited knowledge about them.

It is necessary to point out that the formal assignment of the −I oxidation state to the hydrogen ligand of MH(CO)$_n$ compounds originates from the consideration that the electronegativity of hydrogen is higher than that of any metal.[24] The classification of these compounds as hydrides could be, however, an oversimplification, since the electronegativity of the metal is a function of the oxidation state, of the electronic configuration, and of the type of ligand environment. The problem of charge distribution within the M—H bond is still not completely solved. The reactivity of these compounds is still another problem, which requires further consideration.

MH(CO)$_n$ COMPLEXES AND EARLY TRANSITION ELEMENTS

Vanadium is, among the elements of group V, the most studied as far as carbonyl complexes are concerned. The neutral and the anionic derivatives, V(CO)$_6$ and [V(CO)$_6$]$^-$, are well established (see Reference 25 and references therein). In contrast, for niobium and tantalum, only the hexacarbonylmetalates(−I) have been isolated.[26–30] However, the literature mentions the use of ether solutions of VH(CO)$_6$ (see TABLE 1) for their further reaction, primarily with olefin substrates.[31,32] In view of the low accessibility of seven coordination for vanadium, some doubt existed about a formulation for the compound that would assign a definite coordination position to the hydrogen around the central metal atom. It was thus worth inquiring further about the properties of this compound. Treatment of anhydrous NaV(CO)$_6$ with dry HCl in hydrocarbon solvents was considered to be the most appropriate method to isolate the compound:[25]

$$\text{NaV(CO)}_6 + \text{HCl} \longrightarrow \text{``VH(CO)}_6\text{''} + \text{NaCl} \qquad (3)$$

In contrast to its remarkable inertness toward diluted or even neat CH$_3$COCl or CH$_3$I,[33] NaV(CO)$_6$ reacts promptly with the stoichiometric amount of HCl even at temperatures as low as −80°C. The orange† solutions resulting from this

---

† Hexacarbonylvanadium(0) is a black green solid. In the noncondensed phase [aliphatic hydrocarbon solution, solid solution with Mo(CO)$_6$, gas phase], its color is yellow to orange depending on the concentration.[33]

treatment were examined in the IR carbonyl stretching region, but no carbonyl derivatives other than $V(CO)_6$ ($\bar{v}_{CO}$, 1973/cm, heptane) could be detected. The reaction of $NaV(CO)_6$ with HCl was accompanied by the evolution of H_2, so that the stoichiometry of the reaction can be represented by the following equation:

$$NaV(CO)_6 + HCl \longrightarrow \tfrac{1}{2} H_2 + V(CO)_6 + NaCl \qquad (4)$$

Prior to the formation of $V(CO)_6$, a thermally unstable yellow solid, poorly soluble in saturated hydrocarbons, could be observed in suspension. Attempts to isolate this solid in admixture with NaCl failed due to fast decomposition (gas evolution and formation of $V(CO)_6$) starting from about $-30°C$. About the nature of this precursor to $V(CO)_6$, three tentative suggestions can be made, namely, Formulas 7–9. In view of the experimentally observed low solubility of the precursor and of the already mentioned low accessibility of seven coordination for low-valent vanadium [hexacarbonylvanadium(0) itself is monomeric in the solid state, as shown by results from a recent x-ray investigation],[34] possibility 7 is the least likely one. As far as 8 and 9 are concerned, organometallic compounds containing the $M-CH_2OH$, $M=CHOH$, $M\equiv COH$,[35] and $M-C(O)H$[36,37] groups are of low stability. Structures 8 and 9 would originate from the protonation of $[V(CO)_6]^-$ at the carbonyl carbon and at the carbonyl oxygen, respectively. There are no precedents in the literature of protonation of carbonyl carbon to give formyl derivatives,

$$H-V(CO)_6 \qquad\qquad H(O)C-V(CO)_5 \qquad\qquad H\equiv C-V(CO)_5$$

$$\textbf{7} \qquad\qquad\qquad\qquad \textbf{8} \qquad\qquad\qquad\qquad \textbf{9}$$

while protonation of carbonyl oxygen is well established for metal carbonyl clusters containing triply[38] and doubly[39] bridging carbonyl groups (see below). The usual mode of decomposition[36,37] of formyl derivatives of transition metals is the reverse of the insertion of CO[40] in between the $M-H$ bond (a still rather elusive reactive mode for carbon monoxide‡):

$$M-C(O)H \longrightarrow M(CO)H \qquad\qquad (5)$$

Homolytic cleavage of the $V-H$ bond would then give rise to the observed formation of molecular hydrogen.

Only in the presence of an appropriate Lewis base is the formation of $V(CO)_6$ according to Equation 4 completely suppressed. The most interesting and common base is, of course, water. Acidification of $NaV(CO)_6$ in the presence of a substantially stoichiometric amount of water in diethyl ether leads to a completely different behavior, with formation of $H_3O^+[V(CO)_6]^-$:

$$[V(CO)_6]^- + HCl + H_2O \longrightarrow H_3O^+[V(CO)_6]^- + Cl^- \qquad (6)$$

---

‡ An example of CO insertion in between a metal-hydrogen bond has recently been reported for a rhodium(III) porphyrin complex.[41,42] It is, however, to be pointed out that the $Rh-H + CO \rightarrow Rh-C(O)H$ conversion is probably intermolecular in nature in view of the rigidity of the porphyrin ligand in the tetragonal plane of the pseudooctahedron. If this is not so, this example would violate the general rule that *intramolecular* CO insertions in between M—R bonds in square or octahedral complexes should occur by elementary steps within two coordination *cis* positions, no matter whether the mechanism is CO insertion[43–46] or alkyl migration.[47]

This study has therefore established that (a) "VH(CO)$_6$" is a thermally unstable compound rapidly decomposing to V(CO)$_6$ in the absence of a Lewis base; (b) solutions of "VH(CO)$_6$" should be regarded as containing $H_3O^+[V(CO)_6]^-$, due to the presence of adventitious water; and (c) "VH(CO)$_6$" appears to be an extremely strong acid in aqueous solution:

$$VH(CO)_{6\,(aq)} \longrightarrow H_3O^+_{(aq)} + [V(CO)_6]^-_{(aq)} \tag{7}$$

The latter result is in complete agreement with the earlier finding[48] that the titration curve of aqueous solutions of $H_3O^+[V(CO)_6]^-$ is substantially superimposable on that of HCl and that aqueous solutions of NaV(CO)$_6$ do not undergo hydrolysis to any appreciable extent.[25]

THE ACID STRENGTH OF MH(CO)$_n$ COMPLEXES IN WATER

It is interesting to compare VH(CO)$_6$ with other carbonyl derivatives of the same class, namely, MnH(CO)$_5$ and CoH(CO)$_4$. By taking our most recent results into consideration, the acid strengths in water decrease in the following order:

$$VH(CO)_6 > CoH(CO)_4 > MnH(CO)_5$$

It would be highly desirable to understand the basic reasons for such a trend. The only way to approach this problem from a quantitative point of view is to consider the ionic dissociation, Equation 8 of SCHEME 1, of a given MH(CO)$_n$ complex as the result of the corresponding elementary thermodynamic processes, as shown in SCHEME 1.

$$MH(CO)_{n(aq)} \longrightarrow MH(CO)_{n(g)} \tag{8a}$$

$$MH(CO)_{n(g)} \longrightarrow M(CO)_{n(g)} + H_{(g)} \tag{8b}$$

$$M(CO)_{n(g)} + e^- \longrightarrow [M(CO)_n]^-_{(g)} \tag{8c}$$

$$H_{(g)} \longrightarrow H^+_{(g)} \tag{8d}$$

$$H^+_{(g)} \longrightarrow H^+_{(aq)} \tag{8e}$$

$$[M(CO)_n]^-_{(g)} \longrightarrow [M(CO)_n]^-_{(aq)} \tag{8f}$$

$$MH(CO)_{n(aq)} \longrightarrow H^+_{(aq)} + [M(CO)_n]^-_{(aq)} \tag{8}$$

SCHEME 1. Acid dissociation in water of MH(CO)$_n$ complexes.

The main problem associated with a complete understanding of the thermodynamics of the dissociation process is our limited knowledge of the energetics of Processes 8b and 8c. The situation is particularly unsatisfactory since these are, presumably, the deciding factors as far as the driving force of the whole process is concerned. The hydration energy of $[M(CO)_n]^-_{(g)}$, Process 8f, should be relatively small and, moreover, not largely different from one metal to the other. Acid dissociation will be favored by a largely negative enthalpy change for Process 8c. The latter process corresponds to the electron capture by the 17-electron species resulting from **8b** to give the corresponding 18-electron compound, the electron-

TABLE 2
METAL-HYDROGEN BOND DISSOCIATION ENERGIES

Bond	Dissociation Energy (kcal/mol)	Reference
$Cr^+—H$	35 ± 4	49
$Mn^+—H$	53 ± 3	49
$Co^+—H$	52 ± 4	49
$(OC)_5Mn—H$	51 ± 2	50
$(OC)_4Co—H$	58	51
	54.7	52

ically unsaturated compounds being those of approximate O_h (vanadium, $d^5 \to d^6$), C_{4v} (manganese, $d^7 \to d^8$), and C_{3v} (cobalt, $d^9 \to d^{10}$) symmetries. The simple considerations that the negative charge is dissipated over a larger number of π-back-bonding carbonyl groups and that electron addition is occurring in lower lying orbitals would suggest that Process 8c should be more hexothermic for vanadium than for manganese and cobalt, thus favoring acid dissociation for vanadium, as observed. However, the higher stability of the $V(CO)_6$ radical as compared with $Mn(CO)_5$ and $Co(CO)_4$ makes it hard to decide on the relative magnitude of the energetics of the electron-capture process among the three metals.

The metal-hydrogen bond dissociation energies are of paramount importance in deciding the acid strengths of $MH(CO)_n$ molecules. However, a few such quantities are known in the literature. TABLE 2 reports the available data concerning chromium, manganese, and cobalt. A general conclusion that can be drawn from these data is that M—H bonds of the hydrides of manganese and cobalt are generally speaking strong bonds, requiring about 50 kcal or more to be cleaved. However, the enthalpy contribution to the cleavage of the Co—H bond is slightly higher than for manganese, which appears to be contradictory to the observed higher acidity in water of $CoH(CO)_4$ with respect to $MnH(CO)_5$.

The data for $(OC)_4Co—H$ come from the experimental determinations of the enthalpy change for the reaction:[51,52]

$$Co_2(CO)_8 + H_2 \rightleftharpoons 2\,CoH(CO)_4 \qquad (9)$$

By considering the following sequence and the reported value of 22.5 kcal for Process 10a,[50] relative to $Mn_2(CO)_{10}$, and the value of 51 kcal for Process 10c,

$$M_2(CO)_{2n(g)} \longrightarrow 2\,M(CO)_{n(g)} \qquad (10a)$$

$$H_2 \longrightarrow 2\,H_{(g)} \qquad (10b)$$

$$2\,M(CO)_{n(g)} + 2\,H_{(g)} \longrightarrow 2\,MH(CO)_{n(g)} \qquad (10c)$$

$$M_2(CO)_{2n(g)} + H_{2(g)} \longrightarrow 2\,MH(CO)_{n(g)} \qquad (10)$$

M = Mn, one can calculate the enthalpy change for Reaction 11 to be about 25 kcal. This suggests that it should not be substantially possible to convert

$$Mn_2(CO)_{10} + H_2 \rightleftharpoons 2\,MnH(CO)_5 \qquad (11)$$

$Mn_2(CO)_{10}$ into $MnH(CO)_5$ with molecular hydrogen (the corresponding reaction of $Co_2(CO)_8$ with H_2 is endothermic for 3.2 or 6.6 kcal,[51,52] depending on the literature source). Recent experiments show, however, that Reaction 11 proceeds at 150–180°C under H_2 pressure,[53] thus confirming an earlier report on this subject.[20]

Another uncertainty exists concerning the dependence of the metal-hydrogen bond strength on the metal in a given vertical sequence of elements. Relevant to this point is the observation that while $FeH_2(CO)_4$ and $RuH_2(CO)_4$ are thermally unstable compounds,[54-57] the corresponding osmium derivative can be made in a pure state from $Os(CO)_5$ and H_2 under pressure in a closed system at about 100°C,[58] thus showing that $OsH_2(CO)_4$ is a thermodynamically stable compound with

$$Os(CO)_5 + H_2 \longrightarrow OsH_2(CO)_4 + CO \qquad (12)$$

respect to its recombination with CO or to its decomposition to $Os_3(CO)_{12}$ under the conditions of the experiment. By considering the dissociation energy of H_2 (104 kcal/mol) and the value of the Os—CO bond energy as 52 kcal/mol,[59] the ΔH of Reaction 12 would be zero if the sum of the reorganization energy from $Os(CO)_5$ to $Os(CO)_4$ and the enthalpy change corresponding to the formation of two new Os—H bonds would amount to -156 kcal. This should then set a lower limit for the average dissociation energy of the Os—H bond to about 78 kcal. The observed lower acidity of $OsH_2(CO)_4$ as compared to $FeH_2(CO)_4$ is consistent with a stronger Os—H bond (see Reference 60 and references therein).

From Vanadium to Niobium and Tantalum

In spite of the unsatisfactory situation concerning the thermodynamic data for $MH(CO)_n$ complexes, it is plausible that the observed acid strength of $VH(CO)_6$ is due to a particularly weak V—H bond.§ In view of this and of the preceding considerations concerning the group trends for M—H bonds,¶ it was of interest to study the behavior of $[Nb(CO)_6]^-$ and $[Ta(CO)_6]^-$ toward protons and compare it with $[V(CO)_6]^-$. It has been found that the niobium and tantalum anions behave toward protons in a different way from the corresponding vanadium compound: with anhydrous HCl in an organic solvent (tetrahydrofuran or a saturated aliphatic hydrocarbon) or with a slightly acidic aqueous solution containing Cl^-, the two-electron transfer process of Equation 13 was found to occur:

$$2\,[Nb(CO)_6]^- + 4\,HCl \longrightarrow 2\,H_2 + [Nb_2Cl_3(CO)_8]^- + Cl^- + 4\,CO \quad (13)$$

§ The energetic considerations are not modified by the mode of bonding of hydrogen to the $V(CO)_6$ moiety.

¶ Still insufficient quantitative information is available in the literature concerning the change of steric and electronic properties of a metal-ligand (M—L) combination upon systematic changes of M or L along a group (group trends) and of the metal along a period (period trends). However, the information concerning M—CO and M—R (R = alkyl) bonds[59] appears to point to the general conclusion that the strength of metal-carbon bonds increases along a vertical sequence of transition metals. There may be exceptions, however. One such exception is probably represented by the Ni—Pd—Pt and the Cu—Ag—Au triads. Palladium and silver form presumably weak bonds to carbon monoxide, as suggested by the following experimental observations: (a) nonexistence of $Pd(CO)_4$ under normal laboratory conditions in spite of the *lower* heat of sublimation of the metal with respect to nickel;[61] (b) exceptionally high $\bar{\nu}_{CO}$'s for palladium(II) carbonyl halo complexes,[62] (c) nonexistence of stable solid silver carbonyl molecular complexes.[63-65]

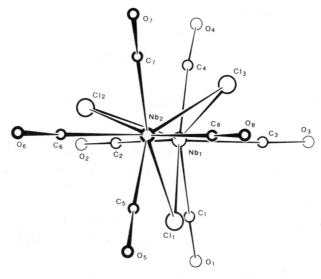

FIGURE 1. A view of the $[Nb_2Cl_3(CO)_8]^-$ anion along an axis slightly inclined with respect to the niobium-niobium vector.

The dimeric anion of d^4 niobium(I) has been isolated as the tetrahydrofuran-stabilized acid $C_4H_8OH^+[Nb_2Cl_3(CO)_8]^-$ and characterized by IR spectroscopy [two $\tilde{\nu}_{CO}$'s at 2007m and 1906s cm^{-1} in tetrahydrofuran corresponding to the C_{4v} local symmetry $(A_1 + E)$ of the $Nb(CO)_4$ moiety] and by x-ray diffraction methods. In the anion, each niobium center is heptacoordinated to four terminal carbonyl groups and to three bridging chlorides (see FIGURE 1). The $[Ta(CO)_6]^-$ anion behaves similarly toward protons.[66]

Not many cases of two-electron transfer chemical processes to protons are known in the literature for organometallic compounds. The possible reaction paths for carbonyl metalates with protons are summarized in SCHEME 2. The occurrence of Reaction 13 shows that the first electron transfer is followed by a faster transfer of the second electron to a proton. This may be due to the formation of a transient hydrogen-containing species $MH(CO)_6$, M = Nb and Ta, which reacts promptly with HCl to give off H_2. This may be taken to suggest that the metal-bonded hydrogen in the transient species is negatively charged with respect to the chlorine-bonded hydrogen of HCl.

THE NATURE OF THE METAL-HYDROGEN BOND IN MH(CO)$_n$ COMPLEXES

Much work has appeared in the literature concerning the polarity of the M—H bond in MH(CO)$_n$ complexes. It is believed that the *ground state polarity* of the bond should be clearly distinguished from the *reactivity* of these species. Physical measurements, such as ^{1}H NMR,[13] infrared spectroscopy in the carbonyl stretching region showing a lower degree of π back donation with respect to the corresponding carbonyl metalate $(MnH(CO)_5$ vs. $[Mn(CO)_5]^-$; $CoH(CO)_4$ vs.

a. Protonation of the metal:

$$[Mn(CO)_5]^- + H^+ \longrightarrow MnH(CO)_5 \text{ (References 19 and 20)}$$

$$[Co(CO)_4]^- + H^+ \longrightarrow CoH(CO)_4 \text{ (References 9–12)}$$

b. One-electron transfer:

$$[V(CO)_6]^- + H^+ \longrightarrow \tfrac{1}{2} H_2 + V(CO)_6 \text{ (Reference 25)}$$

c. Carbonyl metalate–assisted protonation of an added Lewis base:

$$[V(CO)_6]^- + H_2O + H^+ \longrightarrow H_3O^+[V(CO)_6]^- \text{ (Reference 25)}$$

$$[Co(CO)_4]^- + R_3N + H^+ \longrightarrow NR_3H^+[Co(CO)_4]^- \text{ (Reference 67)}$$

d. Protonation of the carbonyl oxygen:

$$[Co_3(CO)_{10}]^- + H^+ \longrightarrow Co_3(CO)_9C\!-\!OH \text{ (Reference 38)}$$

$$[Fe_4H(CO)_{13}]^- + H^+ \longrightarrow Fe_4(CO)_{12}(\mu\!-\!H)(\eta^2\!-\!C\!-\!OH) \text{ (Reference 39)}$$

e. Protonation of the carbonyl carbon:

$$[M(CO)_n]^- + H^+ \longrightarrow M(CO)_{n-1}C(O)H$$

(no examples are known of this kind of reactivity)

SCHEME 2. Reactions of carbonyl metalate anions with protons.

$[Co(CO)_4]^-$), and x-ray photoelectron spectoscopy,[68] tend to suggest that the hydrogen atom is negatively charged, relative to the central metal atom. As far as the reactivity of $MH(CO)_n$ molecules is concerned, they have been found generally to behave as acids,[69,70] according to pattern c of SCHEME 2, the typical example being that of $CoH(CO)_4$. Evidence for radical mechanisms of reactivity for $MH(CO)_n$ complexes has been obtained.[71–74] Moreover, hydridic behavior has been found for $MnH(CO)_5$ with respect to CF_3SO_3H.[75] However, reactivity tells something about the polarity of the metal-hydrogen bond in the presence of a second component, which may alter the polarity of the $M-H$ bond. Relevant to this point is the recent finding concerning the reaction of tertiary amines with hydrocarbon solutions of $CoH(CO)_4$ giving the ion pair resulting from the proton transfer to the amine:[67]

$$NR_3 + CoH(CO)_4 \longrightarrow R_3NH^+ \dots (OC)_3Co(CO)^- \qquad (14)$$

The results of the x-ray investigation show the following facts: (a) the $[Co(CO)_4]^-$ moiety of the ion pair has retained the C_{3v} symmetry of $CoH(CO)_4$ itself;[55] and (b) the proton, attached to the amine nitrogen, interacts with the $Co(CO)_3$ portion of the anion along the threefold axis of symmetry. In the ethyl derivative, $NEt_3H^+[(OC)_3Co(CO)]^-$, the hydrogen atom, which was located, is at distances of 2.85(7) and 2.91(8) Å from the cobalt atom from the symmetry-related carbonyl carbons, respectively. The ion pair situation found in the solid state is maintained in solvents of low polarity such as toluene as shown by the three IR active CO stretching vibrations observed for the amine adducts and for $CoH(CO)_4$, as dictated by the C_{3v} symmetry ($2 A_1 + E$) of the aninonic moiety (see TABLE 3).

TABLE 3
INFRARED CARBONYL STRETCHING VIBRATIONS OF $CoH(CO)_4$ AND ITS
TERTIARY AMINE ADDUCTS*

Compound	$\bar{v}_{CO}$ (cm^{-1})		
$CoH(CO)_4$	2100w	2040s	2010vs
$NMe_3 \cdot CoH(CO)_4$	2015w	1931s	1895vs
$NEt_3 \cdot CoH(CO)_4$	2015w	1934s	1899vs

* Toluene solution.[67]

CONCLUSIONS

This paper points out that the energetics of the metal-hydrogen bond are still largely unknown and that some basic information is necessary in order to discuss the chemical reactivity of metal carbonyl hydrides in a less qualitative way. Work in this laboratory aimed at clarifying the range of stability of carbonyl hydrides of early transition metals and at obtaining information about the strength of metal-hydrogen bonds by chemical methods is in progress.

REFERENCES

1. MASTERS, C. 1979. Adv. Organomet. Chem. 17: 61.
2. OLIVÉ, G. H. & S. OLIVÉ. 1976. Angew. Chem. Int. Ed. Engl. 15: 136.
3. PRUETT, R. L. 1981. Science 211: 11.
4. KAESZ, H. D. & R. B. SAILLANT. 1972. Chem. Rev. 72: 231.
5. PINO, P., F. PIACENTI & M. BIANCHI. 1979. Reactions of CO and hydrogen with olefin substrates: the hydroformylation (OXO) reaction. In Organic Syntheses via Metal Carbonyls. I. Wender & P. Pino, Eds. 2: 43. John Wiley & Sons, Inc. New York, N.Y.
6. WERMER, P., B. S. AULT & M. ORCHIN. 1978. J. Organomet. Chem. 162: 189.
7. ORCHIN, M. 1981. Acc. Chem. Res. 14: 259.
8. HIEBER, W. & F. LEUTERT. 1931. Naturwissenschaften 19: 360.
9. HIEBER, W. 1942. Chemie 55: 7.
10. COLEMAN, G. W. & A. A. BLANCHARD. 1936. J. Am. Chem. Soc. 58: 2160.
11. HIEBER, W. & H. SCHULTEN. 1937. Z. Anorg. Allg. Chem. 232: 29.
12. PINO, P., R. ERCOLI & F. CALDERAZZO. 1955. Chim. Ind. Milan 37: 782.
13. GINSBERG, A. P. 1965. Transition Met. Chem. 1: 112.
14. BAU, R., R. G. TELLER, S. W. KIRTLEY & T. F. KOETZLE. 1979. Acc. Chem. Res. 12: 176.
15. GEOFFROY, G. L. 1980. Prog. Inorg. Chem. 27: 123.
16. TELLER, R. G. & R. BAU. 1981. Struct. Bonding Berlin 44: 1.
17. BEHRENS, H. & R. WEBER. 1957. Z. Anorg. Allg. Chem. 241: 122.
18. DARENSBOURG, M. Y. & J. C. DEATON. 1981. Inorg. Chem. 20: 1644.
19. HIEBER, W. & G. WAGNER. 1957. Z. Naturforsch. Teil B 12: 478.
20. HIEBER, W. & G. WAGNER. 1958. Z. Naturforsch. Teil B 13: 339.
21. STERNBERG, H. W., R. MARKBY & I. WENDER. 1956. J. Am. Chem. Soc. 78: 5704.
22. STERNBERG, H. W., R. MARKBY & I. WENDER. 1957. J. Am. Chem. Soc. 79: 6116.
23. BEHRENS, H. & F. LOHÖFER. 1953. Z. Naturforsch. Teil B 8: 691.
24. COTTON, F. A. & G. WILKINSON. 1972. Advanced Inorganic Chemistry. 3rd Edit.: 115. John Wiley & Sons, Inc. New York, N.Y.
25. CALDERAZZO, F., G. PAMPALONI & D. VITALI 1981. Gazz. Chim. Ital. 111: 455.
26. WERNER, R. P. M. & H. E. PODALL. 1961. Chem. Ind. London: 144.
27. WERNER, R. P. M., A. H. FILBEY & S. A. MANASTYRSKYI. 1964. Inorg. Chem. 3: 298.
28. ELLIS, J. E. & A. DAVISON. 1976. Inorg. Synth. 16: 68.

29. CALDERAZZO, F., G. PAMPALONI & G. PELIZZI. 1982. J. Organomet. Chem. **233:** C41.
30. CALDERAZZO, F., U. ENGLERT, G. PAMPALONI, G. PELIZZI & R. ZAMBONI. 1983. Inorg. Chem. **22:** 1865.
31. SCHNEIDER, M. & E. WEISS. 1974. J. Organomet. Chem. **73:** C7.
32. SCHNEIDER, M. & E. WEISS. 1976. J. Organomet. Chem. **121:** 345.
33. CALDERAZZO, F. & G. PAMPALONI. Unpublished observations.
34. BELLARD, S., K. A. RUBINSON & G. M. SHELDRICK. 1979. Acta Crystallogr. **B35:** 271.
35. VAUGHN, G. D. & J. A. GLADYSZ. 1981. J. Am. Chem. Soc. **103:** 5608.
36. TAM, W., G. Y. LIN & J. A. GLADYSZ. 1982. Organometallics **1:** 525.
37. GLADYSZ, J. A. 1982. Adv. Organomet. Chem. **20:** 1.
38. FACHINETTI, G. 1979. Chem. Commun.: 397.
39. WHITMIRE, K. H. & D. F. SHRIVER. 1981. J. Am. Chem. Soc. **103:** 6754.
40. CALDERAZZO, F. 1977. Angew. Chem. Int. Ed. Engl. **16:** 299.
41. WAYLAND, B. B. & B. A. WOODS. 1981. Chem. Commun.: 700.
42. WAYLAND, B. B., B. A. WOODS & R. PIERCE. 1982. J. Am. Chem. Soc. **104:** 302.
43. PAŃKOWSKI, M. & M. BIGORGNE. 1977. Abstracts, Eighth International Conference on Organomentallic Chemistry, Kyoto, September 12–16: 194.
44. PAŃKOWSKI, M. & M. BIGORGNE. 1983. J. Organomet. Chem. **251:** 333.
45. BIGORGNE, M. Personal communication.
46. BRUNNER, H. & H. VOGT. 1981. Angew. Chem. Int. Ed. Engl. **20:** 405.
47. CALDERAZZO, F. & K. NOACK. 1967. J. Organomet. Chem. **10:** 101.
48. HIEBER, W., E. WINTER & E. SCHUBERT. 1962. Chem. Ber. **95:** 3070.
49. ARMENTROUT, P. B., L. F. HALLE & J. L. BEAUCHAMP. 1981. J. Am. Chem. Soc. **103:** 6501. (An ion-beam experiment.)
50. CONNOR, J. A., M. T. ZAFARANI MOATTAR, J. BICKERTON, N. I. EL SAIED, S. SURADI, R. CARSON, G. AL TAKHIN & H. A. SKINNER. 1982. Organometallics **1:** 1166. (I am grateful to Professor Connor for sending a copy of the manuscript prior to publication.)
51. HUNGVARY, F. 1972. J. Organomet. Chem. **36:** 363.
52. ALEMDAROGLU, N. H., J. M. L. PENNINGER & E. OLTAY. 1976. Monatsh. Chem. **107:** 1043.
53. BIGORGNE, M., F. CALDERAZZO, B. DEMERSEMAN & R. POLI. Unpublished observations.
54. HIEBER, W. & F. LEUTERT. 1931. Chem. Ber. **64:** 2832.
55. MCNEIL, E. A. & F. R. SCHOLER. 1977. J. Am. Chem. Soc. **99:** 6243.
56. FARMERY, K. & M. KILNER. 1970. J. Chem. Soc. A: 634.
57. COTTON, J. D., M. I. BRUCE & F. G. A. STONE. 1968. J. Chem. Soc. A: 2162.
58. L'EPLATTENIER, F. & F. CALDERAZZO. 1967. Inorg. Chem. **6:** 2092.
59. CONNOR, J. A. 1977. Top. Curr. Chem. **71:** 71.
60. WALKER, H. W., C. T. KRESGE, P. C. FORD & R. G. PEARSON. 1979. J. Am. Chem. Soc. **101:** 7428.
61. DASENT, W. E. 1965. Nonexistent Compounds. Marcel Dekker. New York, N.Y.
62. CALDERAZZO, F. & D. BELLI DELL'AMICO. 1981. Inorg. Chem. **20:** 1310.
63. MANCHOT, W. & J. KÖNIG. 1927. Chem. Ber. **60:** 2183.
64. SOUMA, Y. & H. SANO. 1976. Bull. Chem. Soc. Jpn. **49:** 3296.
65. BACKÉN, W. & R. VESTIN. 1980. Acta Chem. Scand. **A34:** 73.
66. CALDERAZZO, F., G. PAMPALONI & P. F. ZANAZZI. 1982. Chem. Commun.: 1304.
67. CALDERAZZO, F., G. FACHINETTI, F. MARCHETTI & P. F. ZANAZZI. 1981. Chem. Commun.: 181.
68. CHEN, H. W., W. L. JOLLY, J. KOPF & T. H. LEE. 1979. J. Am. Chem. Soc. **101:** 2607.
69. STERNBERG, H. W., I. WENDER, R. A. FRIEDEL & M. ORCHIN. 1953. J. Am. Chem. Soc. **75:** 2717.
70. NISHIHARA, H., T. MORI, Y. TSURITA, K. NAKANO, T. SAITO & Y. SASAKI. 1982. J. Am. Chem. Soc. **104:** 4367.
71. SWEANY, R. L. & J. HALPERN. 1977. J. Am. Chem. Soc. **99:** 8335.
72. SWEANY, R. L., D. S. COMBERREL, M. F. DOMBOURIAN & N. A. PETERS. 1981. J. Organomet. Chem. **216:** 57.
73. NALESNIK, T. E. & M. ORCHIN. 1982. Organometalics **1:** 222.
74. NALESNIK, T. E., J. H. FREUDENBERGER & M. ORCHIN. 1982. Organomet. Chem. **236:** 95.
75. TROGLER, W. C. 1979. J. Am. Chem. Soc. **101:** 6459.

LIGAND DESULFURIZATION IN OSMIUM CARBONYL CLUSTER COMPOUNDS*

Richard D. Adams, Zain Dawoodi, Donald F. Foust, and Li-Wu Yang

Department of Chemistry
Yale University
New Haven, Connecticut 06511

INTRODUCTION

Much of our recent research has focused on the reactions of hydride-containing metal carbonyl cluster compounds with heteronuclear unsaturated small molecules.[1] While our initial interests were in determining the modes of hydrogen transfer from the metal atoms to the unsaturated molecules, we have recently discovered that these clusters are very effective in abstracting sulfur atoms from sulfur-containing ligands. The sulfur atoms invariably remain bonded to the clusters as sulfido ligands. Herein will be described the nature of the desulfurization of N-arylthioform-amido- and arylthiolato- ligands by triosmium clusters.

Some of the resultant sulfido osmium carbonyl clusters have been found to contain unusual structures, bonding, and reactivity. Some examples of these compounds and their reactivity properties will also be described.

RESULTS AND DISCUSSION

Desulfurization of N-Arylthioformamido Ligands

The reaction of the cluster compound $H_2Os_3(CO)_{10}$ with N-arylisothiocyanate (RN=C=S) molecules yields the compounds $HOs_3(CO)_{10}[\mu\text{-}\eta^2\text{-SC(H)=NAr}]$, Ar = C_6H_5 and $p\text{-}C_6H_4F$, which are formed by the addition of one mole of the arylisothiocyanate to the cluster and the transfer of one hydride ligand to the cyanocarbon atom.[2] Details of the molecular structure of **I**, Ar = $p\text{-}C_6H_4F$, were established by a single-crystal x-ray diffraction analysis. An ORTEP (Oak Ridge thermal ellipsoid plot) diagram of its molecular structure is shown in FIGURE 1. The molecule consists of a triangular cluster of three metal atoms containing 10 linear terminal carbonyl ligands. The metal-metal internuclear separations, Os(1)-Os(2) = 2.861(1) Å, Os(1)-Os(3) = 2.870(1) Å, and Os(2)-Os(3) = 2.868(1) Å, are similar to those found in $Os_3(CO)_{12}$,[3] 2.877(3) Å, and are indicative of Os—Os single bonds. The N-p-fluorophenylthioformamido ligand symmetrically bridges one edge of the cluster via the sulfur atom. The carbon-sulfur internuclear separation, C(17)-S, in the thioformamido ligand is 1.782(12) Å. This is similar to that of a carbon-sulfur single bond, 1.80 Å. The carbon-nitrogen distance, C(17)-N at 1.279(13) Å, is similar to that of a carbon-nitrogen double bond. It is believed

* This research was supported in part by the Division of Chemical Sciences of the U.S. Department of Energy and the National Science Foundation. R. D. Adams wishes to thank the Alfred P. Sloan Foundation for a fellowship.

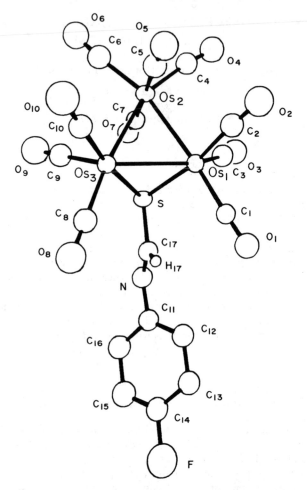

FIGURE 1. An ORTEP diagram of $HOs_3(CO)_{10}[\mu-\eta^1\text{-}SC(H)\!=\!N\text{-}p\text{-}C_6H_4F]$ showing 50% thermal motion probability ellipsoids. (Reproduced with permission from Reference 2. Copyright 1982, American Chemical Society.)

that this bonding arrangement can be attributed to a localization of the π-electron density of the thioformamido ligand between the carbon and nitrogen atoms. This is probably a consequence of the dinuclear coordination of the sulfur atom, which localizes electron density of the sulfur atom in metal-sulfur bonds and in turn prevents its use for carbon-sulfur π-bond formation. **I** also contains one hydride ligand (not observed crystallographically), which is believed to bridge the Os(1)-Os(3) edge of the cluster on the opposite side of the Os_3 plane from the sulfur atom.

Upon photolysis, **I** loses one mole of carbon monoxide to yield the product $HOs_3(CO)_9[\mu_3-\eta^2\text{-}SC(H)\!=\!NAr]$, **II**, $Ar = p\text{-}C_6H_4F$. The molecular structure of **II** was established by a single-crystal x-ray diffraction analysis, and an ORTEP

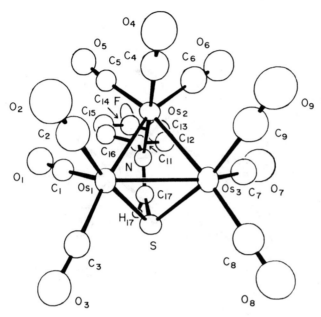

FIGURE 2. An ORTEP diagram of $HOs_3(CO)_9[\mu_3\text{-}\eta^2\text{-}SC(H)=N\text{-}p\text{-}C_6H_4F]$ showing 50% thermal motion probability ellipsoids. (Reproduced with permission from Reference 2. Copyright 1982, American Chemical Society.)

diagram of its structure is shown in FIGURE 2. This molecule also consists of a triangular cluster of three osmium atoms with three metal-metal bonds, $Os(1)\text{-}Os(2) = 2.817(1)$ Å, $Os(1)\text{-}Os(3) = 2.860(1)$ Å, and $Os(2)\text{-}Os(3) = 2.824(1)$ Å. Each metal atom contains three linear terminal carbonyl ligands. The thioformamido ligand is a triple bridge with the sulfur atom bridging the $Os(1)\text{-}Os(3)$ edge of the cluster and the nitrogen atom coordinated to the third metal atom, $Os(2)$. The C—S and C—N distances in the thioformamido group are 1.774(15) Å and 1.250(17) Å respectively, and imply that the single and double bond character observed in **I** persists in the structure of **II**. This can be explained in the same way as for **I**. One hydride ligand (not observed crystallographically) is believed to bridge the $Os(1)\text{-}Os(3)$ bond on the opposite side of the cluster from the sulfur atom.

When heated to 125°C (refluxing octane solution), **II** is converted into an isomer $HOs_3(CO)_9(\mu_3\text{-}S)(\mu\text{-}HC=N\text{-}p\text{-}C_6H_4F)$, **III**. The molecular structure of **III** was established by x-ray crystallographic methods, and an ORTEP diagram of its structure is shown in FIGURE 3. This molecule consists of an "open" cluster of three osmium atoms. The $Os(1)\text{-}Os(2)$ and $Os(2)\text{-}Os(3)$ distances at 2.836(1) Å and 2.988(1) Å, respectively, are similar to those of osmium-osmium single bonds.[3] However, the large $Os(1)\text{-}Os(3)$ distance at 3.779(1) Å suggests the absence of a direct metal-metal bond. As in **II** each atom contains three linear terminal carbonyl groups. There is a triply bridging sulfido ligand and an $N\text{-}p$-fluorophenyl-formimidoyl ligand ($HC=N\text{-}p\text{-}C_6H_4F$), which bridges the "open" edge of the cluster. The hydride ligand (unobserved) is believed to bridge the $Os(2)\text{-}Os(3)$ bond.

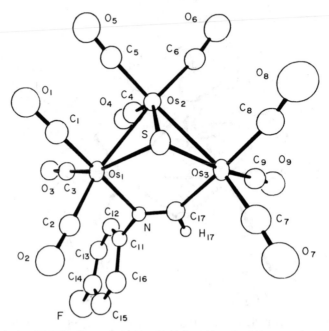

FIGURE 3. An ORTEP diagram of $HOs_3(CO)_9(\mu_3\text{-}S)(\mu\text{-}HC{=}N\text{-}p\text{-}C_6H_4F)$ showing 50% thermal motion ellipsoids. (Reproduced with permission from Reference 2. Copyright 1982, American Chemical Society.)

The formation of **III** is believed to have occurred through the cleavage of the carbon-sulfur bond in the thioformamido ligand and formation of a metal-carbon bond on one of the sulfur-bridged metal atoms (see SCHEME 1). An intermediate product, **IV**, containing sulfido and formimidoyl ligands bridging different edges of the triangular cluster might be formed initially. The formation of **III** would be completed by donation of a pair of electrons on the sulfido ligand to the third metal atom. The formation of such a new sulfur-metal donor should induce cleavage of a metal-metal bond, as observed.

The characterization of the sequence of Products I–III reveals certain important

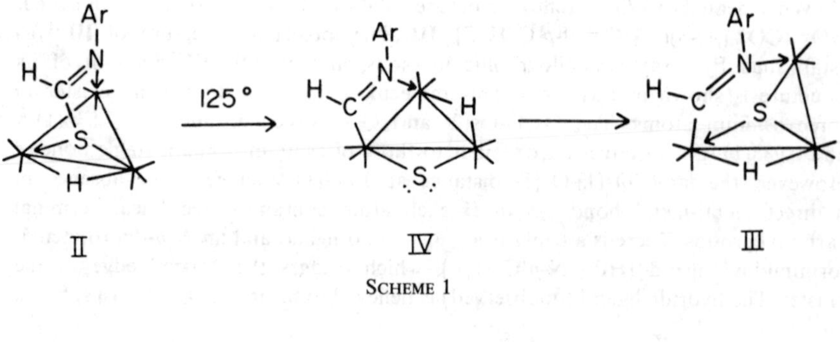

SCHEME 1

features about the manner in which the cluster acting as a unit induces the desulfurization of the thioformamido ligand. A principal effect is the localization of π-electron density in the thioformamido ligand between the carbon and nitrogen atoms. This reduces the C—S bond order to approximately unity. The second important effect is realized upon decarbonylation of **I**. This leads to the coordination of the nitrogen atom to the third metal atom and positions the thioformamido ligand in a way that more readily permits the formation of the bridging formimidoyl ligand, perhaps through a concerted process of carbon-sulfur bond cleavage and metal-carbon bond formation. The important conclusion is that it is a combination of effects *promoted by the multinuclear coordination* of the thioformamido ligand that produces overall the desulfurization process.

Desulfurization of Arylthiolato Ligands and the Synthesis of Sulfido Osmium Carbonyl Cluster Compounds

In an effort to test the scope of the desulfurization reaction about multinuclear sites, we investigated the reactivity of the arylthiolato complexes $HOs_3(CO)_{10}(\mu\text{-}SAr)$, **V**, $Ar = C_6H_5$ or C_6F_5. It was observed that the appropriate arene (ArH) was formed in high yield when nonane solutions of the compounds **V** were heated to reflux (150°C) under a nitrogen atmosphere for 2–6 hours.[4] Clearly the phenyl carbon-sulfur bond was cleaved, but unexpectedly the phenyl group combined with the hydride ligand to form benzene, which was then eliminated. The fate of the sulfido metal carbonyl moiety was established by thin-layer chromatography (TLC) product separation of the resultant reaction mixture and single-crystal x-ray diffraction analyses of the various components. A variety of sulfido-osmium carbonyl cluster compounds have been characterized and are listed in SCHEME 2. Line drawings of the structures are shown below the corresponding chemical formulas.

One of the most interesting of these is $Os_4(CO)_{12}(\mu_3\text{-}S)_2$, **VI**. An ORTEP diagram of **VI** is shown in FIGURE 4. This molecule consists of a "butterfly" tetra-hedron of four osmium atoms with two triply bridging sulfido ligands bridging the open triangular faces of the cluster. Five osmium-osmium distances are sufficiently short to imply significant direct metal-metal interactions, but the Os(1) --- Os(3) and S(1) --- S(2) distances at 3.551(1) Å and 3.191(2) Å, respectively, seem to be too large to permit significant direct bonding interactions. The nature of the metal-metal bonding is unclear since electron-counting procedures indicate that the cluster contains 64 electrons, assuming that the sulfido ligands each serve as four electron donors, but an electron "precise" cluster with 64 electrons should have only four metal-metal bonds.[5] Likewise, an "electron precise" tetranuclear cluster with five metal-metal bonds should have 62 electrons. Thus, it appears that **VI**, which has five metal-metal bonds and 64 electrons, contains two too many electrons. For this reason we have chosen to describe **VI** as electron rich.[4] However, a second way of rationalizing the electronic configurations of cluster compounds is through the skeletal electron pair (SEP) theory.[6] According to this theory, the Os_4S_2 cluster of **VI** is a *nido*-form of the pentagonal bipyramidal polyhedron. Cluster bonding for such a polyhedron would require eight skeletal electron pairs. Electron counting according to this method shows that **V** does indeed contain eight skeletal electron pairs. Thus, this cluster appears to conform to the SEP theory, which does predict the existence of five direct metal-metal interactions. However, close

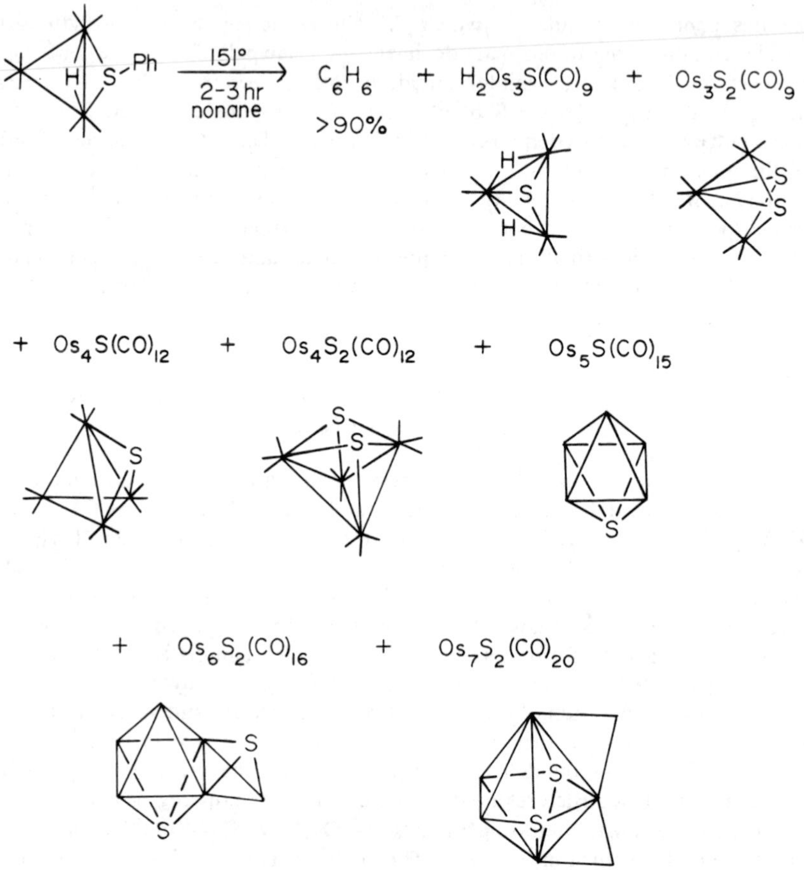

SCHEME 2

inspection of the five metal-metal bonds in **VI** shows that there are significant differences among them. In particular, the Os(1)-Os(4) and Os(2)-Os(3) distances at 3.091(1) Å and 3.002(1) Å respectively are both significantly longer than the remaining three, Os(1)-Os(2) = 2.914(1) Å, Os(2)-Os(4) = 2.935(1) Å, and Os(3)-Os(4) = 2.940(1) Å. It is believed that the increased lengths of two of these five bonds is evidence for selective bond weakening, perhaps, caused by the two "extra" electrons in the cluster core (see above). It seemed possible that such weakening of the metal-metal bonds could have influences on the reactivity of the molecule. Thus, efforts to find and identify unusual reactivity were undertaken. Some results of these studies are given in the next section.

Addition Reactions to $Os_4(CO)_{12}(\mu_3\text{-}S)_2$, Compound VI

Many metal complexes can catalyze reactions between selected small molecules. The reactivity of these small molecules is modified through the formation of

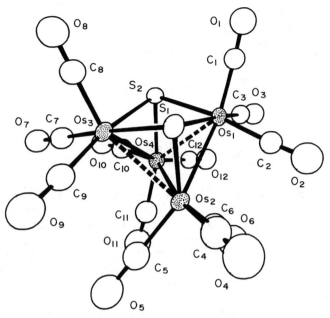

FIGURE 4. An ORTEP diagram of $Os_4(CO)_{12}(\mu_3\text{-}S)_2$, **VI**, showing 50% thermal motion of probability ellipsoids. (Reproduced with permission from Reference 4. Copyright 1982, American Chemical Society.)

bonds between the metal atoms and the small molecules. The two principal ways in which these bonds are formed are by nucleophilic addition and oxidative addition reactions. Via the former process, a donor-acceptor bond is formed between the molecule or ligand and the metal atom. Via the latter, the small molecule is added to the metal atom and a degree of electron transfer from the metal atom to the molecule produces "formally" an oxidation of the metal atom. Because of the importance of these addition reactions to the subject of catalysis, Compound **VI**, described in the previous section, was investigated specifically for its reactivity toward addition reactions.

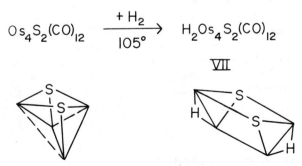

SCHEME 3. Oxidative addition.

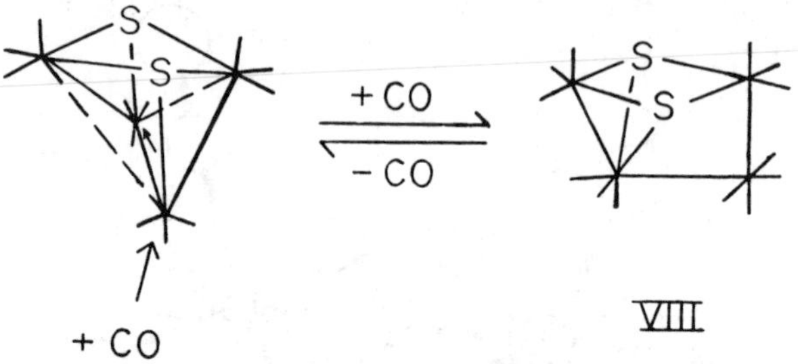

SCHEME 4. Nucleophilic addition.

Compound VI was found to add one mole of hydrogen at 105°C in hydrocarbon solvent to yield the product $H_2Os_4(CO)_{12}(\mu_3\text{-}S)_2$, **VII** (see SCHEME 3). The six atoms, Os_4S_2, in the cluster of **VII** are arranged in the form of a trigonal prism. (See the line structure shown in SCHEME 3.) An x-ray crystal structure analysis of the homologue of **VII**, $H_2Os_4(CO)_{12}(\mu_3\text{-}Se)_2$, was recently reported by Johnson et al.[7] Compound VII contains only three metal-metal bonds. Thus, it is electron precise, which means that each metal atom obeys the 18-electron rule. Two of the three metal-metal bonds contain bridging hydride ligands. Overall, the addition of hydrogen to **VI** resulted in the cleavage of the two weak metal-metal bonds. Formally, this is an *oxidative addition* reaction, since two of the four metal atoms have increased their oxidation states by one unit.

Under an atmosphere of carbon monoxide (1 atm) Compound VI was found to add one mole of CO per mole of cluster at room temperature in hydrocarbon solvent (see SCHEME 4). The product, $Os_4(CO)_{13}(\mu_3\text{-}S)_2$, **VIII**, contains a planar arrangement of four osmium atoms with a triply bridging sulfido ligand on each side of the Os_4 plane.[8] A line drawing of its structure is shown in SCHEME 4. Compound VIII contains only three metal-metal bonds; and as with the formation of Compound VII, the formation of Compound VIII involves a cleavage of both of the weak metal-metal bonds in Compound VI. Like Compound VII, Compound VIII is electron precise; mechanistically, it appears that **VIII** has been formed via a *nucleophilic addition* of a carbon monoxide molecule to one of the "hinge" metal atoms of the butterfly tetrahedron in **VI**. This evidently induces a shift of the sulfido ligand bonded to that atom to the adjacent "hinge" metal atom. A simultaneous cleavage of the two weak metal-metal bonds yields Compound VIII directly. Interestingly, the reaction is fully reversible such that Compound VI is reformed quantitatively when hexane solutions of **VIII** are refluxed.

In summary Compound VI does readily engage in the selective addition of small molecules. The establishment of ligand combination processes and the development of catalytic reaction cycles seem to be reasonable goals of future research. These studies are in progress.

REFERENCES

1. ADAMS, R. D. 1983. Acc. Chem. Res. **16**: 67.
2. ADAMS, R. D. & Z. DAWOODI. 1981. J. Am. Chem. Soc. **103**: 6510.

3. CHURCHILL, M. R. & B. G. DEBOER. 1977. Inorg. Chem. **16:** 878.
4. ADAMS, R. D. & L. W. YANG. 1982. J. Am. Chem. Soc. **104:** 4115.
5. CARTY, A. J. 1982. Pure Appl. Chem. **54:** 113.
6. JOHNSON, B. D. G., Ed. 1980. Transition Metal Clusters: Chapter 3. John Wiley and Sons. New York, N.Y.
7. JOHNSON, B. F. G., J. LEWIS, P. G. LODGE, P. R. RAITHBY, K. HENRICK & M. MCPARTLIN. 1979. J. Chem. Soc. Chem. Commun.: 719.
8. ADAMS, R. D. & L. W. YANG. 1983. J. Am. Chem. Soc. **105:** 235.

ORGANOMETALLIC REACTIONS OF PLATINUM INVOLVING METALLIC HYDRIDES*

George M. Whitesides, Robert H. Reamey, Robert L. Brainard,
Alan N. Izumi, and Thomas J. McCarthy

Department of Chemistry
Harvard University
Cambridge, Massachusetts 02138

INTRODUCTION

Hydridoalkylmetal compounds are intermediates in a large number of reactions—both homogeneous and heterogeneous—that are catalyzed by transition metals. We have maintained a program for some years designed to examine the mechanisms of elementary reactions in which carbon-hydrogen bonds are broken by oxidative addition to platinum, and formed by reductive elimination from it. Our work has focused on platinum for two reasons. First, many soluble platinum organometallic compounds are stable and easily characterized, and thus suitable for detailed mechanistic study. Second, platinum is a useful heterogeneous catalyst, with catalytic activity in reforming and olefin hydrogenation.[1-4]

One objective of our work is to compare the chemistry of homogeneous alkyls with that of metal surface alkyls. This work is intended to help clarify several central problems in current research in catalysis. Does the catalytic activity displayed by aggregated metals (either bulk metal or metal clusters) reflect special electronic features associated with this aggregation, or is heterogeneous catalysis by aggregated metals simply localized surface organometallic chemistry (FIGURE 1)? Do the arrays of adjacent metal centers in a heterogeneous catalyst participate in reactions in some cooperative way, or do these metal atoms act singly?

It will be difficult to resolve these questions fully. The approach we have taken to these problems lies in comparison of the reactivities of alkyl groups attached to single metal atoms in soluble organometallic compounds with those of alkyl groups attached to metal surfaces. We suggest that if the reactions observed for alkyl groups in soluble organometallic complexes closely resemble those of alkyl groups on metal surfaces, then we may be able to rationalize the heterogeneous catalytic reactions in terms of single-center or highly localized chemistry. If, on the other hand, the reactions observed for heterogeneous metal alkyls have no counterpart in soluble organometallic chemistry, then we must consider carefully the possibility that aggregated metals possess unique reactivity.

Carbon-hydrogen bond activation provides one reaction that may be used to explore this problem. This reaction occurs readily over a wide variety of metal clusters and bulk metals,[1,2] but has been observed only infrequently for single metal atoms in soluble organometallic complexes.[5-7] Why is there this difference?

* This work has been supported by the National Science Foundation, most recently by Grant CHE-82–05143, and by grants to the Harvard Materials Research Laboratory.

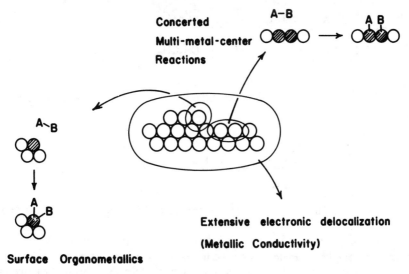

Concerted

Multi-metal-center

Reactions

A–B

A B

A̅B

A B

Surface Organometallics

Extensive electronic delocalization

(Metallic Conductivity)

FIGURE 1. Why are bulk metals (or metal clusters) catalytically active? Is catalysis by metals a single metal atom phenomenon, one requiring cooperating metal atoms, or one requiring extensive electronic delocalization over many centers?

RESULTS AND DISCUSSION

Mechanisms

Most of our work on the mechanisms of reactions of soluble organometallic compounds has been concerned with bis(trialkylphosphine)dialkylplatinum(II) complexes. The techniques used to study the transformations of these compounds are an unexceptional mixture of kinetics, isotopic labeling, and thermodynamics. All of these methods are straightforward in principle (although often demanding in practice), with the possible exception of the thermodynamics. Obtaining accurate kinetic data (on which the thermodynamic parameters rest) in organometallic systems may be difficult. Rates can be influenced by precipitation of zero-valent metals, by trace impurities, by air and water, and by the surfaces of reaction vessels. Further, the interpretation of the thermodynamic parameters—especially the activation entropy—requires caution. We will not review these methods and problems in this paper, and instead refer to the several published papers that contain details.[8–15] Here we only summarize relevant results.

For most thermal decomposition reactions of complexes having the structure L_2PtR_2, a critical step is an initial dissociation of a phosphine ligand (although occasional exceptions to this apparent requirement for initial dissociation are known). A representative scheme is that for the β-hydride elimination reaction observed in bis(tri-*n*-butylphosphine)dibutylplatinum(II)[8] and bis(triethylphosphine)diethylplatinum(II).[9] A reaction mechanism for the latter is outlined in FIGURE 2. The coordinately unsaturated complex $LPtR_2$ appears to be the essential reactive species in this and many similar reactions. We suggest that the creation of a vacant coordination site on this four-coordinate soluble

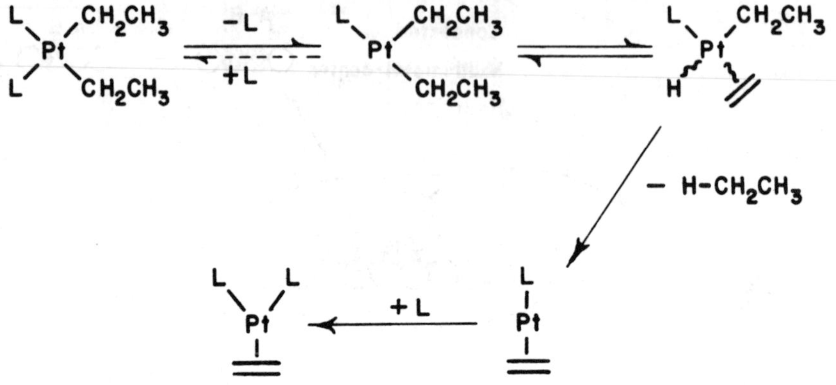

FIGURE 2. β-Hydride elimination. Mechanism of thermal decomposition of $(Et_3P)_2PtEt_2$.

platinum complex is analogous to the formation of a reactive site on the surface of a phosphine-poisoned heterogeneous catalyst. Once this vacant, reactive site has been formed, a number of subsequent reactions appear to occur rapidly.

Activation of β Carbon-Hydrogen Bonds

As a generalization, this type of C—H bond cleavage reaction is one that occurs most rapidly for those platinum complexes having accessible β carbon-hydrogen bonds. The β-hydride elimination reaction is, however, usually not the overall rate-limiting step: in some instances, the initial phosphine dissociation is rate limiting; in others, a subsequent reductive elimination of alkane with concomitant carbon-hydrogen bond formation seems to be slow. In certain cases, reductive elimination of alkane is not possible (as in the decomposition of complexes having the structure L_2PtClR) and olefin loss or, perhaps, β-hydride activation is the rate-limiting step.

Activation of γ Carbon-Hydrogen Bonds

This reaction has been of particular interest to us because it provides an intramolecular model for the reaction of greatest interest as a probe for comparisons of soluble and surface metal alkyls, that is, *intermolecular* activation of unactivated carbon-hydrogen bonds. The reaction is also of considerable interest as an important step in the many processes that form metallacycles by internal cyclometalation.[16]

The mechanism of a typical reaction—that of thermal decomposition of bis(triethylphosphine)dineopentylplatinum(II)—is shown in FIGURE 3. This reaction is very similar to that observed for β-hydride activation for L_2PtR_2 complexes: an initial dissociation of phosphine is followed by reversible γ carbon-hydrogen activation, and the overall rate-limiting step is reductive elimination of neopentane with concomitant carbon-hydrogen bond formation. We have studied this reaction

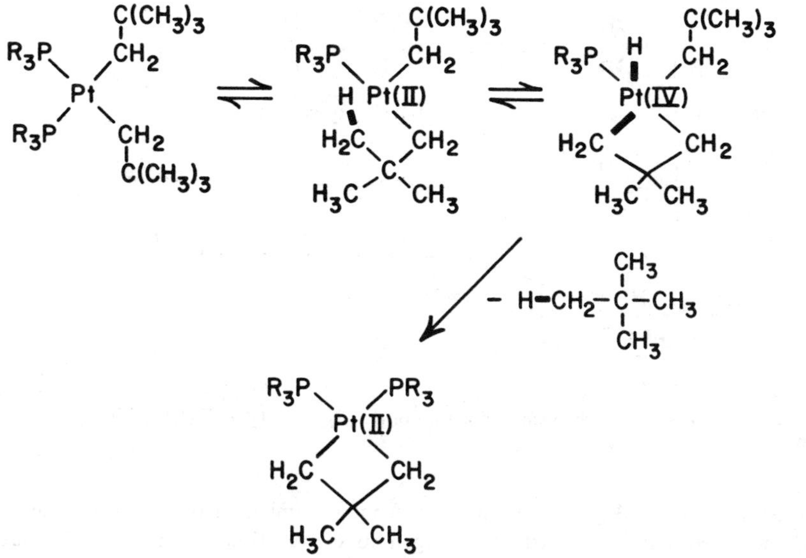

FIGURE 3. γ Carbon-hydrogen bond activation. Mechanism of thermal decomposition of $(Et_3P)_2Pt(CH_2C(CH_3)_3)_2$.

in considerable detail,[10] trying to understand why it proceeds so readily while the corresponding intermolecular reaction proceeds unobservably slowly. Although this analysis is not yet complete, it is sufficiently advanced to exclude a number of possible reasons for the difference in the intra- and intermolecular reactions. Differences in bond energies are probably not dominant, since the bonds formed in both inter- and intramolecular reactions are similar. It is unlikely that electronic effects due to differences in bond angles are sufficient to explain the difference. There seems to be no particular stability associated with the formation of platinacyclic rings.[14] Hence, it seems probable that the intramolecular reaction proceeds in preference to the intermolecular reaction because of some combination of differences in reaction entropy and differences in the amount of intramolecular nonbonded steric strain. At present, the relative contribution of these two factors is not known. Probably both are significant.

Although the reason for the facility of the intramolecular activation reactions is not well understood, the qualitative fact that they occur readily is important. There is, thus, little question that an unactivated carbon-hydrogen bond will react readily with a platinum atom present in the ligand environment provided by one phosphine and two alkyl groups: adjacent or aggregated platinum centers are not required for reactivity toward unactivated C—H bonds.

Intermolecular Reactions

We have never observed authentic intermolecular activation of carbon-hydrogen bonds by oxidative addition of alkanes to the platinum(II) center in complexes of

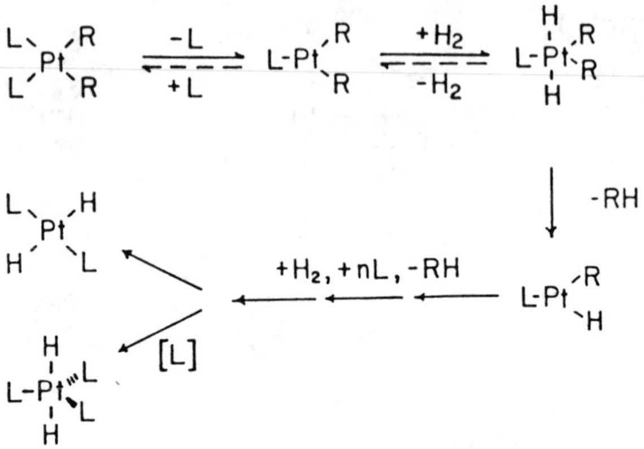

FIGURE 4. Mechanism of reaction of $(Et_3P)_2Pt(CH_2C(CH_3)_3)_2$ with H_2.

structure $LPtR_2$, although recent work from a number of other research groups has provided examples of carbon-hydrogen bond activation by reaction with soluble metal centers in solution.[5-7] The contrast between the homogeneous and heterogeneous reactivity of platinum toward carbon-hydrogen bonds is striking. Intermolecular reactions of dihydrogen with coordinatively unsaturated platinum centers is, however, a facile reaction, and occurs by a mechanism very similar to that of the intramolecular reaction that cleaves carbon-hydrogen bonds[15] (FIGURE 4). Why does dihydrogen react so much more readily than compounds containing carbon-hydrogen bonds? By how much do the two reaction types differ in rate?

We have made a qualitative estimate for these rate differences by comparing thermodynamic parameters for intramolecular carbon-hydrogen activation and intermolecular hydrogen-hydrogen activation (FIGURE 5). Our estimate was calculated by assuming that intermolecular carbon-hydrogen bond activation of a solvent molecule (assumed to be cyclohexane) would proceed with the *activation energy* of the *intramolecular* reaction and the *activation entropy* of the *intermolecular* reaction with dihydrogen.† The result of this calculation is that intermolecular activation of carbon-hydrogen bonds of solvent would be expected to occur 10 times faster than intramolecular activation of carbon-hydrogen bonds. Thus expectation is not in accord with experiment, and is thus based on one or more incorrect premises. What are they? We do not presently know, and will not be certain until we have been able to examine and analyze authentic intermolecular carbon-hydrogen bond-cleaving reactions. We note, however, that the estimation summarized in FIGURE 5 explicitly neglects substantial differences between nonbonded interactions for addition of the carbon-hydrogen bond of cyclohexane to L_2PtR_2 and those for addition of (the much smaller) H_2. There may also be significant differences in the entropies of the transition states for these reactions, reflecting restrictions to bending or torsional motions.

† To facilitate comparison, concentration of H_2 was set equal to the concentration of C—H bonds in pure cyclohexane (110 M).

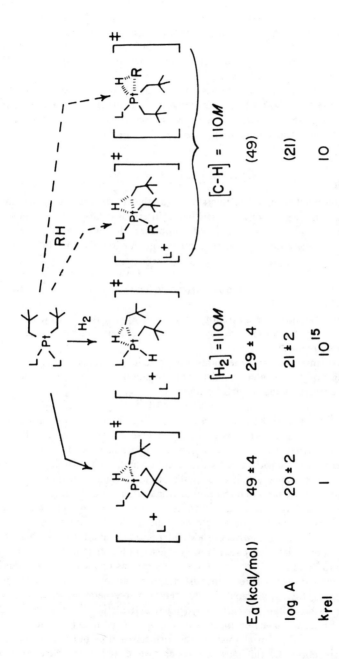

FIGURE 5. Comparisons of kinetic parameters for several organometallic reactions, and estimates of the relative rate of inter- and intramolecular reactions breaking carbon-hydrogen bonds.

Whatever the origin of the failure of the calculation summarized in FIGURE 5, several useful facts emerge from it. First, the relative rates of intra- and intermolecular cleavage of comparable carbon-hydrogen bonds by addition to an LPtR$_2$ fragment are probably not greatly different. Second, the slower rate of the intermolecular reaction is not due exclusively to the difference in entropy arising from the necessity of combining two particles into one: the effective concentration of C—H bonds in liquid hydrocarbon is very high. Third, the most probable (in our opinion) origin for the slowness of the intermolecular reaction is steric congestion in the transition state. This congestion might be reflected in either E_a (that is, in nonbonded steric strain) or in log A (restricted rotation and bending in a crowded transition state).

Heterogeneous Reactions

The original problem posed in this study was that of understanding why bulk platinum metal activates carbon-hydrogen bonds intermolecularly with considerable facility, while platinum ions in soluble complexes do not do so. We have suggested that there are three broad types of rationalizations for this difference.

1. The bulk metal has characteristic electronic features reflecting cooperative interactions between metals or extensive electronic delocalization that give it unique catalytic activity.

2. The bulk metal provides no more than an array of individual metal surface atoms, but these metal atoms are, by their nature, coordinatively unsaturated and hence highly reactive.

3. Catalysis involves some mechanism intermediate between 1 and 2, that is, localized surface organometallic chemistry, but organometallic chemistry requiring cooperation of several adjacent metal centers.

In order to distinguish between these possibilities, it is important to develop techniques for studying heterogeneous catalytic reactions over metals. There is general agreement that metal surface alkyls are intermediates in many of these catalytic systems, and considerable evidence concerning their structures (at least under certain circumstances). There is, however, little agreement concerning the factors that influence the rates of formation or transformation of these intermediates. For mechanistic studies in this area, we would like to be able to prepare alkyls on the surfaces of catalytically interesting transition metals by routes independent of those normally used in catalysis, and under circumstances in which carbon skeletons and isotopic labeling patterns are (at least initially) well defined. Ideally, we would like to be able to carry out these preparations under conditions similar to those encountered during heterogeneous catalytic reactions.

Our efforts in this area have so far centered on heterogeneous catalytic hydrogenation, rather than on carbon-hydrogen activation. The reason for this choice is pragmatic: we have developed a new type of experimental system that is applicable to studies of the mechanism of the former reaction, but not (yet) to that of the latter reaction. In this new approach, we examine the heterogeneous platinum-catalyzed reduction by dihydrogen of soluble (bisolefin)dialkylplatinum(II)

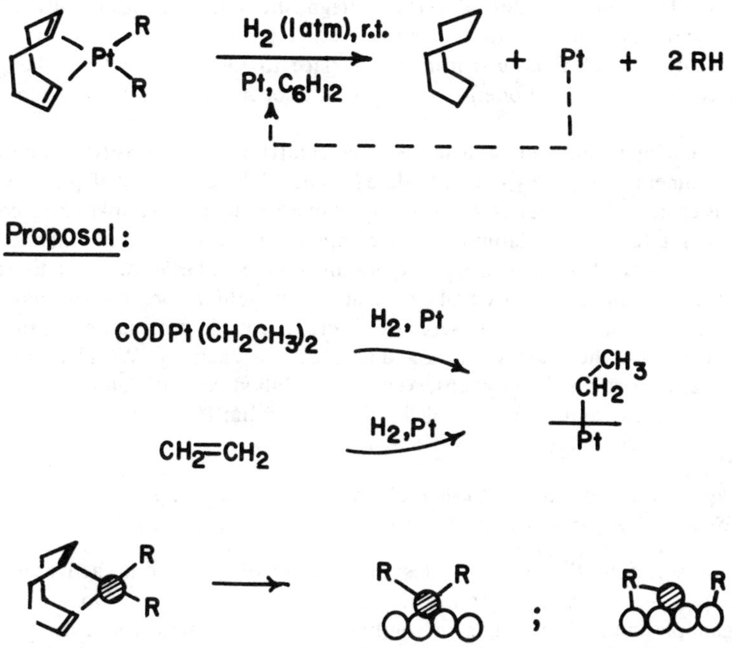

Proposal:

FIGURE 6. Heterogeneous platinum-catalyzed hydrogenation of (COD)PtR$_2$.

complexes (FIGURE 6).[18] We believe that this reaction proceeds by initial adsorption of the organoplatinum compound on the catalyst surface, with concomitant dissociation into adsorbed diolefin and adsorbed "dialkylplatinum." Once the alkyl group originally present in the dialkyl platinum moiety has been transferred to the platinum surface, we believe that it loses its memory of its origin, and becomes indistinguishable from corresponding alkyl groups formed by reaction of olefins with hydrogen on the metal surface. Evidence supporting the proposed reaction mechanism is summarized elsewhere.[18] The essential features of this evidence are:

1. The kinetics of the platinum-catalyzed reduction of (cycloocta-1,5-diene)-dimethylplatinum(II) by dihydrogen are very similar to those for the reduction of typical olefins by dihydrogen using the same catalyst system.

2. Hydrogen interchange occurs between the olefin and the alkyl groups during reduction. This exchanging hydrogen can, in turn, exchange with deuterium present in deuterated solvents.‡ We infer that the exchanging hydrogen is present as mobile platinum surface hydride.

3. The platinum metal originally present in the soluble organoplatinum complex plates or precipitates on the surface of the catalyst. This process constantly renews the surface of the heterogeneous catalyst and effectively prevents its poisoning. The system is thus a particularly easy one to study kinetically.

---

‡ Methanol-d$_4$ or diethyl ether saturated with D$_2$O was used in these experiments, using procedures outlined in Reference 18.

4. The relative rates of certain diagnostic reactions (especially hydrogen-deuterium exchange between adsorbed alkyl groups and solvent and reductive elimination of alkyl groups from the surface to form alkanes) are the same for the catalytic reduction of ethylene and of (cyclooctadiene)diethylplatinum.

What have studies of this heterogeneous catalytic reaction involving homogeneous organometal complexes shown to date? Some of the chemistry of the alkyl groups transferred to the catalyst surface via soluble (diolefin)dialkylplatinum complexes is similar to that of platinum alkyl complexes in solution, but other features of the chemistry of these species are quite different. Similarly, the catalytic reduction of the diolefin group originally present in the soluble organoplatinum complex is similar to that observed in separate heterogeneous hydrogenation of the diolefin, but here also there are significant differences in reactivity. We give one example of characteristic reactivity observed for the alkyl and diolefin moieties in the heterogeneous reduction of (diolefin)dialkylplatinum(II) complexes.

Relative Rates of α and β Carbon-Hydrogen Bond Activation in Soluble Complexes and on Surfaces

In solution, the relative rates of carbon-hydrogen bond activation are, in general, $\beta > \gamma > \alpha$ (α carbon-hydrogen bond activation is seldom observed for organoplatinum compounds in solution). On the platinum surface, the relative rates of these reactions can be inferred by determining the relative rates of deuterium transfer to cyclooctadiene from, e.g., $PtCH_2CD_3$ and $PtCD_2CH_3$. These results indicate qualitatively that the relative rates of C—H bond-cleavage reactions are $\alpha > \beta$, γ. This result is significant in that it suggests either a bridging alkylidine species 1 or a surface carbene 2 is readily formed on the surface.

Stereochemistry of Olefin Hydrogenation

Deuterogenation of norbornadiene over supported platinum yields almost entirely the *exo* deuterated product. Deuterogenation of (norbornadiene)dimethylplatinum yields a mixture of deuterated species, containing a significant quantity of *endo* deuterated material (the ratio of *endo*- to *exo*-deuterium is approximately 3:1) (Equation 1). The *exo* stereochemistry observed for the deuterogenation of norbornadiene itself is consistent with initial adsorption of the olefin on its least hindered side. Rationalization of the stereochemistry of hydrogenation of (norbornadiene)dimethylplatinum is less clear because of the complexity of the product mixture formed. The result is consistent with (but does not demand) an

$$(1)$$

initial adsorption of NBDPtMe$_2$ at platinum, followed by transfer of the diolefin to the metal surface on its *endo* face and subsequent reduction at this face.

Exploration of this new catalytic heterogeneous reaction has just begun. It is, however, one that offers promise of a range of opportunities for the preparation and study of otherwise inaccessible platinum surface alkyls, especially those—phenyl, CF$_3$, CH$_3$—that would be impossible to form from olefinic precursors by hydrogenation.

CONCLUSION

What do we conclude from these studies concerning the origin of the apparent difference in the ability of bulk platinum, and of individual soluble platinum atoms present in organometallic complexes of structure LPtR$_2$, to break carbon-hydrogen bonds? Our tentative conclusion is that, in fact, the intrinsic reactivity of platinum in these two different environments toward carbon-hydrogen bonds may not be very different. In solution, however, for reasons that are not yet entirely clear, intramolecular carbon-hydrogen bond activation involving the alkyl platinum moieties or the phosphines, or some other reaction involving the phosphines, is sufficiently rapid that it essentially precludes intermolecular carbon-hydrogen bond activation. If it were possible to suppress this (these) intramolecular reaction, it should then be possible to observe intermolecular carbon-hydrogen activation. The technical key to further progress in this area thus seems to be the design and synthesis of phosphines and alkyl ligands that are capable of maintaining the platinum (or other transition metal) centers in soluble, reactive form, but are themselves unreactive in cyclometalation, intramolecular carbon-hydrogen activation, or other reactions.

SUMMARY

This paper discusses mechanisms of a number of elementary reactions of organo-platinum compounds in solution, with particular emphasis on reactions that activate carbon-hydrogen bonds. It also summarizes a new reaction—the platinum-catalyzed heterogeneous hydrogenation of soluble (diolefin)dialkylplatinum(II) compounds—which provides a bridge between the chemistry of soluble and surface platinum alkyls.

ACKNOWLEDGMENTS

A number of individuals other than those listed as coauthors have made important contributions to this work. The names of these individuals will be found in the references listed in the text.

REFERENCES

1. WEBB, G. 1978. *In* Catalysis **2:** 145–175. The Chemical Society. London, England.
2. KUZNETSOV, B. N., Y. I. YERMAKOV, M. BOUDART & J. P. COLLMAN. 1978. J. Mol. Catal. **4:** 49–55.
3. CSICSERY, S. M. 1979. Adv. Catal. **28:** 293–321.
4. VASSILIEV, Y. B., V. S. BAGOTZKY, O. A. KHAZOVA, V. V. CHERNY & A. M. MERETSKY. 1979. J. Electroanal. Chem. **98:** 253–272.
5. JANOWICZ, A. H. & R. G. BERGMAN. 1982. J. Am. Chem. Soc. **104:** 352–354.
6. CRABTREE, R. H., M. F. MELLEA, J. M. MIHELCIC & J. M. QUIRK. 1982. J. Am. Chem. Soc. **104:** 107–113.
7. HOYANO, J. K. & W. A. G. GRAHAM. 1982. J. Am. Chem. Soc. **104:** 3723–3725.
8. WHITESIDES, G. M., J. F. GAASCH & E. R. STEDRONSKY. 1972. J. Am. Chem. Soc. **94:** 5258–5270.
9. MCCARTHY, T. J., R. G. NUZZO & G. M. WHITESIDES. 1981. J. Am. Chem. Soc. **103:** 3396–3403.
10. FOLEY, P., R. DiCOSIMO & G. M. WHITESIDES. 1980. J. Am. Chem. Soc. **102:** 6713–6725.
11. NUZZO, R. G., T. J. MCCARTHY & G. M. WHITESIDES. 1981. J. Am. Chem. Soc. **103:** 3404–3410.
12. IBERS, J. C., R. DiCOSIMO & G. M. WHITESIDES. 1982. Organometallics **1:** 13–20.
13. DiCOSIMO, R. & G. M. WHITESIDES. 1982. J. Am. Chem. Soc. **104:** 3601–3607.
14. DiCOSIMO, R., S. S. MOORE, A. F. SOWINSKI & G. M. WHITESIDES. 1982. J. Am. Chem. Soc. **104:** 124–133.
15. REAMEY, R. H. & G. M. WHITESIDES. J. Am. Chem. Soc. (Submitted.)
16. WEBSTER, D. E. 1977. Adv. Organomet. Chem. **15:** 147–188.
17. PARSHALL, G. W. 1975. Acc. Chem. Res. **8:** 113–121.
18. MCCARTHY, T. J., Y.-S. SHIH & G. M. WHITESIDES. 1981. Proc. Nat. Acad. Sci. USA **78:** 4649–4651.

IRIDIUM HYDRIDES WITH DI(TERTIARY PHOSPHINE) BRIDGES AND CHELATES*

Richard Eisenberg and Barbara J. Fisher

Department of Chemistry
University of Rochester
Rochester, New York 14627

INTRODUCTION

The activation of substrate is an essential step in catalysis, and generally involves a weakening or breaking of bonds within the substrate. The effectiveness of transition metal complexes as catalysts is closely related to their ability to perform this process. Substrate activation is often accomplished by donation of electron density from filled metal d orbitals into vacant antibonding orbitals of the substrate, thus perturbing its electronic structure, and resulting in an oxidative addition reaction if a substrate bond is cleaved. Complexes of electron-rich metals are particularly effective in activating substrates in this manner, and within this genre, no set of complexes has been more vigorously studied over the past two decades than those of Rh(I) and Ir(I). Complexes of these d^8 ions possess a rich oxidative addition chemistry, and are active as catalysts for a variety of reactions including hydrogenations, hydroformylation, and carbonylations (see Reference 1 for examples).

While complexes of Ir(I) are often not as catalytically active as analogous Rh(I) systems, the electron richness of Ir(I) frequently yields more stable substrate adducts and oxidative addition products (see Reference 2 and references therein). In this context, Vaska's complex, $IrCl(CO)(PPh_3)_2$, is especially notable, undergoing reactions with numerous substrates including H_2, HX (X = Cl, Br, or I), MeI, BzBr, RCOX, and R_3SiH among others.[3-5] In this reaction chemistry and that of closely related analogues, the phosphine ligands L generally maintain their *trans* disposition and yield stable adducts having Structures 1 or 2 depending on the substrate XY and the mechanism of adduct formation.

1

2

In this paper, we describe our studies on iridium complexes containing di(tertiary phosphine) ligands. These ligand systems may either chelate a single metal center or bridge two Ir ions. The former leads to a *cis* stereochemistry of phosphine donors different from that observed in most adducts of Vaska's complex

* This research was supported in part by the National Science Foundation and by the Office of Naval Research.

and its analogues, while the latter produces two metal centers in close proximity for the binding and activation of substrates.

The relative tendency of the di(tertiary phosphine) ligand system $Ph_2P(CH_2)_nPPh_2$ to bridge or chelate has been addressed by Sanger, who synthesized a monomer only when $n = 2$ and dimers when $n = 1$ and 3 via Equation 1.[6] Compound 3, which was first reported by Vaska, forms because of the favorable driving force of five-membered chelate ring formation.[7] Complexes 4 and 5, on the other hand, maintain what seems to be the electronically favorable disposition of *trans* P donors with the creation of face-to-face dimers of Ir(I) having ligand sets similar to that found in Vaska's complex.

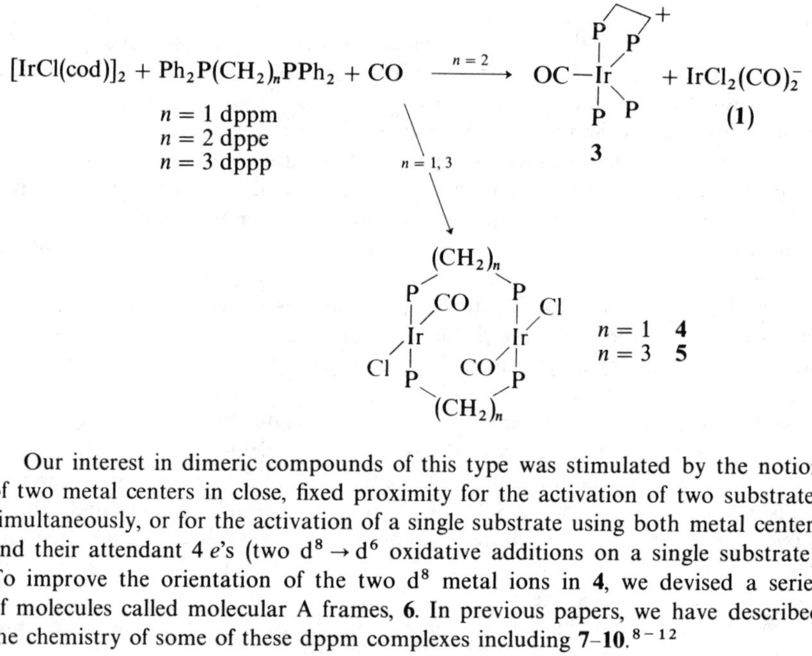

$$[IrCl(cod)]_2 + Ph_2P(CH_2)_nPPh_2 + CO \xrightarrow{n=2} \quad OC-Ir \begin{array}{c} \\ \\ \end{array} \quad + IrCl_2(CO)_2^- \quad (1)$$

$n = 1$ dppm
$n = 2$ dppe
$n = 3$ dppp

$$n = 1, 3$$

$$\begin{array}{c} n = 1 \quad 4 \\ n = 3 \quad 5 \end{array}$$

Our interest in dimeric compounds of this type was stimulated by the notion of two metal centers in close, fixed proximity for the activation of two substrates simultaneously, or for the activation of a single substrate using both metal centers and their attendant 4 e's (two $d^8 \rightarrow d^6$ oxidative additions on a single substrate). To improve the orientation of the two d^8 metal ions in **4**, we devised a series of molecules called molecular A frames, **6**. In previous papers, we have described the chemistry of some of these dppm complexes including **7–10**.[8-12]

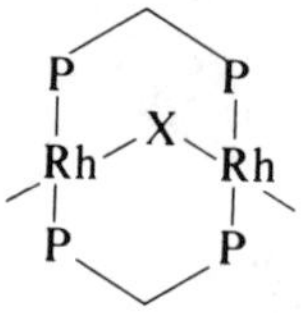

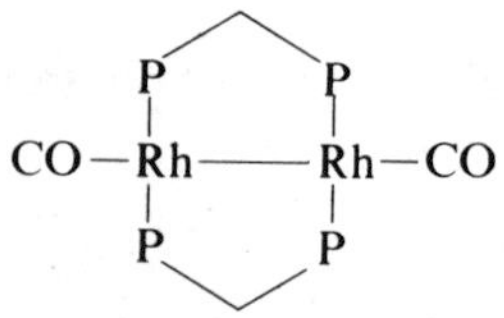

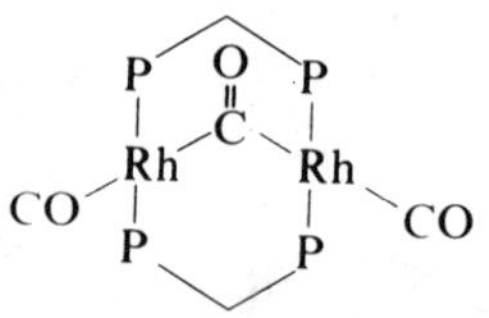

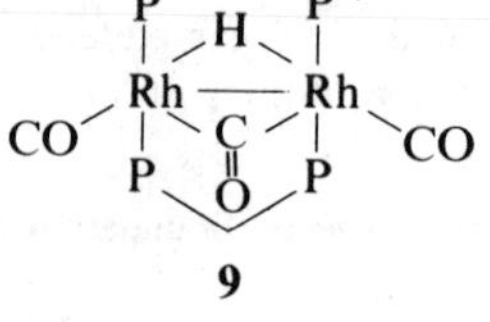

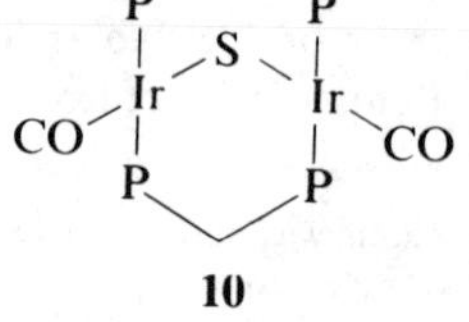

Complex 9 is modestly active as a catalyst for the water gas shift reaction, Equation 2, but its catalyst lifetime is relatively short. Complex 10 forms reversible adducts

$$CO + H_2O = CO_2 + H_2 \qquad (2)$$

with CO and with H_2 but not with both simultaneously. Further studies on these and related A-frame systems are in progress.[13]

While the dppm ligand keeps the two bridged metal centers in close proximity, the dppp ligand allows the binuclear complexes to be more flexible with metal-metal distances ranging from ~ 3.5 Å to >6 Å. The chemistry of **5** has recently been explored by Pignolet and Wang, who find that **5** oxidatively adds H_2 to form a mixture of the dihydride $[Ir_2H_2(CO)_2Cl_2(dppp)_2]$, **11**, and the tetrahydride $[Ir_2H_4(CO)_2Cl_2(dppp)_2]$, **12**, in Equation 3.[14]

The structural assignments of **11** and **12** are supported by crystallographic evidence. Loss of H_2 from **12** appears facile, and the increased steric bulk at one Ir center after the first oxidative addition appears to inhibit reactivity at the second metal center.

The studies described in this paper use the work of Sanger and Pignolet as a starting point. Because the bromo and iodo analogues of Vaska's complex were known to be more reactive than the parent chloro system $IrCl(CO)(PPh_3)_2$, we commenced studies on the bromo and iodo analogues of **5**. The context in which our studies were undertaken was the development of H_2 reduction catalysts for CO_2, a goal that still remains to be reached. Based on Herskovitz' work, it was known that electron-rich Ir(I) centers are capable of reacting with CO_2.[15] We envisioned that the presence of nearby hydrides on a second metal center would facilitate the desired reduction of bound CO_2.

The investigations that we outline here include studies of binuclear dppp complexes and the hydrides that they form, the cleavage of these dimers into mononuclear species, and the formation and reaction chemistry of previously unknown mononuclear complexes containing only one dppe ligand. One of these dppe systems upon irradiation activates arene C—H bonds and promotes the formation of benzaldehyde and benzyl alcohol from benzene and synthesis gas. This reaction represents an important example of C—H bond functionalization.

THE SYNTHESIS AND CHARACTERIZATION OF NEW IRIDIUM HYDRIDES CONTAINING DI(TERTIARY PHOSPHINE) LIGANDS

The iridium(I) anion $[Ir(CO)_2Br_2]^-$ as its n-Bu$_4$N$^+$ salt serves as the convenient starting material for the preparation of the mono- and binuclear complexes of iridium reported here.

dppp Complexes

The room temperature reaction of $(n\text{-}Bu_4N)[Ir(CO)_2Br_2]$ with dppp in acetone under N_2 leads to the evolution of CO and the essentially quantitative production of the pale yellow complex $[Ir_2Br_2(CO)_2(dppp)_2]$, **13**. The dimeric structure of **13** is assigned based on elemental analyses, a singlet at δ 19.03 in the ^{31}P NMR

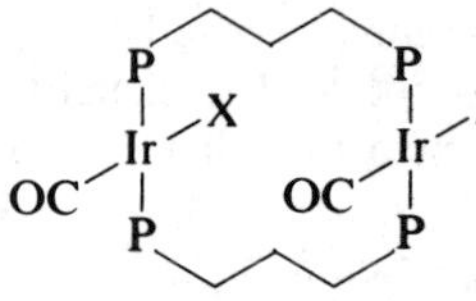

13 X = Br
14 X = I

spectrum of the complex, and by analogy with the chloro complex **5** reported by Sanger. The CO ligands in **13** are shown in a *cis* orientation because of two v_{CO}'s at 1,944 and 1,915 cm^{-1}. The diiodo complex **14** is prepared by metathesis using a 100-fold excess of LiI and a slurry of the chloro complex **5** in benzene and, based on the single v_{CO} at 1,950 cm^{-1}, is assigned a structure with CO ligands in the *trans* orientation as found in **5**.

The oxidative addition of H_2 to **13** and **14** yields binuclear hydride complexes. Under 1 atm H_2 in CH_2Cl_2, **13** and **14** form the tetrahydride complexes **15** and **16**. The formation of **16** in tetrahydrofuran (THF) at 25°C is essentially complete within 1 hour compared with 24 hours for the formation of **15** and only incomplete conversion of the chloro complex **5** to its tetrahydride **12** under the same conditions.

$$[IrX(CO)(dppp)]_2 + 2H_2 \longrightarrow \qquad\qquad (4)$$
$$X = Br, I$$

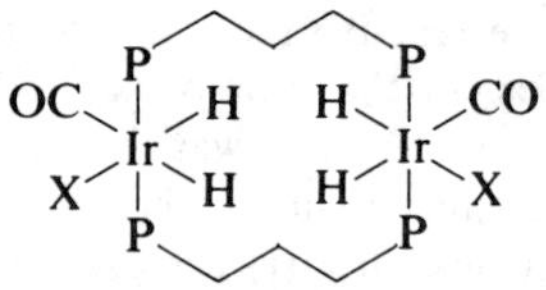

15 X = Br
16 X = I

This observation is consistent with the notion of increased reactivity with halide ligand in the order Cl < Br < I. Oxidative addition of only a single molecule of H_2 to **13** to yield the Ir(I)-Ir(III) dihydride **17** can be accomplished by using only one equivalent of H_2. In acetone, the reaction of **13** with H_2 yields a mixture of **15** and **17** because of the inhomogeneous nature of the reaction.

The binuclear hydrides **15–17** were characterized spectroscopically, and relevant

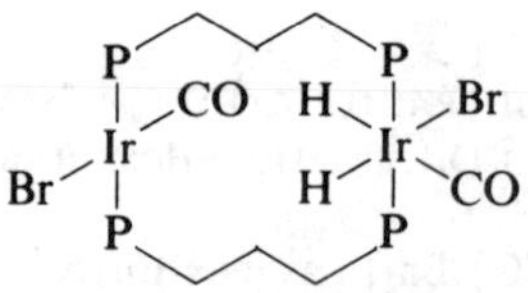

17

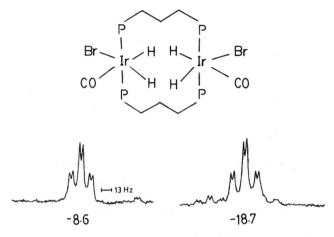

FIGURE 1. Hydride region of **15**.

data are presented in TABLE 1. The hydride region of the ^{1}H NMR spectrum, as illustrated in FIGURE 1, is particularly informative, showing for each species two triplets of doublets separated by ~9 ppm indicating chemically different hydride ligands. The triplet splitting at each chemical shift is 13 Hz, and is due to coupling to two equivalent P nuclei. The doublet splitting is 3 Hz and is due to J_{HH} between the two hydrides. Homonuclear decoupling of each triplet results in the loss of doublet splitting at the other. These observations are consistent with "micro" structure **A**, in which the phosphine donors are *trans*.

Additional support for our interpretation of the ^{1}H NMR spectral data is obtained by analogy to the known mononuclear hydride complexes IrH$_2$X(CO)(PPh$_3$)$_2$, which possess geometry **A**. The ^{1}H NMR spectra of these complexes are essentially the same as we observed for **15–17**.

A very different hydride pattern is obtained, however, when a reaction solution of Ir(CO)$_2$I$_2^-$ + dppp is treated with H$_2$. Prior to treatment with H$_2$, a complex may be isolated that exhibits ν_{CO} at 2,040 and 1,955 cm^{-1} and a singlet in the ^{31}P NMR spectrum at δ −28.5. The ^{1}H NMR spectrum obtained upon the addition of H$_2$ is shown in FIGURE 2. The main features of the hydride pattern are two groups of resonances separated by ~6 ppm, one of which is a broad doublet of doublets and the other a more complicated multiplet. The spectrum is consistent with chemically different P nuclei, and the magnitude of the larger doublet-of-doublets splitting (P$_{P-H}$ = 120 Hz) suggests that one of the hydrides is *trans* to a phosphine donor. "Micro" structure **B** is consistent with the ^{1}H NMR results. The second hydride is located *trans* to iodide based on its chemical shift.

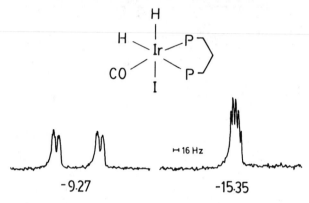

FIGURE 2. Hydride region of **19**.

The spectroscopic data provide the basis for a consistent interpretation of the reactions involving $Ir(CO)_2I_2^-$ + dppp and H_2 as shown in Equation 5. The initial product having ν_{CO} of 2,040 and 1,955 cm^{-1} is mononuclear, unlike the other dppp complexes formed in analogous reactions, and the two carbonyl stretches indicate a *cis* disposition of CO ligands in the complex. This complex, $IrI(CO)_2(dppp)$, is assigned structure **18** with a chelating di(tertiary phosphine) ligand. The singlet in the ^{31}P NMR spectrum of **18** may be ascribed to stereochemical nonrigidity of this five-coordinate d^8 species. The reaction of **18** with H_2 leads to loss of CO, which is detected in the gas phase above the solution, and the formation of **19**,

$$Ir(CO)_2I_2^- + dppp \longrightarrow \quad \underset{\textbf{18}}{I-\overset{\overset{\displaystyle P}{\displaystyle |}}{\underset{\underset{\displaystyle CO}{\displaystyle |}}{Ir}}\overset{P}{\underset{CO}{}}} \quad \xrightarrow[-CO]{+H_2} \quad \underset{\textbf{19}}{\overset{H}{\underset{H}{}}\overset{\overset{\displaystyle P}{\displaystyle |}}{\underset{\underset{\displaystyle CO}{\displaystyle |}}{Ir}}\overset{P}{\underset{I}{}}} \qquad (5)$$

which possesses "micro" structure **B**. Alternative binuclear formulations of **18** and **19** with bridging dppp ligands in *cis* positions of the coordination sphere are ruled out based on highly unfavorable steric interactions between neighboring diphenyl-phosphino groups and based on parallel reaction chemistry observed using dppe, which is discussed below.

Cleaving dppp-Bridged Binuclear Hydrides

When CH_2Cl_2 solutions of the binuclear hydride complex **17** are refluxed for extended times or heated in the presence of NEt_3 for shorter times (two to five hours), a striking change occurs in the hydride region of the ^1H NMR spectrum. Specifically, the resonances characteristic of arrangement **A** are replaced by those consistent with arrangement **B**. The change from *trans* to *cis* P donors occurs with a cleavage of the binuclear hydride system to the mononuclear structure **20** as shown in Equation 6.

$$
\underset{\textbf{15}}{\overset{\displaystyle \text{Br}_{\diagdown}|_{\diagup}\text{H} \quad \text{H}_{\diagdown}|_{\diagup}\text{CO}}{\underset{\text{OC}^{\diagup}|^{\diagdown}\text{H} \quad \text{H}^{\diagup}|^{\diagdown}\text{Br}}{\text{Ir} \qquad \text{Ir}}}} \quad \xrightarrow{\;\Delta\;} \quad \underset{\textbf{20}}{\overset{\displaystyle \text{H}_{\diagdown}|_{\diagup}\text{P}}{\underset{\text{H}^{\diagup}|^{\diagdown}\text{Br}}{\text{Ir}}}} \qquad (6)
$$

The lack of integrity of **17** as a binuclear species was surprising since Pignolet and Wang had employed harsher conditions in their hydrogenation experiments with [IrCl(CO)(dppp)]$_2$, **5**, and had not observed any evidence of dimer cleavage. Moreover, reactions of similar complexes having dppm bridges under a variety of conditions have revealed no tendency of these dimers to break apart into monomeric species. Our observation of dimer cleavage provides an important caveat to studies based on using di- and poly(tertiary phosphine) ligands to hold two or more metal atoms together in systems having structural integrity.

dppe Complexes

In order to provide additional support for the structural assignments of **18–20** as mononuclear species, the analogous complexes with dppe in place of dppp were synthesized and characterized. The ligand dppe shows a much greater tendency toward chelation than does dppp, and only rarely forms a bridge between two metals. The new mononuclear mono(dppe) complexes Ir(CO)X(dppe), **21** (X = Br) and **22** (X = I), were prepared by reacting (n-Bu$_4$N)[Ir(CO)$_2$X$_2$] with dppe in refluxing THF or benzene according to Equation 7. Formation of **3** in this reaction cannot be avoided, but the two different products can be easily separated. Complexes **21** and **22** have very similar spectroscopic properties

$$
(n\text{-Bu}_4\text{N}^+)[\text{Ir(CO)}_2\text{X}_2]^- + \text{dppe} \longrightarrow \quad \underset{\substack{\textbf{21 } X = Br \\ \textbf{22 } X = I}}{\overset{\displaystyle \text{X}_{\diagdown} \quad _{\diagup}\text{P}}{\underset{\text{CO}^{\diagup} \quad {}^{\diagdown}\text{P}}{\text{Ir}}}} \qquad (7)
$$

(see TABLE 1). The single ν_{CO} of 1,980 cm^{-1} for **21** and 1,980 cm^{-1} for **22** and the two doublets in the ^{31}P NMR spectrum of each complex provide convincing evidence that **21** and **22** are mononuclear mono(dppe) species.

Solutions of the orange, square planar Ir(I) complexes **21** and **22** in THF or CH$_2$Cl$_2$ react extremely rapidly with H$_2$ to form the mononuclear dihydride species **23** and **24**, respectively. These complexes are readily isolated as colorless

$$
\underset{\substack{\textbf{23} \quad X = Br \\ \textbf{24} \quad X = I}}{\overset{\displaystyle \text{H}_{\diagdown}|_{\diagup}\text{P}}{\underset{\text{H}^{\diagup}|^{\diagdown}\text{X}}{\underset{\text{CO}}{\text{Ir}}}}}
$$

TABLE 1

SPECTROSCOPIC DATA OF IRIDIUM dppp AND dppe COMPLEXES

Compound	Infrared (cm⁻¹)*		³¹P NMR	¹H NMR (hydride region only)¶
	ν_{CO}	ν_{Ir-H}		
[Ir₂Br₂(CO)₂(dppp)₂] **13**	1915, 1944		19.03(s)‡	
[Ir₂I₂(CO)₂(dppp)₂] **14**	1950		14.0(s)‡	
[Ir₂H₄Br₂(CO)₂(dppp)₂] **15**	1945, 1980	2100, 2190	0.04(s), 7.47(s)‡	−8.6 (t of d), −18.7 (t of d), $J_{PH\text{-}cis}$ = 13 Hz, J_{HH} = 3 Hz
[Ir₂H₄I₂(CO)₂(dppp)₂] **16**	1980	2090, 2160		−9.74 (t of d), −16.84 (t of d) (CDCl₃) $J_{PH\text{-}cis}$ = 13 Hz, J_{HH} = 3 Hz
[Ir₂H₂Br₂(CO)₂(dppp)₂] **17**	1945, 1980	2100, 2190		−8.4, −8.8, −18.2, −18.8 (all t of d) (CDCl₃) $J_{PH\text{-}cis}$ = 13 Hz, J_{HH} = 3 Hz
IrI(CO)₂(dppp) **18**	1955, 2040		−28.49(s)‡	
IrH₂I(CO)(dppp) **19**	2042	2105	21.9(d), 29.9(d)‡ $J_{P\text{-}P}$ = 30.52 Hz	−9.27 (d of d, $J_{PH\text{-}trans}$ = 120 Hz, $J_{PH\text{-}cis}$ = 16 Hz), −15.35 (m) (C₆D₆)
IrH₂Br(CO)(dppp) **20**	2043	2220		−9.27 (d of d, $J_{PH\text{-}trans}$ = 120 Hz, $J_{PH\text{-}cis}$ = 16 Hz), −17.99 (m) (C₆D₆)
IrBr(CO)(dppe) **21**	1980		43.7(d), 47.9(d)§ $J_{P\text{-}P}$ = 14 Hz	

Compound	IR	IR	^{31}P NMR	^{1}H NMR
IrI(CO)(dppe) **22**	1980		64.3(d), 62.21(d)§ $J_{P-P} = 9.5$ Hz	
IrH$_2$Br(CO)(dppe) **23**	2030	2195	33.8(d), 26.6(d)§	−9.05 (d of d of d, $J_{PH\text{-}cis} = 17$ Hz, $J_{PH\text{-}trans} = 130$ Hz, $J_{HH} = 4.5$ Hz), −18.26 (m) (CDCl$_3$)
IrH$_2$I(CO)(dppe) **24**	2040	2160	28.06(d), 20.62(d)‡ $J_{P-P} = 7$ Hz	−9.92 (d of d of d, $J_{PH\text{-}cis} = 17$ Hz, $J_{PH\text{-}trans} = 128$ Hz, $J_{HH} = 4.5$ Hz), −16.27 (m) (acetone$_{d-6}$)
IrBr(CO)$_2$(dppe) **25**	1940, 2040		53.51(s)§	
IrI(CO)$_2$(dppe) **26**	1950, 2040		32.0(s)§	
IrH$_3$(CO)(dppe) **27**	2010	1940, 2000, 2060	30.8(s)§	−9.48 (d of d of d, $J_{PH\text{-}trans} = 124$ Hz, $J_{PH\text{-}cis} = -12.2$ Hz, $J_{HH} = 4.5$ Hz), −10.86 (t, $J_{PH\text{-}cis} = 19$ Hz) (C$_6$D$_6$)
IrD$_3$(CO)(dppe) **28**	2030	†		
IrH(CO)$_2$(dppe) **29**	1913, 1966	2000	33.8(s)§	−10.36 (t, $J_{PH\text{-}cis} = 41$ Hz) (C$_6$D$_6$)

* IR spectra were recorded on a Perkin-Elmer 467 grating infrared spectrophotometer. All spectra were taken of KBr pellets except for **18**, **25**, and **26**, which where in benzene solution.

† $\nu_{IR\text{-}D}$ not observed.

‡ Measured in 10-mm tubes on Jeol PFT-100 spectrometer at 41.25 MHz.

§ Measured in 5-mm tubes on Bruker WH-400 at 162 MHz. Positive chemical shifts are downfield from H$_3$PO$_4$ (external).

¶ All proton NMR spectra were recorded on a Bruker WH-400 at 400.134 MHz. Positive chemical shifts are downfield from tetramethylsilane (TMS).

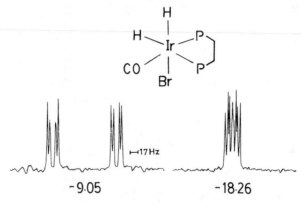

FIGURE 3. Hydride region of **23**.

crystals by the addition of EtOH and removal of solvent. Complexes **23** and **24** are spectroscopically similar to Complexes **19** and **20** (see TABLE 1). The hydride region of the ^{1}H NMR spectrum of **23** is shown in FIGURE 3. The "downfield" hydride resonance is a doublet of doublets of doublets ($J_{H-Ptrans} = 130$ Hz; $J_{H-Pcis} = 17$ Hz; $J_{H-H} = 4.5$ Hz), and is separated from the upfield" hydride multiplet by 9 ppm. Irradiation of the upfield multiplet results in loss of the smallest doublet splitting in the downfield hydride resonances, leaving a doublet-of-doublets pattern. The observation of hydride-hydride coupling for **23** and **24** represents the only difference in their ^{1}H NMR spectra from those observed for the analogous dppp complexes, **19** and **20**.

The addition of H_2 to **21** and **22** is reversible, as is the addition of CO to form the five-coordinate dicarbonyl complexes **25** and **26**, which are analogous to **18** (Equation 8). Refluxing THF or benzene solutions of these adducts under N_2 leads to loss of the addend molecule and regeneration of **21** or **22**. The

$$\text{IrX(CO)(dppe)} \xrightleftharpoons{\text{CO}} \text{X}-\underset{\underset{\text{CO}}{|}}{\overset{\overset{\text{P}}{|}}{\text{Ir}}}\overset{\text{P}}{\underset{\text{CO}}{\diagdown}} \tag{8}$$

25 X = Br
26 X = I

dicarbonyl complexes **25** and **26** exhibit two ν_{CO}'s at 1,940 and 2,040 cm^{-1}, and at 1,950 and 2,040 cm^{-1}, respectively, in close parallel with **18**.

Preparation and Characterization of $IrH_3(CO)(dppe)$

The trihydride complex IrH$_3$(CO)(dppe), **27**, is a particularly interesting compound, which was prepared according to Equation 9 by the reaction of NaBH$_4$ in ethanol with **22** in CH$_2$Cl$_2$ under H$_2$. This complex was isolated as a tan, air-stable powder and, when recrystallized, is colorless. The hydride region of the ^{1}H NMR spectrum of **27** is shown in FIGURE 4, and based on the splitting pattern observed, a facial configuration for the hydrides can be assigned unambiguously. The apical hydride, H$_a$, is unique and is cis to the 2 P donor atoms of dppe, giving rise to the triplet at $\delta = -10.86$ ppm ($J_{P-H} = 19$ Hz). The equatorial

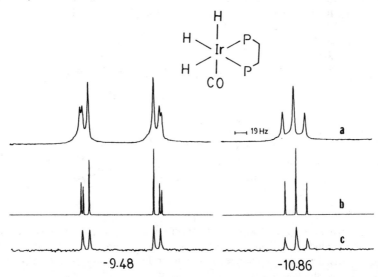

$$\underset{\textbf{22}}{\underset{\text{I}}{\overset{\text{OC}}{}}\text{Ir}\underset{\text{P}}{\overset{\text{P}}{}}\Bigg]} + BH_4^- + H_2 \longrightarrow \underset{\textbf{27}}{\underset{H_b}{\overset{H_bH_a}{}}\text{Ir}\underset{\text{CO}}{\overset{P}{}}\Bigg]} \qquad (9)$$

hydrides, H_b, are chemically equivalent (δ -9.27 ppm) and are split by a *trans* P (J_{P-H} = 124 Hz), a *cis* P (J_{P-H} = -12.2 Hz), and each other (4.5 Hz). A computer simulation of the hydride region confirms these assignments and is shown in FIGURE 4.

We found that Complex 27 loses H_2 both thermally and photochemically, with the rate of photogenerated loss much greater.† This was demonstrated by photolyzing 27 in benzene-d_6 solution under D_2 and CO and monitoring the ¹H NMR spectral changes with time. In reactions under D_2 and CO, only the hydride region of the spectrum was affected. After short photolysis times (20 minutes) or longer thermolysis times (2 hours), the outermost doublets of the -9.27 ppm resonance were observed to lose their hydride-hydride coupling as shown in FIGURE 4, and the integrated intensity of the hydride resonances decreased by $\sim 50\%$ relative to the dppe resonances of the complex. After 2.5 hours of photolysis, only traces of the hydride resonances remained, indicating that deuterium incorporation was essentially complete, converting 27 to $IrD_3(CO)(dppe)$, 28. This conversion requires 8 hours thermally.† The IR spectrum of the isolated product (see TABLE 1) together with the ¹H NMR data supports the formulation

FIGURE 4. Hydride region of 27. (a) Experimental spectrum. (b) Simulated spectrum. (c) Partial deuteration after thermolysis of 27 under D_2 for eight hours.

† These exchange experiments have been redone under more dilute conditions which decrease the rate of recombination between thermally or photolytically released H_2 or CO and the four-coordinate metal complex $IrH(CO)(dppe)$, making productive rates of thermal and photochemical exchange more favorable. Under these conditions, the thermal and photochemical exchange reactions proceed at comparable rates and occur more rapidly than the observations reported here. See Reference 16.

of Complex 28 as $IrD_3(CO)(dppe)$. When **28** dissolved in C_6H_6 is photolyzed under H_2, it is converted back to the trihydride, **27**.

Photolysis or thermolysis of **27** in benzene under CO leads to the rapid appearance of a new hydride resonance at $\delta -10.36$ ppm (t) accompanied by new resonances in the methylene and phenyl regions of the spectrum. After 2.5 hours of irradiation, all resonances of **27** are replaced by ones associated with the new triplet at -10.36 ppm. (The thermal reaction is complete in 8 hours.) The phenyl and methylene regions in the ^{1}H NMR spectrum of this material are nearly identical to those of the five-coordinate Ir(I) complex $IrI(CO)_2(dppe)$, **26**, which forms by CO addition to **22**. A larger scale photolysis of **27** under CO allows isolation of this new material, which we identify as $IrH(CO)_2(dppe)$, **29**, based on NMR and IR spectral data. Both **26** and **29** exhibit only sharp singlets in their $^{31}P\{^1H\}$ NMR spectra, indicating equivalence of the two dppe P donors in each complex at room temperature. Complex **29** may also be prepared from **26** and BH_4^- under a CO atmosphere in absolutely dry solvents.

The formation of **28** and **29** by photolysis or thermolysis of the trihydride **27** under D_2 and CO, respectively, is consistent with the reductive elimination of H_2 from **27**, generating the reactive four-coordinate species $IrH(CO)(dppe)$, **30**, which then adds D_2 or CO as shown in Equation 10. The formation of the trideuteride **28** requires at least two passes through the reductive elimination/ oxidative addition sequence in Equation 10 with production of one equivalent of HD. The photochemically promoted reductive elimination of H_2 from metal polyhydrides is now well documented, and has been found in a number of cases to generate highly reactive species (see Reference 17 for examples).

$$(10)$$

The proposed four-coordinate Ir(I) intermediate, **30**, is reactive to other substrates including benzene. When a benzene-d_6 solution of **27** is photolyzed under N_2 or vacuum, a change in the hydride spectrum similar to that seen under D_2 as shown in FIGURE 4 is observed after 20 minutes, along with a corresponding loss in the integrated intensity of the hydride resonances relative to those of dppe. This result indicates that deuterium incorporation into **27** is taking place with the solvent serving as the deuterium source.

Although a phenyl hydride species **31** corresponding to the oxidative addition product of benzene to $IrH(CO)(dppe)$ is not seen directly, the observed H/D exchange is most readily explained by its intermediacy.

$$\begin{array}{c} H(D) \\ H{\diagdown}\,|\,{\diagup}P \\ {\hspace{0.6em}}Ir{\hspace{0.6em}} \\ (d_5)Ph{\diagup}\,|\,{\diagdown}P \\ CO \end{array}$$

31

Carbonylation of Benzene to Benzaldehyde

The proposed existence of **31** stimulated further experiments to determine if CO insertion and elimination of carbonylated product could be seen. This was indeed the case. In all photolyses of **27** in C_6D_6 under CO or CO/H_2 mixture, a new resonance in the ^1H NMR spectrum at δ 9.63 ppm was observed, which is assignable to benzaldehyde. The only other observed change in the ^1H NMR spectrum was complete conversion of **27** to **29** as noted above. Experiments were done to confirm the formation of benzaldehyde. Photolysis of **27** in C_6H_6 under 600 torr CO followed by gas chromatographic (GC) analysis of the volatiles showed benzaldehyde present. Treatment of the volatiles with semicarbazide yielded a white crystalline material, which was shown to be benzaldehyde semicarbazone. In all cases, the amounts of benzaldehyde detected by GC analysis were small (5–8 mM after eight hours of photolysis). The carbonylation reaction does not appear to go thermally. When **27** in C_6H_6 under 600 torr CO is heated at 90°C for five days, no evidence of benzaldehyde formation is found by either GC or NMR methods.

The formation of benzaldehyde when **27** is photolyzed in benzene under CO indicates the occurrence of Equation 11, and represents an important example of C—H functionalization. The small amounts of benzaldehyde formed may be the

$$\text{(l)} \quad + \; CO \; = \; \text{CHO} \quad \text{(l)} \tag{11}$$

result of unfavorable thermodynamics for Equation 11. We calculate $\Delta G°$ and K_{eq} at 298°C for Equation 11 as $+1.7$ kcal/mol and 5.9×10^{-2} atm^{-1}, respectively. It should be noted that uncertainties in entropy values for liquid benzene and benzaldehyde permit a range of ΔG and K values to be calculated, but in all cases the reaction is thermodynamically unfavorable. This notion is confirmed when the trihydride **27** is photolyzed in C_6D_6 in the presence of C_6H_5CHO under vacuum. Within five minutes, the formation of the dicarbonyl species **29** was noted, and after eight hours, the conversion of **27** to **29** was complete and the benzaldehyde had decreased to a small but relatively steady value with a concomitant increase in the benzene resonance. It thus appeared that the equilibrium of Equation 11 was being approached from either direction.

Attempts were then made to determine K_{eq} and ΔG of Equation 11 experimentally. However, GC analysis of solutions from prolonged irradiation (>36 hours) revealed a new product, which has been identified by GC and NMR techniques to be benzyl alcohol. This result may prove to be highly significant since the reduction of benzaldehyde to benzyl alcohol, Equation 12, is thermodynamically favorable (ΔG is between -5 and -8 kcal/mol depending on the entropy values used).

$$\text{(I)} \quad \text{C}_6\text{H}_4\text{CHO} \quad + \quad \text{H}_2 \quad = \quad \text{C}_6\text{H}_4\text{CH}_2\text{OH} \quad \text{(I)} \tag{12}$$

The reaction may thus serve as a convenient drain for Equilibrium 11 so that benzene carbonylation can proceed productively. Longer term photolyses of **27** in C_6H_6 under 1:1 $CO:H_2$ (600 torr), however, reveal only the formation of ~ 2 equivalents of benzyl alcohol. While the only Ir complex detectable appears to be **29**, which can reenter the catalytic cycle by dissociation of CO, additional products are formed in these longer term experiments (125 hours), as evidenced by the development of yellow and orange colors in the reaction solution.

Further studies to examine the trihydride **27** and the dicarbonyl hydride **29** as catalysts for arene C—H bond functionalization are continuing.

SUMMARY

In this paper we have examined the synthesis and characterization of iridium hydrides having di(tertiary phosphine) ligands as bridges and chelates. Binuclear dppp bridge complexes of formula [IrH$_2$X(CO)(dppp)]$_2$ where X = Br or I possess the phosphine donors in *trans* disposition as shown by ^{1}H NMR spectroscopy. Upon heating, these dimers cleave into monomeric species of the same stoichiometry. The dppe complexes are all mononuclear and contain a chelated di(tertiary phosphine) ligand. The hitherto unreported IrX(CO)(dppe) X = Br or I complexes have been described, as has their reaction chemistry to form reversible adducts with CO and H$_2$. The trihydride complex IrH$_3$(CO)(dppe) in C_6H_6 under CO leads to the formation of benzaldehyde in possibly thermodynamically limited amounts. An intriguing observation under continuing study is the subsequent conversion of benzaldehyde to benzyl alcohol. These observations represent an important example of C—H bond functionalization.

ACKNOWLEDGMENTS

We wish to thank the Johnson Mathey Co., Inc. for a generous loan of iridium salts. We also wish to acknowledge valuable discussions with Prof. William D. Jones, Prof. Jack A. Kampmeier, Dr. Curtis Johnson, and Mr. Frank Feher.

REFERENCES

1. PARSHALL, G. W. 1980. Homogeneous Catalysis. John Wiley & Sons, Inc. New York, N.Y.
2. MILSTEIN, D. & J. C. CALABRESE. 1982. J. Am. Chem. Soc. **104:** 3773.
3. VASKA, L. & J. W. DiLUZIO. 1962. J. Am. Chem. Soc. **84:** 679.
4. CHOCK, P. B. & J. HALPERN. 1966. J. Am. Chem. Soc. **88:** 3511.
5. DEEMING, A. J. & B. L. SHAW. 1971. J. Chem. Soc. A: 1802.

6. SANGER, A. S. 1977. J. Chem. Soc. Dalton Trans.: 1971.
7. VASKA, L. & D. L. CATONE, 1966. J. Am. Chem. Soc. **88**: 5324.
8. KUBIAK, C. P. & R. EISENBERG. 1980. Inorg. Chem. **19**: 2726.
9. KUBIAK, C. P. & R. EISENBERG. 1980. J. Am. Chem. Soc. **102**: 3637.
10. KUBIAK, C. P., C. WOODCOCK & R. EISENBERG. 1982. Inorg. Chem. **21**: 2119.
11. KUBIAK, C. P., C. WOODCOCK & R. EISENBERG. 1980. Inorg. Chem. **19**: 2733.
12. KUBIAK, C. P. & R. EISENBERG. 1977. J. Am. Chem. Soc. **99**: 6129.
13. WOODCOCK, C. & R. EISENBERG. 1982. Organometallics **1**: 886.
14. PIGNOLET, L. & H. H. WANG. 1980. Abstract No. 168, Fall Meeting of the American Chemical Society, Las Vegas, Nev., September 1981.
15. HERSKOVITZ, T. & L. J. GUGGENBERGER. 1976. J. Am. Chem. Soc. **98**: 1615.
16. FISHER, B. J. & R. EISENBERG. 1982. Organometallics **2**: 764.
17. GEOFFROY, G. L. & M. S. WRIGHTON. 1979. Organometallic Photochemistry. Academic Press, Inc. New York, N.Y.

SURFACE HYDROGEN IN HETEROGENEOUS CATALYSIS*

H. Saltsburg and M. Mullins

*Department of Chemical Engineering
University of Rochester
Rochester, New York 14627*

INTRODUCTION

In recent years there has been an increasing effort to relate the catalytic properties of the transition metal complexes that have been studied in homogeneous systems to the properties of transition metals involved in heterogeneous catalysis.[1,2] Significant progress has been made in the understanding of the molecular details of homogeneous catalytic reactions due in no small part to the array of sophisticated spectroscopic and structural tools available to study these systems and the relative ease of data interpretation. By comparison, some of the most powerful structural techniques available for the study of surface species are not functional under typical industrial catalytic reaction conditions, and intrinsically present interpretive problems even when they can be used.[3] Techniques that utilize the properties of electrons or ions either as probes or as signals are restricted to vacuum operation. Photon-based *in situ* techniques are suitable but suffer from interpretive difficulties. Furthermore, although hydrogen is of major significance in many catalytic reactions, direct *in situ* observation of this surface species is difficult. Neutron scattering, although it can be used for this purpose, requires a substantial facility.[4]

Since the raison d'etre of catalysis is to cause reactions to selectively produce a desired product, in addition to the structural identification techniques of the surface scientist, there is a need for procedures that operate under more severe reaction conditions and that intrinsically permit one to examine the reaction components on the surface. For example, one of the difficulties in establishing the mechanism and kinetic behavior of heterogeneous catalytic reactions has been that, typically, the experimental observables are the concentrations of reactants and products in the gas phase rather than those on the surface where the reaction actually occurs. Data derived only from bulk gas phase properties are, therefore, incomplete. Direct measurement of appropriate properties of surface species would help elucidate details of surface processes.

In this study a new technique is described that permits one to measure the thermodynamic activity of chemisorbed hydrogen under reaction conditions. The method utilizes a solid state electrochemical device, similar in principle to that previously applied to the study of surface oxygen involved in the catalytic oxidation of SO_2.[5,6] Other applications of the oxygen sensor have been reported.[7,8]

In the following, the procedure for measuring the thermodynamic activity of

* Initial support for these studies came from the Office of Naval Research under Contract N00014–76–C–0001. Fellowship support of the E. H. Hooker Foundation to M. Mullins is gratefully acknowledged.

82

hydrogen at a gas-solid interface will be described. A summary of the results of a study of the Pt-catalyzed hydrogenation of ethylene and preliminary results of a study of the (dissociative) adsorption of propane, butane, and cyclohexane will be presented. For the hydrogenation reaction: (a) hydrogen on the surface of platinum can be observed quantitatively at atmospheric pressure and elevated temperatures during reaction; (b) during the ethylene hydrogenation reaction, gaseous hydrogen is not equilibrated with adsorbed hydrogenic species; (c) the species involved in the steady-state reaction are related by both the rate expressions and a newly discovered steady-state isotherm, which relates the gaseous pressures of hydrogen and ethylene to the activity of surface hydrogen; (d) the stoichiometric implications of the proposed mechanism are consistent with currently proposed models for the structure of ethylene adsorbed on platinum; and (e) although the hydrogen is a limiting reactant, a single rate-limiting step cannot be identified. For paraffin adsorption: (f) saturated hydrocarbons can exhibit dissociative adsorption without steady production of gaseous hydrogen; and (g) the energetics of the dissociative adsorption process suggest a stabilization of the surface hydrocarbon and hydrogenic species, which may be related to resonant bond stabilization of the surface species.

SURFACE HYDROGEN ACTIVITY

Theory

Wagner first discussed the possible use of solid electrolyte cells as a means of observing the reaction of atomic species during a heterogeneous catalytic reaction.[9] The technique was first demonstrated by Vayenas and Saltsburg using an oxygen-ion-conducting solid electrolyte to investigate the catalytic oxidation of SO_2.[5,6] The present study utilizes a similar electrochemical technique to study reactions involving surface hydrogen.

Thomas and White reported the rapid conduction of protons within a beta″-alumina electrolyte.[10] Using this electrolyte and platinum film electrodes, a room temperature gaseous hydrogen concentration cell was demonstrated by Lundsgaard and Brook.[11] Although the absolute electromotive force (EMF) was not that predicted, voltage changes across the cell were found to be in accord with the Nernst equation if the appropriate electrode reaction (half cell) at the three-phase interface could be written:

$$H_2 = 2\,H^+ + 2\,e^- \tag{1}$$

If one assumes that this reaction is the dominant exchange current reaction, and is at equilibrium at the three-phase boundary of gas, metal, and electrolyte, then the open circuit potential difference across the electrolyte can be described in terms of an appropriate equilibrium. The conducting ion being the proton, its activity is equal on both sides of the electrolyte. The resulting chemical potential balance is:

$$\mu_{H_2}^1 - \mu_{H_2} = 2(\mu_e^1 - \mu_e) = 2\,FE \tag{2}$$

where $\mu_{H_2}^1$ refers to a reference electrode, μ_{H_2} to the catalyst/electrode surface, E is the electric potential (EMF), and F is the Faraday constant. Defining the hydrogen activity on the platinum as $a_H^2(Pt)$, the chemical potential for the surface species, μ_{H_2}, may be written either in terms of $a_{H_2}(Pt)$ or $a_H^2(Pt)$ since equilibrium among surface hydrogenic species is assumed. One obtains:

$$\mu_{H_2} = \mu_{H_2(gas)}^0 + RT \ln a_H^2(Pt) \tag{3a}$$

$$= \mu_{H_2(gas)}^0 + RT \ln a_{H_2}(Pt) \tag{3b}$$

If one electrode is designated as a reference, the other as a reactor, and the reference activity is identified with the gas phase hydrogen partial pressure (assuming gas surface equilibrium exists on the reference side), the EMF of the cell may be described by:

$$EMF = (RT/2F) \ln (a_H^2/P_{H_2}) \tag{4}$$

Using pure hydrogen at one atmosphere as the reference gas, the measured EMF becomes a simple function of the hydrogen activity on the surface at the three-phase boundary within the reaction cell:

$$EMF = (RT/F) \ln a_H \tag{5a}$$

$$= (RT/2F) \ln a_H^2 \tag{5b}$$

Experimental

The essential component of the experimental apparatus (FIGURE 1) is a closed-end, gas tight, solid electrolyte type which is used as the reaction vessel. The tube is formed from isostatically pressed beta″-alumina (8.85% Na_{2O}–0.75% Li_{2O}), and has a volume of approximately 20.0 cm^3. A conducting, porous, planar layer of metallic platinum was produced on both the inside and outside of the closed end by coating the tube surfaces with Engelhardt Liquid Bright Platinum No. 05x and baking at 500°C with excess oxygen for one hour. The inner Pt surface functioned both as catalyst and electrode. The outer Pt surface served as a reference electrode. The entire exterior of the electrolyte was enclosed in a stainless vessel containing the continuously flowing reference gas. Platinum lead wires led to the external measuring circuit. The balance of the system was constructed of stainless steel with Teflon pressure seals. The superficial area of the platinum film was approximately 1.5 cm^2, with a total surface area of $\sim 2{,}200$ cm^2 as determined by both ethylene titration and acetylene adsorption.[12] Potential changes across the cell were measured under conditions precluding circuit loading. A precision of 1 to 2 mV was typical.

For both adsorption and reaction studies, the device was modeled as a continuously stirred tank reactor (CSTR) and typically operated in a steady flow mode. Reactant gases (c.p.) were used without further purification and mixed to the desired compositions on line with total flow rates set to eliminate bulk mass transfer effects.[13] Effluents were analyzed gas chromatographically. Feed stream gases were preheated using a noncatalytic stainless steel coil. The reactor system temperature was controlled and monitored by a thermocouple within the reference cell enclosure. A six-inch section of the reactor tube, including the closed end, was heated, but the Pt catalyst/electrode surfaces were in a plane in the center of the heated region within 0.15 cm of each other. During the course of

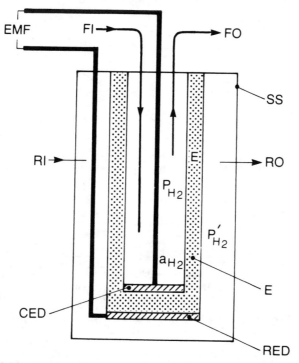

FIGURE 1. Schematic of reaction system: E, beta″-alumina electrolyte; CED, Pt catalyst electrode; RED, Pt reference electrode; RI, RO, reference gas flow in and out; FI, FO, feed gas flow in and out; SS, reference gas enclosure. Electrical leads shown as heavy lines.

experiments, the temperature variation was within 1.5°C for the cell, gases, and catalyst surface.

RESULTS

Hydrogen Activity in the Absence of Reaction

In the absence of other chemical reactions, gaseous hydrogen is in equilibrium with the surface. This activity of surface hydrogen, when expressed as a partial pressure of hydrogen, is equal to the gas phase partial pressure. The performance of the beta″-alumina/Pt concentration cell was tested under those conditions. The changes in cell EMF are consistent with those predicted by the Nernst equation over a wide range of temperatures and hydrogen partial pressures. Typical data are shown in FIGURE 2. The cell functions as a gaseous "pH" sensor.

Hydrogen Activity during Steady-State Ethylene Hydrogenation

Although in the absence of chemical reaction the hydrogen activity at the surface of the Pt film was equal to the partial pressure of gas phase hydrogen,

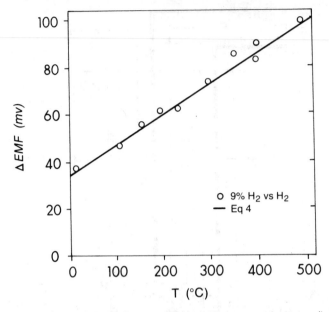

FIGURE 2. Dependence of the EMF on temperature. Nine percent hydrogen (in argon) vs. hydrogen at 1 atm. Solid line: Equation 4.

the corresponding measurement during the steady-state catalytic hydrogenation of ethylene showed that the activity was not equal to the gas phase partial pressure. Instead, the activity was found to be a temperature-dependent function of both hydrogen and ethylene partial pressures. This electrochemically measured activity is interpreted as that of hydrogenic species on the surface satisfying Equation 1, regardless of source. The magnitude represents the hydrogen gas pressure that would be required to satisfy a gas surface equilibrium if such an independent equilibrium existed. Failure to develop thermodynamic equilibrium between gaseous and surface hydrogen is a consequence of the catalytic hydrogenation reaction: removal of the ethylene from the feed stream led eventually to an activity characteristic of the hydrogen partial pressure in the feed stream.

Two functions were required to describe the isothermal behavior of the activity as a function of the gas species during the reaction, one for each of two temperature regimes. The dividing temperature was in the range of 450–500 K. The data may be represented by the relationship:

$$P_{H_2}/a_H^n = K_{Et(n)} P_{Et} + 1 \qquad (6)$$

where below 450 K, $n = 1$, and

$$K_{Et(1)} = 320 \, (500/T - 1) \, \text{atm}^{-1/2} \qquad (7)$$

$$P_{Et} = \text{partial pressure of ethylene}$$

Above 500 K, $n = 2$ and

$$K_{Et(2)} = 221 \, \exp \, (-3468/RT) \, \text{atm}^{-1} \qquad (8)$$

Ethylene Hydrogenation: Reaction Studies

The steady-state rate of ethane production from Pt-catalyzed ethylene-hydrogen mixtures was measured over a range of temperatures (298–700 K) and reactant partial pressures (0.001–1.0 atm) while maintaining the total pressure at 1 atmosphere. Conversions were kept below 5% to simplify the reactor analysis. The reaction rate reached a maximum between 450–500 K, with the rate increasing rapidly from room temperature to the maximum temperature and exhibiting a slight decline thereafter. This observation is in agreement with that of most other investigators including Sinfelt,[14] zur Strassen,[15] and Rideal.[16] The rate of ethane production, $\Gamma_{C_2H_6}$, also passes through a maximum as the pressure of ethylene is increased, but is increased monotonically with increasing hydrogen pressure.

It is possible to describe both the low- and high-temperature reaction rates analytically using two coefficients determined experimentally: K_r, a global rate coefficient, was derived from the initial reaction rate data and K_{Et} from the steady-state isotherms. Again, two temperature regimes need to be represented: above and below 450–500 K. The rates may be described by the following functional forms:

$$\Gamma_{C_2H_6} = \frac{K_r(m) P_{Et} P_{H_2}}{\left(1 + K_{Et(n)} P_{Et}\right)^m} \tag{9}$$

where $m = 2$ ($n = 1$) for the low-temperature and $m = 1$ ($n = 2$) for the high-temperature range.

The temperature dependence of the rate coefficients indicates an exothermic process at high temperature, while at low temperature an endothermic process is dominant. The two rate coefficients converge at a temperature of 505 K, close to the experimentally observed maximum reaction temperature of 450–500 K.

Hydrocarbon Adsorption: Ethylene

The dissociation of ethylene on platinum surfaces into a stoichiometrically specified hydrogen-deficient hydrocarbon residue and atomic hydrogen has been suggested by many authors (e.g., Beeck).[17] Direct observation of this dissociation process can be made using the hydrogen-sensitive reactor system by injecting sufficiently small pulses of ethylene into an inert carrier gas flowing slowly through the reaction cell. Even at room temperature dissociation occurred: hydrogen was observed in the gas phase effluent together with a small amount of ethane (possibly due to self-hydrogenation). The surface activity also changed showing that surface hydrogen was produced. If the pulse contained more ethylene than could be completely adsorbed, the subsequent flushing of the cell by the pure carrier gas following the pulse rapidly removed the gaseous hydrogen and hydrocarbons, but the activity of surface hydrogen did not vanish showing that most of the surface hydrogen was not immediately removed. When finally only inert carrier gas was flowing, ethylene could be detected in the gas phase and appeared at a rate proportional to the change in atomic surface hydrogen activity, a_H. That process must represent the reversal of the dissociation process, which involved the formation of surface hydrogen and the hydrogen-deficient hydrocarbon residue. During this slow desorption process, reaction could be interrupted with oxygen and restored with hydrogen showing that the process was not simple evaporation. The extent of ethylene dissociation was only slightly dependent upon temperature below 450 K. Although the dissociative adsorption of ethylene and the recombination of an ethylene residue with surface hydrogen is an important component of the reaction sequence, this process does not lead directly to product since ethane was not detected during this desorption.

Dissociative Adsorption of Propane, Butane, and Cyclohexane

Propane, butane, and cyclohexane exhibited dissociative adsorption in the sense that a temperature-dependent hydrogen activity became measurable upon exposure of the Pt film to the steady flow of pure hydrocarbon at a total pressure of one atmosphere. With propane, surface hydrogen was detected, beginning at approximately 275°C and, as the temperature increased, the hydrogen activity decreased until a minimum value was reached at about 325°C. At that point the activity abruptly began to increase with further increase in temperature. The behavior of butane and cyclohexane was similar but occurred over different temperature ranges. The data for these hydrocarbons are shown in FIGURES 3 and 4 in terms of the temperature dependence of the log of the activity (i.e., EMF). Neither methane nor ethane was observed to undergo dissociative adsorption on platinum at temperatures below 450°C.

Below the temperature of minimum hydrogen activity, hydrogen was not detected in the gas phase effluent on a steady-state basis. Above that temperature, both hydrogen and dehydrogenation products were readily and continuously observed. Furthermore, the activity at temperatures beyond the minimum was a function of the hydrocarbon flow rate. At high flow rates through the reactor, the activity increased with increased flow, while the hydrogen content of the gas phase asymptotically approached some limiting value. At very low flow rates, the activity was found to be equal to the partial pressure of gaseous hydrogen. The products detected during the dehydrogenation were principally propylene from propane, butadiene and butene from butane, and cyclohexadiene and benzene from cyclohexane. The complexity of the high-temperature reactions, particularly for the butane and cyclohexane, made it difficult to close the mass balance. Cracking, hydrogenolysis, and coking became apparent and complete analysis of the gaseous products was not carried out. Hydrogen production from both propane and butane appeared to be first order in hydrocarbon at low flow but approached zero order at the highest flow rates.

The temperature dependence of the EMF in the high-temperature region also was dependent upon flow except for the limiting regions of very slow and very fast flow. Assuming that the temperature dependence of the EMF is a representation of a ΔH for the process that is signaled by the changing hydrogen activity, the low-temperature data yield a value of −9.2 kcal/mol with propane, −12.8 kcal/mol with butane, and −28.8 kcal/mol with cyclohexane. The higher temperature range yields 28.5 kcal/mol with propane, 30.5 kcal/mol with cyclohexane, and two values with butane: at very low flow rate 61 kcal/mol and, when affected by the flow rate, 31.5 kcal/mol.

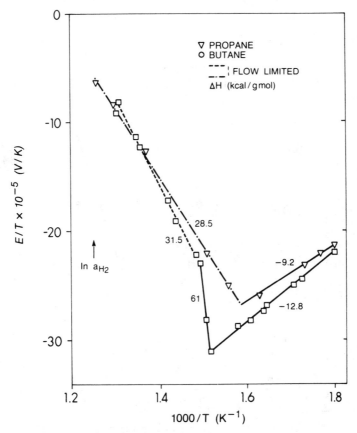

FIGURE 3. Dependence of EMF on temperature for propane and butane interacting with Pt. H for each segment in kcal/gmol. Broken lines: flow limited.

DISCUSSION

Adsorption of Hydrogen during Ethylene Hydrogenation

Direct observation of surface hydrogen activity, a_{H_2}, shows that when Pt is exposed to ethylene under hydrogenation conditions, a_{H_2} (surface) always is less than P_{H_2} (gas). Since an independent species equilibrium requires that $a_{H_2} = P_{H_2}$, the existence of steady-state relationships between a_{H_2}, P_{H_2}, and $P_{C_2H_4}$, as represented by Equation 6, requires that gaseous hydrogen and surface hydrogen not be in thermodynamic equilibrium during the reaction. This possibility had been considered previously by Halsey who pointed out that if reaction follows the adsorption of reactants, the independent equilibrium between reactants and corresponding surface species will be disturbed.[18] This in turn will lead to a steady-state relationship (isotherm) between gas and surface species characteristic of the reaction and not of an independent species equilibrium. He showed that

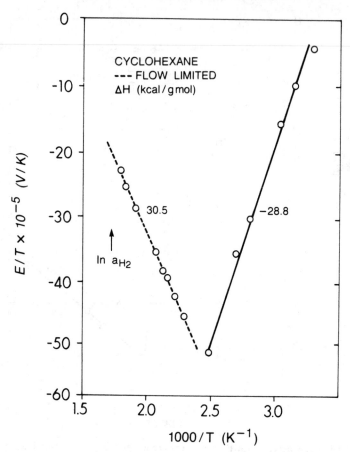

FIGURE 4. Dependence of EMF on temperature for cyclohexane interacting with Pt. H for each segment in kcal/gmol. Broken lines: flow limited.

using the mechanism proposed by Beeck, without independent species equilibrium, an appropriate steady-state surface mass balance led to a different steady-state reaction rate expression and a different (steady-state) isotherm could be derived. In such a coupled system, no one step need be rate limiting. Halsey's proposal was not confirmed by the available experimental data but, since direct measurement confirms the nonequilibrium conjecture, the lack of agreement appears to be in the proposed mechanism. Similar steady-state considerations have been used by Halsey in an analysis of CO oxidation on Pd.[19]

Elementary Processes in the Ethylene Hydrogenation Reaction

Since ethylene hydrogenation was first reported in 1897,[20] the large number of studies of the reaction suggest that most of the possible elementary reaction processes have been considered. Some concensus exists as to their interrelationship in the form of the Horiuti-Polanyi mechanism, which invokes a reaction between nondissociatively adsorbed ethylene and adsorbed hydrogen.[21] The discovery of the steady-state isotherm and the results of the kinetic studies provided experimental information suggesting that the nature of the steps that actually occur is more complex.

While equilibrium dissociative adsorption of hydrogen has been observed often, during these reaction studies, gaseous hydrogen was never observed to be in simple equilibrium with the surface species. The extent to which ethylene dissociates upon adsorption was less clear, but the electrochemical studies showed that the dissociation process does occur. Further, at room temperature, surface hydrogen was observed to react in the presence of gas phase ethylene to product ethane and the reaction continued until the surface hydrogen was consumed. This observation suggested an initial gas-surface reaction, although adsorption followed by reaction could not be ruled out by that observation alone. The inverse process was not observed at low temperature: the ethylene residue formed by adsorption was observed to recombine with surface hydrogen, but this process did not yield ethane.

The possibility of a direct gas-surface reaction has been debated over the years. The data obtained electrochemically, together with other experimental observations, suggest that such a process is important but it is easily masked by the overwhelming adsorption of ethylene at low temperature. The interpretive problem is complicated since the gas surface step, which removes free surface hydrogen to form a hydrogen-addition species, is accompanied by another (dissociative) adsorption process which produces hydrogen and a hydrogen-deficient residue. The latter residue is not an ethane precursor. Farkas and Farkas, using deuterium exchange, showed that the rates for hydrocarbon exchange (recombination) and hydrogenation were comparable and functions of gas phase composition and temperature.[22] Since the rate of hydrogen-deuterium exchange was found to be slower than the hydrogenation rate, they discussed the possibility that the rate-limiting step was hydrogen adsorption.

That the residues resulting from dissociative adsorption are relatively inert during the course of the ethylene hydrogenation reaction (in the sense that they were ethane precursors) was shown by Taylor in ^{14}C-ethylene hydrogenation studies.[23] Electrochemically derived information regarding the structural form of the residue species is inconclusive: the experimental kinetic and thermodynamic data cannot be used to distinguish between adsorbed species of a given form. The stoichiometry, however, must be reflected in the steady-state isotherm (Equation 5). The data are not inconsistent with the formation of a sigma- or pi-bonded species of the vinylic or ethylidyne type, such as suggested by Demuth.[24] The existence of such a species, which is deficient in one hydrogen atom, is also indicated by deuterium exchange experiments: Twigg, Kimball, and Bond showed that the principal form of residual ethylene recovered was a mono-deuterated species,[25,26,27] a result that is consistent with the recombination step observed in this study. The principal deuteroethane product obtained is a dideuterated species, also consistent with a primary gas surface hydrogenation step. Soma has proposed that at 203 K there exist, simultaneously, a nondissociatively adsorbed ethylene species which undergoes hydrogenation, and a relatively inert dissociatively pi-bonded species.[28] However, at the temperatures used in this study no evidence for nondissociatively adsorbed ethylene was obtained.

As a consequence of these observations, four major reactions affecting surface hydrogen were proposed.[29] They are similar to those suggested by Farkas and Farkas.[22] In detail: (I) dissociative hydrogen adsorption to produce surface hydrogen, which, however, is not in equilibrium with the gas; (II) dissociative ethylene adsorption to produce surface hydrogen and a residue that is not a direct ethane precursor; (III) reaction of gaseous ethylene with surface hydrogen to produce a hydrogen-addition compound; (IV) kinetically irreversible reaction of the addition compound to form gaseous ethane. The inverses of Processes I, II, and III are kinetically permitted. These elementary processes are described by Equations 10a–d:†

$$\text{(I)} \qquad\qquad H_2(gas) + 2^* = 2\,H^* \qquad\qquad \text{(10a)}$$

$$\text{(II)} \qquad\qquad C_2H_4(gas) + 2^* = C_2H_3^* + H^* \qquad\qquad \text{(10b)}$$

$$\text{(III)} \qquad\qquad C_2H_4(gas) + H^* = C_2H_5^* \qquad\qquad \text{(10c)}$$

$$\text{(IV)} \qquad\qquad C_2H_5^* + H^* = C_2H_6(gas) \qquad\qquad \text{(10d)}$$

Since an independent chemical potential balance on gaseous and surface hydrogen is inappropriate for this system, the less restrictive surface mass balance suggested by Halsey must suffice. Three surface species, hydrogen, ethylene residue formed from the dissociative adsorption, $C_2H_3^*$, and the hydrogen addition compound $C_2H_5^*$, which is the ethane precursor, were included in the analysis. Only hydrogen and ethylene gas were considered since ethane is produced in an irreversible step. It was further assumed that mass action kinetics provide an appropriate description of the kinetic behavior. Finally, some relationship between the thermodynamic activity of the surface species and the surface concentration is needed to utilize mass action kinetics. This problem has not yet been resolved experimentally but the use of a Langmuir-like adsorption "isotherm," in which the activity appears in place of the pressure, was proposed by Wagner[9] and was used in this analysis.

Since a physically adsorbed precursor state could not be distinguished from the gas phase in these studies, "gas" will include both possibilities. With that proviso, using these steps in a steady-state mass balance on all surface species, it was possible to derive the functional forms of the nonequilibrium isotherms and, with appropriate assumptions, the rate expressions. In the high-temperature regime, the work of Farkas and Farkas suggests that Process II is equilibrated,[22] and that assumption was included.

The mechanism proposed by Equations 10 includes Process III, the direct attack of surface hydrogen by a nonsite blocking ethylene species, which is at variance with the usual discussion. The evidence that the direct process is appropriate, and that the surface adsorption mechanism is not, is based upon the following arguments. The consensus mechanism invoking the reaction between adsorbed ethylene and adsorbed hydrogen as the initial step leading to ethane, when examined in terms of the steady-state analysis, does not lead to the observed rate and isotherm expressions. Consideration of the structural and electrochemical

† Asterisks denote adsorption sites.

evidence requires the existence of a hydrogen-deficient species that is not an ethane precursor. It may be formed either directly from the gas phase or via an adsorbed intermediate since inclusion of such an adsorbed intermediate in the proposed mechanism does not alter the results of the analysis in predicting the rate and isotherm forms. If Process III in Equation 10 is replaced by a process in which the initial reaction leading to ethane is the reaction between the adsorbed species, the solution of the coupled differential equations does not permit the existence of multiple steady states. Such multiplicity is predicted by Equations 10, and two isothermal states have been observed during the course of this research.[30]

It is interesting that the rate of arrival of gaseous hydrogen is much larger than the rate of utilization of the hydrogen by the reaction to produce ethane. This is a result of the relative rates of the processes involved. The surface reactions are sufficiently fast and the surface is sufficiently covered with nonreactive residues to make hydrogen adsorption kinetically difficult. Thus, although the limiting reactant for the overall process is hydrogen, no single step solely determines the surface concentrations and hence no one rate-limiting step can be identified as being responsible for the global reaction rates. It is the balance among the steps of the entire reaction network that must be considered.

Dissociative Adsorption of Propane, Butane, and Cyclohexane

Each of the three hydrocarbons propane, butane, and cyclohexane showed evidence of dissociative adsorption on Pt: significant steady-state surface hydrogen activity was observed in the temperature regime below the temperature of the activity minimum but no steady production of any gaseous product, hydrogen, or hydrocarbon was observed. The temperature dependence of the activity provides an important clue as to the origin of this behavior. If one assumes that the hydrocarbon is adsorbed dissociatively on the Pt surface, then both the hydrogen and the hydrocarbon residue that are formed are stable. Furthermore, it is reasonable to expect that a precursor to a dehydrogenation product will be formed, particularly as the temperature is increased toward the point at which the dehydrogenation reaction can be observed. One possible origin of that precursor stability could be the ability of the hydrocarbon residue to form appropriate resonant structures with the metal. Thus, for propane, the initially adsorbed species might be propyl radical and hydrogen, but it is unlikely that this system is stable. The resonance stabilization can be estimated if one assumes that the allyl radical is the stable species: the extra stabilization energy should be of the order of the difference in energy between propyl and allyl, which has been estimated to be 9 kcal/mol.[31] Since the stability of the hydrogen is associated with the hydrocarbon, the hydrogen activity should also reflect this stability. The value of ΔH derived from the temperature dependence of that activity, assuming that the steady-state EMF can be treated by equilibrium theory, is essentially the stabilization energy. A similar estimate for butyl conjugate stabilization is 13 kcal/mol. Although no estimate has been made for the stability of a conjugated cyclohexane residue, it is not unreasonable to use, as a first approximation, the stabilization conferred upon benzene relative to cyclohexatriene, which is 36 kcal/mol. These values are in surprising agreement with the observed ΔH's. There are few data on precursors, but during the adsorption of cyclohexene on Ni, at 150°C, even though no gaseous products were observed, magnetic measurements indicated that significant bonding to the metal had occurred. In addition, the number of bonds formed corresponded to the number of exchangeable hydrogens. Only at temperatures in excess of 230°C were the products, cyclohexadiene and benzene, observed. Presumably at 230°C benzene and cyclohexadiene could desorb and precursor formation is suggested.[32]

The lack of reactivity of methane and ethane is in accord with the notion of the need for a stabilized precursor. Since neither can form resonant species, they do not dissociate readily upon adsorption and are invisible to the hydrogen sensor.

If the metal provides the opportunity for stabilization of the adsorbed dehydrogenation precursor species it must also prevent the simple recombination of surface hydrogen to form the saturated hydrocarbon or gaseous hydrogen. One possibility is that a complex with Pt is formed as part of an equilibrium structure. The decrease in activity with increasing temperature would reflect this equilibrium behavior and must be associated with some decrease in the coverage of hydrogen. The appearance of a minimum activity, which corresponds to nearly the same value of ΔG for each of the three hydrocarbons, approximately 9 kcal/mol, suggests that the limit is due to a property of the hydrogen metal system. Although the magnitude of ΔG is suggestive of the formation of a simple hydride, platinum hydride is not known as a stable independent species. Since the stabilization of the system seems to be determined primarily by the hydrocarbon, the stability of the hydrogen must be associated also with the adsorbed hydrocarbon. The sharp transition between the dissociative adsorption and the dehydrogenation regions could be also the result of a structural change induced by the variation in hydrogen coverage. Such structural effects have been reported for Pt during the oxidation of CO.[33] At this point, however, no clear explanation for the minimum in activity can be offered.

As the temperature increases beyond the point of minimum activity, dehydrogenation begins and both unsaturated hydrocarbons and hydrogen appear in the gas phase. Near the activity minimum, where the extent of reaction is small, and at low flow rates through the reactor, gaseous hydrogen is in equilibrium with surface hydrogen and the magnitude of ΔH is that which corresponds to the appropriate hydrogenation-dehydrogenation equilibrium. The temperature dependence reflects the equilibrium origin of the hydrogen. Thus, for the propane-propylene-hydrogen equilibrium, the value of ΔH is 30 kcal, which is the value derived from the temperature dependence of the activity. Butane-butadiene is also represented by the observed ΔH, 61 kcal/mol. In the high-temperature range the dependence of the activity upon the flow rate becomes apparent and, in FIGURE 3, a change in the value of ΔH is indicated for butane. At a given temperature, increased flow results in increased surface hydrogen activity even though gas analysis shows that a saturation limit for hydrogen in the gas has been reached. Interpreting the magnitude of ΔH in that region is less clear. It is a value commonly reported for the desorption energy of hydrogen, and for the formation of single double bonds from saturated species. The former suggests that hydrogen desorption has become rate limiting. The dependence of hydrogen production on hydrocarbon pressure is in accord with this suggestion. At high flow, hydrocarbon is being brought to the surface and decomposed faster than hydrogen desorption can occur and the hydrogen activity then increases on the surface.

CONCLUSIONS

The use of the beta″-alumina/Pt electrochemical cell has made it possible to study the behavior of surface hydrogen during the Pt-catalyzed hydrogenation of ethylene and the adsorption of propane, butane, and cyclohexane.

During the course of steady-state ethylene hydrogenation, surface hydrogen is not in thermodynamic equilibrium with gaseous hydrogen. The activity, a_{H_2}, is less than the partial pressure, P_{H_2}, and is a function of the gaseous ethylene and hydrogen pressures. Two quantitative relationships (steady-state isotherms) have been found, one of which is applicable below 450–500 K, the other being appropriate above that temperature range.

The rate of ethane formation increased with increasing temperature, and the well-known maximum rate was observed at approximately 450–500 K. Two different rate expressions were found, one of which was applicable above, the other below, the rate maximum. An overall rate parameter and the coefficient of the steady-state isotherm are the relevant parameters of the rate expressions.

As a result of qualitative studies of the behavior of surface hydrogen, several of the processes important to the hydrogenation reaction have been observed directly. Dissociative adsorption of ethylene occurs on platinum to form a species deficient in one hydrogen atom as well free surface hydrogen. The hydrogen-deficient species is not a precursor of ethane. The hydrogenation of ethylene to form ethane proceeds initially by the interaction of adsorbed atomic hydrogen and "gaseous" ethylene to form an adsorbed addition species followed by a surface addition of hydrogen to form ethane gas.

The failure of the equilibrium assumption for the adsorption of hydrogen during the reaction suggests that a steady-state mass balance on surface species may be the only limitation that can be imposed a priori. The solution of these equations, formulated for the reaction scheme outlined above, does yield the form of both the steady-state isotherms and rate expressions. The Horiuti-Polanyi mechanism does not yield these forms as a result of a steady-state analysis nor does it permit the existence of the observed multiple steady states. Although hydrogen plays the role of the limiting reagent in this process, as a result of the balance between competitive reactions, the isolation of a single rate-determining step is not possible.

Finally, the reaction of propane, butane, and cyclohexane with Pt exhibits two characteristic reaction patterns. At low temperature, there is evidence for dissociative adsorption but no reaction products appear on a continuous basis. This is the region in which olefin hydrogenation is catalyzed. The energetics of the dependence of the activity upon temperature are suggestive of a stabilization of an unsaturated hydrocarbon radical by resonant structures involving the metal. As a result, analogies with cluster complexes may not be appropriate. Such resonance effects could provide the basis for empirical observations relating the ease of hydrogenation to the substituent structure of the olefin (e.g., Lebedev's rule).[34] Such stabilized structures may be the origin of the carbonaceous layers observed on Pt.[35] At the higher temperature the expected equilibrium dehydrogenation is observed and there is evidence that, under some circumstances, hydrogen desorption may be a rate-limiting process. It appears that hydrogen again may be the limiting reagent, but whether a rate-limiting step can be identified remains an unanswered question.

The minimum hydrogen activity separating the hydrogenation from the dehydrogenation regime is unexplained.

REFERENCES

1. MUETTERTIES, E. L. 1977. Science **196**: 839.
2. MUETTERTIES, E. L., R. R. BURCH & A. M. STOLZENBERG. 1982. Annu. Rev. Phys. Chem. **33**: 89.
3. SOMORJAI, G. 1981. Chemistry in Two Dimensions: Surfaces. Chapter 2. Cornell University Press. Ithaca, N.Y.
4. KELLY, R. D., J. J. RUSH & T. E. MADEY. 1979. Chem. Phys. Lett. **66**: 159.
5. VAYENAS, C. G. 1977. Ph.D. Dissertation. University of Rochester. Rochester, N.Y.
6. VAYENAS, C. G. & H. M. SALTSBURG. 1979. J. Catal. **57**: 296.
7. STOUKIDES, M. & C. G. VAYENAS. 1981. J. Catal. **69**: 18.
8. HAALAND, D. 1980. J. Electrochem. Soc. **127**: 796.
9. WAGNER, C. 1970. Adv. Catal. **21**: 323.
10. THOMAS, J. M. & A. J. WHITE. 1972. J. Mater. Sci. **7**: 838.
11. LUNDSGAARD, J. S. & R. J. BROOK. 1974. J. Mater. Sci. **9**: 2061.
12. MULLINS, M. 1983. Ph.D. Dissertion. University of Rochester. Rochester, N.Y.
13. SATTERFIELD, C. 1970. Mass Transfer in Heterogeneous Catalysis: 155. MIT Press. Cambridge, Mass.
14. SINFELT, J. H. 1964. J. Phys. Chem. **68**: 856.
15. ZUR STRASSEN, H. 1934. Z. Phys. Chem. **A169**: 81.
16. RIDEAL, E. K., A. FARKAS & L. FARKAS. 1934. Proc. R. Soc. **A146**: 630.
17. BEECK, O. 1950. Discuss. Faraday Soc. **8**: 118.
18. HALSEY, G. D. 1963. J. Phys. Chem. **67**: 2038.
19. HALSEY, G. D. 1977. Surf. Sci. **64**: 681.
20. SABATIER, P. & J. B. SERENDEN. 1897. C. R. **124**: 1358.
21. HORIUTI, J. & K. MIYAHURA. 1968. Hydrogenation of Ethylene on Metallic Catalysts. NSRDs-NBS 13.
22. FARKAS, A. & L. FARKAS. 1938. J. Am. Chem. Soc. **60**: 22.
23. TAYLOR, G. F., et al. 1968. J. Catal. **12**: 191.
24. DEMUTH, J. E. 1980. Surf. Sci. **93**: L82–L88.
25. TWIGG, G. H. 1950. Discuss. Faraday Soc. **8**: 152.
26. KEMBALL, C. 1956. J. Chem. Soc. **146**: 735.
27. BOND, G. C. & P. B. WELLS. 1964. Adv. Catal. **15**: 92.
28. SOMA, Y. 1979. J. Catal. **59**: 239.
29. MULLINS, M. & H. SALTSBURG. (In preparation.)
30. SALTSBURG, H. & M. MULLINS. 1983. In Spillover of Adsorbed Species. G. M. Pajonk, S. J. Teichner & J. E. Germain, Eds.: 295. Elsevier. Amsterdam, Oxford, New York & Tokyo.
31. MORRISON, R. T. & R. N. BOYD. 1973. Organic Chemistry. 3rd edit.: 211. Allyn and Bacon. New York, N.Y.
32. SELWOOD, P. W. 1962. Adsorption and Collective Paramagnetism. Academic Press. New York, N.Y.
33. ERTL, G., P. R. NORTON & J. RUESTIG. 1982. Phys. Rev. Lett. **49**: 177.
34. GERMAIN, J. E. 1969. Catalytic Conversion of Hydrocarbons: 64. Academic Press. New York, N.Y.
35. SOMORJAI, G. & F. ZAERA 1982. J. Phys. Chem. **86**: 365.

ORGANOMETALLIC ELECTRON, H-ATOM, AND HYDRIDE RESERVOIRS; SINGLE ELECTRON OR H-ATOM TRANSFERS IN THE REDUCTION OF IRON SANDWICH COMPLEXES BY HYDRIDES OR $H_2/Pd/C$; REDUCTION OF ISOLOBAL CARBONYL COMPLEXES, A MODEL SYSTEM FOR THE FISCHER-TROPSCH SYNTHESIS*

Pascal Michaud, Claude Lapinte, and Didier Astruc†

Laboratory of Organometallic Chemistry
ERA CNRS No. 477
University of Rennes
35042 Rennes Cedex, France

Pairwise mechanisms in organic and organometallic reactions are one of the fundamental principles set in the chemistry of this century and are still currently accepted. Main group and transition metal hydrides are representative substrates for this concept in various areas (Equations 1–4). In particular, their role in catalysis is crucial, as exemplified by the interest developed in the present conference. This role is derived from the ability of dihydrogen to effect oxidative addition to a transition metal species in which the coordination sphere of the metal bears at most 16 valence electrons. We propose to focus here on a different approach to the reactivity of metal hydrides using model complexes in which the coordination sphere of the transition metal is saturated or oversaturated, e.g., bears 18, 19, or even 20 valence electrons. Iron, the most common metal and that operating in Fischer-Tropsch catalysis, is being used throughout these studies. The two major results of this approach are (1) reactions of metal hydrides may proceed by nonpairwise mechanisms (single electron and/or H-atom transfers); and (2) organometallic complexes may play the role of "electron reservoirs," "H-atom reservoirs," or "hydride reservoirs" in reactions involving metal hydrides. This dichotomy between the classical and the present approaches will be detailed using the following classical examples.

Main group hydrides are popular for their use in the reduction of carbonyl and other organic unsaturated functions.[1–4] Stoichiometric reactions of common main group hydrides such as $NaBH_4$ with transition metal carbonyl complexes are frequently used as models for the Fischer-Tropsch synthesis[5–8] (Equations 1 and 2). Stoichiometric reduction of hydrocarbon fragments by the same hydrides in cationic organometallic complexes is a useful application of transition metals to synthesis[9,10] (Equation 3). In both the organic and organometallic areas, the transfer of a hydride (H^-) from the metal to the substrate has been adopted as a classical mechanistic feature (Equations 1–3). Another one is the

* Organometallic Electron Reservoirs. 11. For part 10, see Reference 54.

† D. Astruc thanks the CNRS for a fellowship (1978–1982) and for ATP Grant No. 9812, the DGRST for a predoctoral fellowship to P. Michaud (1980–1982), and B.A.S.F. (Ludwigshafen, Federal Republic of Germany) for a gift of iron carbonyl.

concerted addition of both atoms of dihydrogen from two adjacent transition metal atoms to an olefin in heterogeneous catalysis[11] (Equation 4).

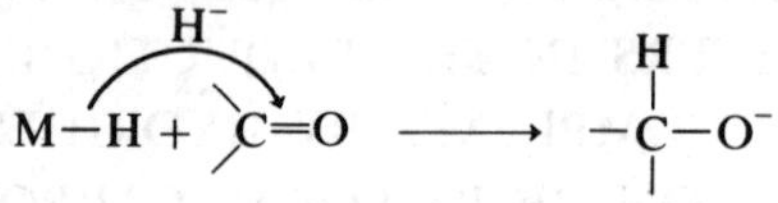

$$\overset{\displaystyle H^-}{\overset{\frown}{M-H}} + \overset{}{\diagdown}C=O \longrightarrow -\overset{\displaystyle H}{\underset{\displaystyle |}{\overset{|}{C}}}-O^- \tag{1}$$

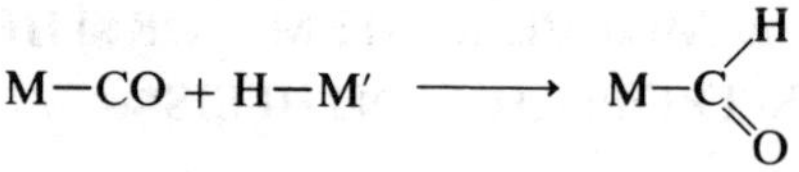

$$M-CO + H-M' \longrightarrow M-C\overset{\displaystyle H}{\underset{\displaystyle O}{\diagup}} \tag{2}$$

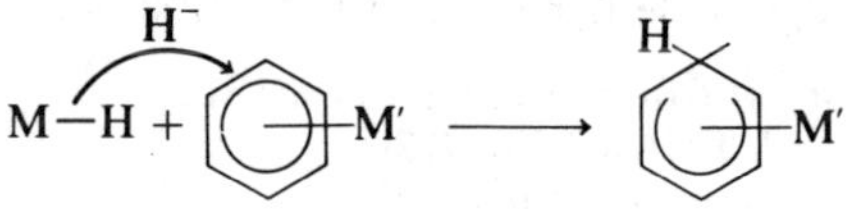

$$M\overset{\frown}{-H} + \underset{}{\bigcirc}-M' \longrightarrow \underset{}{\bigcirc}-M' \tag{3}$$

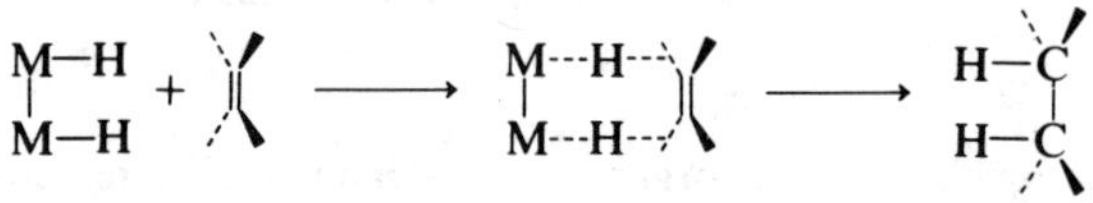

$$\begin{matrix} M-H \\ | \\ M-H \end{matrix} + \diagup\!\!\!\!\diagdown \longrightarrow \begin{matrix} M\cdots H\cdots \\ | \\ M\cdots H\cdots \end{matrix}\diagup\!\!\!\!\diagdown \longrightarrow \begin{matrix} H-C \\ | \\ H-C \end{matrix} \tag{4}$$

We have compared the stoichiometric reductions of the isolobal complexes $CpFe^+L_1L_2L_3$ [L_1, L_2, $L_3 = (CO)_3$,[13] 2^+, $(CO)_2(PPh_3)$, 3^+, arene,[14] 1^+, $Cp = C_5H_5$, C_5Me_5] with the hydrides $NaBH_4$, $LiAlH_4$, $MoH_4(dppe)_2$,[15] and $C_5Me_5Fe(CO)_2H$. Catalytic hydrogenation of the olefins $CpFeC_6Me_5{=}CH_2$,[16] 4 ($Cp = C_5H_5$ or C_5Me_5) has been carried out using Pd/C.

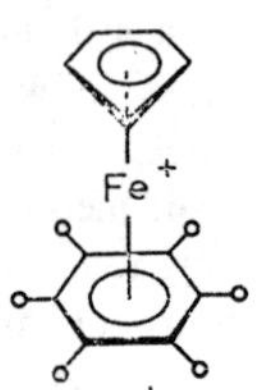

1^+

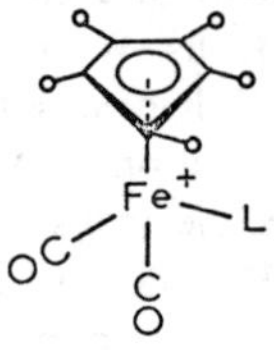

2^+: L=CO

3^+: L = PPh$_3$

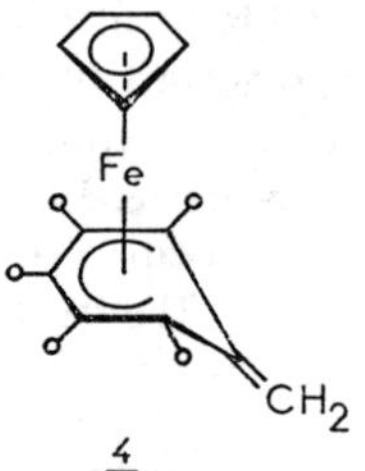

4

It should be noted that nonpairwise organometallic mechanisms have been emphasized by Kochi.[12]

STEPWISE DOUBLE H-ATOM TRANSFER IN THE Pd/C-CATALYZED HYDROGENATION OF OLEFINS: "H-ATOM RESERVOIR INTERMEDIATES"[17]

Although concerted dihydrogen transfer is the currently accepted mechanism for the heterogeneously catalyzed hydrogenation of olefins, the intermediacy of semihydrogenated species has been proposed in some cases on the basis of the stereochemistry of the reduction products;[11] however, such species were never isolated or even characterized spectroscopically.

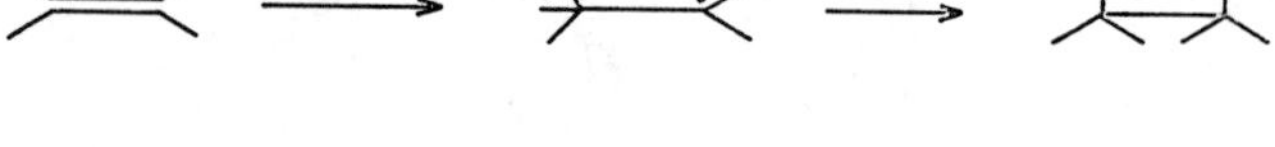

$$\diagup\!\!\!\!\diagdown\overset{\displaystyle H_2/Cata.}{\longrightarrow}\overset{\displaystyle H}{\diagup}\!\!\!\diagdown\cdot\longrightarrow\overset{\displaystyle H \quad H}{\diagdown\!\!\!\diagup\!\!\!\diagdown} \tag{5}$$

The red olefin **4** has been characterized by an x-ray crystal structure study,[18] and its Pd/C-catalyzed hydrogenation in toluene at 20°C and 1 atm gives the expected orange reduction product **5**, the reaction being complete in two hours[18] (Equation 6).

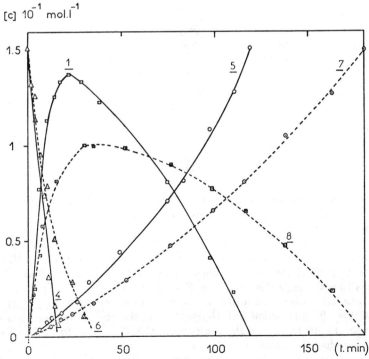

$$\underline{4} \qquad\qquad\qquad\qquad \underline{5} \qquad (6)$$

After 15 minutes of reaction, the reaction medium becomes deep green; subsequent workup gives a black green powder, the Mössbauer spectra of which indicate the formation of 62.5% of the 19-electron d^7 complex $C_5H_5Fe^I(C_6Me_6)$, **1** [isomer shift, 0.73 mm s^{-1}, quadrupole splitting, 0.50 mm s^{-1}; the % conversion is deduced from the Mössbauer absorption factors of **5**, **4**, and **1** ($\cong 0.1$)].

The decrease of the C=C band for **4** (1,600 cm^{-1}) and the increase of the exo C—H band for **5** (2,750 cm^{-1}) in the infrared spectra allow monitoring of the reaction (FIGURE 1) and indicate that the maximum concentration of **4** is

FIGURE 1. Pd/C hydrogenation of $C_5H_5Fe^{II}\eta^5$-$C_6Me_5CH_2$ (**4**) to $C_5H_5Fe^{II}\eta^5$-C_6Me_6H (**5**) via $C_5H_5Fe^I C_6Me_6$ (**1**) (solid line) and of $C_5Me_5Fe^{II}\eta^5$-$C_6Me_5CH_2$ (**6**) to $C_5Me_5Fe^{II}\eta^5$-C_6Me_6H (**7**) via $C_5Me_5Fe^I C_6Me_6$ (**8**) (dashed line) at 20°C, followed by IR (two reproducible runs).

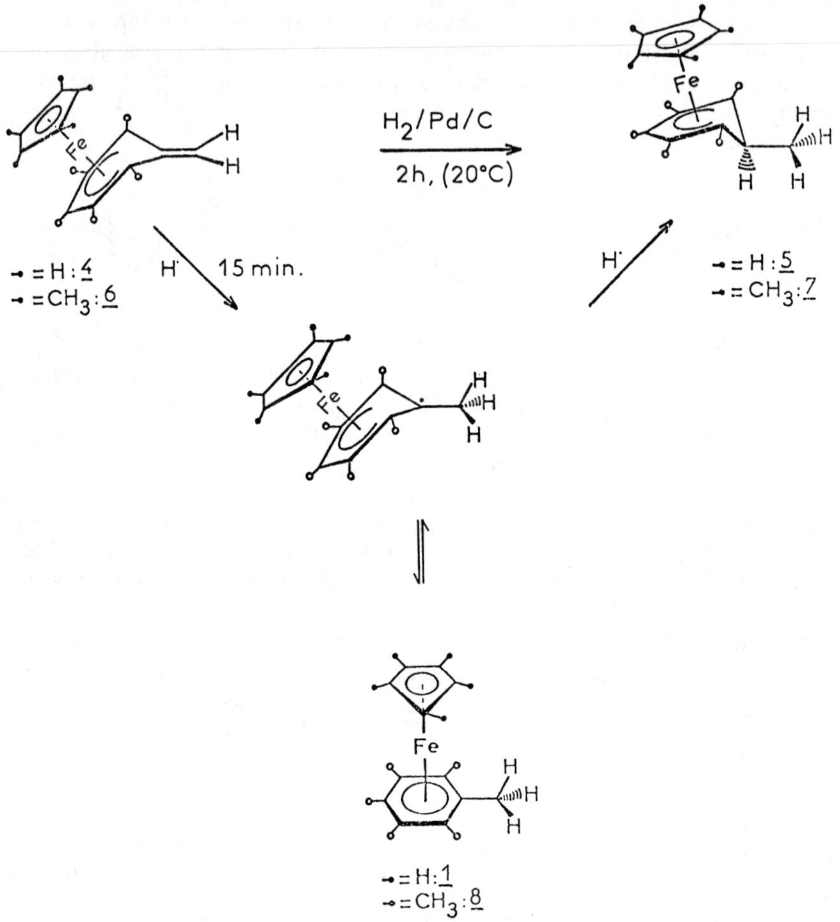

$H_2/Pd/C$

2h, (20°C)

→ = H : 4
→ = CH₃ : 6

H· 15 min.

H·

→ = H : 5
→ = CH₃ : 7

→ = H : 1
→ = CH₃ : 8

SCHEME 1

reached after 20 minutes of reaction. Starting from a pure sample of **1**, it is possible to effect solely the second H-atom transfer from Pd/C/H₂ under the same reaction conditions.

The related olefinic complex $C_5Me_5Fe^{II}\eta^5-C_6Me_5CH_2$,[19] **6**, shows the same behavior, giving the double H-atom transfer complex $C_5Me_5Fe^{II}\eta^5-C_6Me_6H$, **7**, via the intermediate 19-electron d^7 complex $C_5Me_5Fe^IC_6Me_6$, **8**, (SCHEME 1); the second step is slower in this latter case (FIGURE 1).

In one case where the olefin is more sterically crowded, e.g., $C_5H_5Fe^{II}$-C_6Et_5CHMe, **9**, the reaction with H₂/Pd/C under the same conditions gives only the simple H-atom transfer 19-electron complex $C_5H_5Fe^IC_6Et_6$, **10**, as the ultimate reaction product (Equation 7).

$$CpFe^{II}C_6Et_5CHMe \xrightarrow[\text{1 atm, 20°C}]{Pd/C/H_2} CpFe^IC_6Et_6 \qquad (7)$$

        9   **10**

These experiments show that it is possible to transfer one H-atom at a time to olefins from H_2 and a heterogeneous transition metal catalyst when the semihydrogenated olefin is thermally stable.

ELECTRON-TRANSFER AND HYDROGEN-ATOM-TRANSFER PATHS IN THE REDUCTION OF d^6 AND d^7 ORGANO IRON CATION BY HYDRIDES: "ELECTRON-RESERVOIR" INTERMEDIATES[20]

The hydride reduction of unsaturated hydrocarbon fragments coordinated to transition metals is now a classic transformation in organometallic chemistry.[9] In organic chemistry there are a few reports mentioning electron-transfer paths in the reduction of diarylketones,[21] polyaromatics,[22] polyacetylene,[23] and N-hetero-cycles.[24] Some other reactions between transition metal hydrides and acceptors have also been reported.[25-27]

We have carried out a mechanistic study of the well-known reduction of CpFe$^+$(arene) cations with $NaBH_4$ and $LiAlH_4$. Davies, Green, and Mingos have rationalized the site of hydride attack in terms of charge control[9] (there are also a few examples for which the reaction is not under kinetic control).[9,28] We find that, in many cases, these reactions proceed by an electron-transfer path; the electron-transfer products were isolated at room or low temperature, quantitatively in some cases or as more transient intermediates in others.

The study was extended to d^6 bis (arene) iron dications, to d^6 (arene) (cyclohexadienyl) iron monocations, and to d^7 (arene)$_2$ iron monocations.

The reaction between yellow CpFe$^+$(C_6H_6)PF$_6^-$, **11$^+$**, with $LiAlH_4$ in tetrahydrofuran (THF) gives orange CpFe(η^5-C_6H_7), **12**, at 20°C,[30,31] but when the reaction is carried out at −60°C, the forest-green complex CpFe(C_6H_6) **11**, forms and, after 1/2 hour, can be extracted in 80% yield and characterized by electron paramagnetic resonance (EPR) and its temperature-dependent Mössbauer quadrupole doublet.[14] Otherwise, if the reaction is allowed to continue, transformation into the cyclohexadienyl complex occurs upon warming the solution, which becomes orange at −33°C (2 minutes) (SCHEME 2).

The rate of electron transfer depends on the amount of $LiAlH_4$ used for the reaction. The second step, H-atom transfer, can proceed either with AlH_3 [formed in stoichiometric reaction between d^6 CpFe$^+$(C_6H_6)PF$_6^-$ and $LiAlH_4$] or with $LiAlH_4$ and a pure sample of the d^7 neutral complex.

This reaction path with $LiAlH_4$ in THF or DME is general for the series of CpFe$^+$ (arene) cations with a variety of arene ligands (C_6Me_6, **1$^+$**, C_6Et_6, **10$^+$**, C_6H_5F, **13$^+$**, C_4Me_4S, **14$^+$**). In the two latter cases (**13$^+$** and **14$^+$**), it affords the synthesis (at low temperature) of d^7 complexes that were not accessible by Na/Hg reduction. The solvent greatly influences the mechanism since d^7 electron transfer intermediates were not found when the reactions were carried out in ether. (With carbanions, the opposite trend was found: electron transfer proceeds in ether and nucleophilic attack in THF). The second H-atom transfer generally proceeds with attack on the arene ligand to give cyclohexadienyl complexes, except with **10$^+$**, which gives η^4-$C_5H_6Fe(0)(C_6Et_6)$, **15**.

$NaBH_4$ also reacts with CpFe$^+$(C_6Me_6)PF$_6^-$, **1$^+$**, via the electron-transfer (ET) neutral d^7 complex CpFe^I(C_6Me_6), **1**, but the latter cannot be obtained in

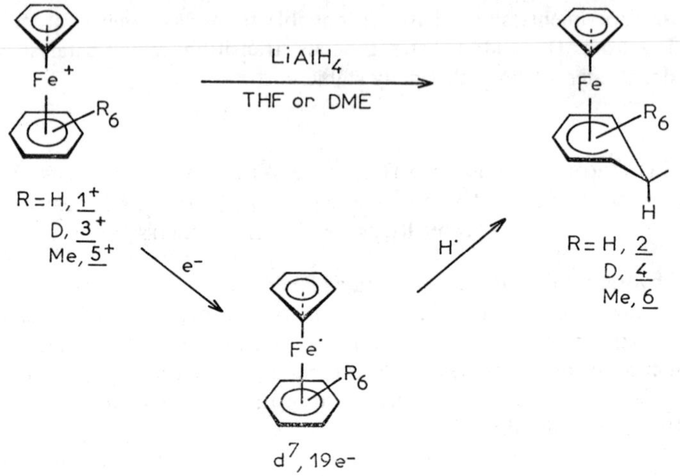

R = H, $\underline{1}^+$
D, $\underline{3}^+$
Me, $\underline{5}^+$

LiAlH$_4$
THF or DME

R = H, $\underline{2}$
D, $\underline{4}$
Me, $\underline{6}$

d^7, 19 e$^-$

SCHEME 2

good yield by this route. Its intermediacy has been demonstrated by its EPR and UV-visible spectra.

However, the reduction of 10^+ by NaBH$_4$ gives the d^7 complex 10 as the final reaction product (SCHEME 3).

Reaction of pink $(C_6Me_6)_2Fe^{++}(PF_6^-)_2$, 16^{++}, with NaBH$_4$ or LiAlH$_4$ in THF gives orange $(C_6Me_6)Fe(C_6Me_6H_2)$, 17. The first step is a hydride transfer giving dark red $(C_6Me_6)Fe^+(\eta^5\text{-}C_6Me_6H)PF_6^-$, 18^+, and the purple d^7 electron-transfer product $(C_6Me_6)_2Fe^+PF_6^-$, 16^+, is not found in the reaction.

Reaction of this d^7 19-electron complex 16^{+33} with NaBH$_4$ or LiAlH$_4$ gives the ET product, e.g., the d^8 20-electron complex 16.[32] A further intermediate in the reaction of NaBH$_4$ and LiAlH$_4$ is the ivory-brown 19-electron d^7 complex $(C_6Me_6)Fe^I(\eta^5\text{-}C_6Me_6H)$, 18, resulting from electron transfer to 18^+ and formed in virtually quantitative spectroscopic yield (SCHEME 4).

Complex 18 was fully characterized by its UV-visible and Mössbauer spectra, elemental analysis, and electrochemical data; its Mössbauer parameters confirm

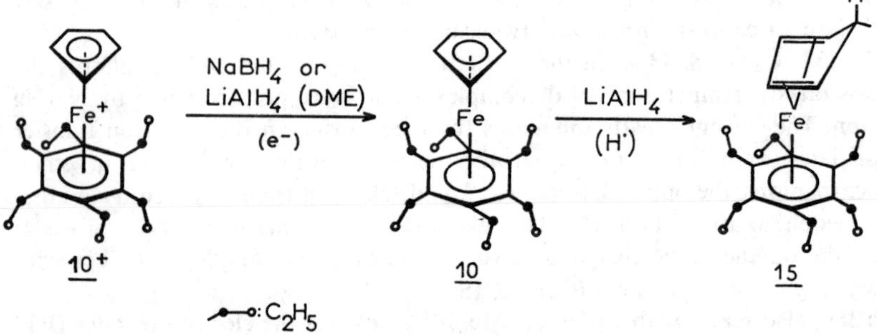

10^+

NaBH$_4$ or
LiAlH$_4$ (DME)
(e$^-$)

$\underline{10}$

LiAlH$_4$
(H$^{\cdot}$)

$\underline{15}$

$\bullet\!\!-\!\!\circ : C_2H_5$

SCHEME 3

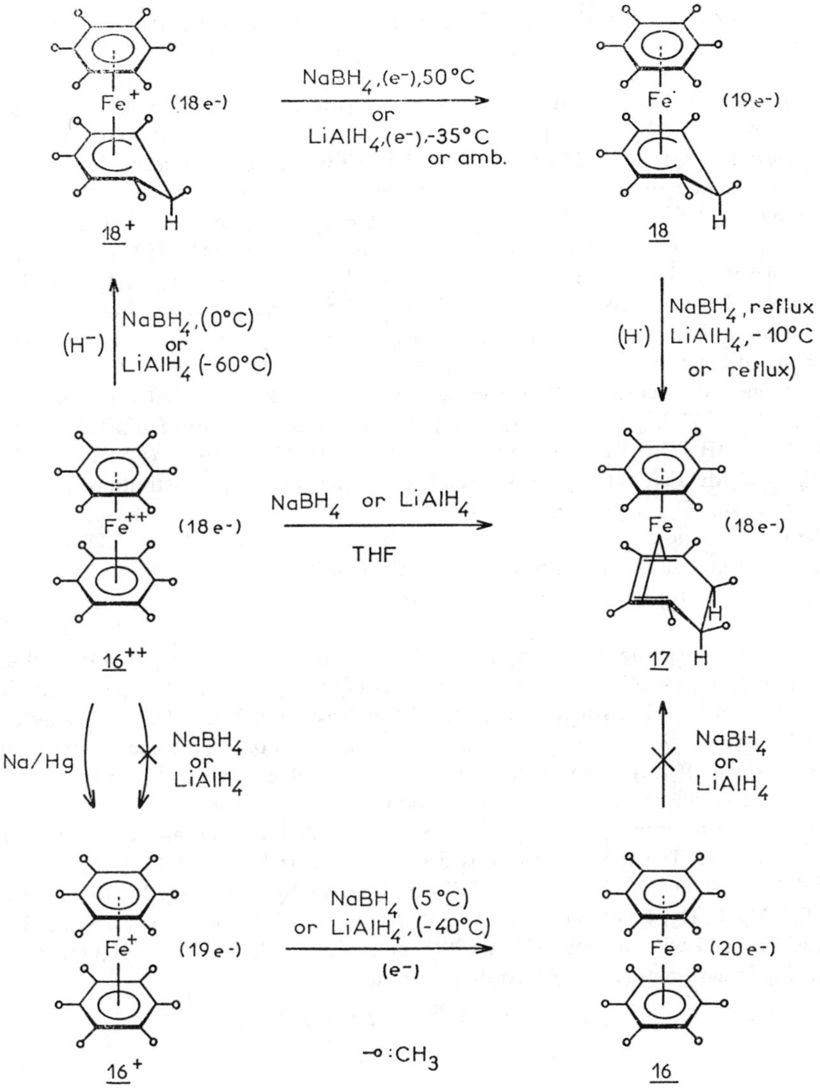

SCHEME 4

the Fe^I oxidation state, but the temperature dependence of the quadrupole splitting is less marked than for the Fe^I complexes of perfect sandwich structure.[14,33] [**18**: I.S. (mm s^{-1} vs Fe) 0.46 (293 K); 0.58 (77 K); Q.S. (mm s^{-1}) 0.91 (293 K); 1.04 (77 K)]. Complex **18** can also be synthesized in high yield by Na/Hg reduction of **18**$^+$ in DME at 20°C. Thus (arene)Fe^+(cyclohexadienyl) complexes behave like $CpFe^+$ (arene) complexes toward hydride reduction. It is probable that many other organometallic cations follow the same trend (the study of reactions with transition metal hydrides is under way).

Reduction by B, Fe, and Mo Hydrides of CO Coordinated to Fe in
$C_5Me_5Fe^+(CO)_3PF_6^-$

In view of the ET paths found for the hydride reduction of $CpFe^+$(arene) complexes, it was of interest to investigate the reaction of isolobal $C_5Me_5Fe^+(CO)_3PF_6^-$, **2$^+$**, and $C_5Me_5Fe^+(CO)_2L\ PF_6^-$ complexes (L = PPh$_3$; **3$^+$**). Reduction of coordinated CO has been successfully effected using rhenium complexes.[33-45] The first formyl [Fe(CO)$_4$CHO]$^-$ complex was synthesized by Collman et al. and many other transition metal formyl[46] (mostly anionic) complexes are known. However no "[Fe—CHO], [Fe—CH$_2$OH], or [Fe—CH$_3$]" complex has ever been characterized upon the reduction of an iron carbonyl complex, although Cutler isolated the alkoxymethyl complex $CpFe(CO)_2CH_2OCH_3$ upon reduction of $CpFe^+(CO)_3$ by NaBH$_3$CN.[47]

Pioneering studies by Green et al. afforded Fp$_2$ [Fp = CpFe(CO)$_2$] via FpH from (FpCO)$^+$PF$_6^-$ and NaBH$_4$ and η^4-C$_5$H$_6$Fe(CO)$_2$PPh$_3$ from (FpPPh$_3$)$^+$PF$_6^-$ and NaBH$_4$.[48] The reduction of (Fp'CO)$^+$PF$_6^-$ and (Fp'PPh$_3$)$^+$PF$_6^-$ [Fp' = C$_5$Me$_5$Fe(CO)$_2$] was envisaged because the permethyl substitution sterically protects the Cp ring from H$^-$ attack and stabilizes potential intermediates of the CO reduction.

The degree of reduction of CO by hydrides in **2$^+$** depends dramatically on the Lewis acid properties of the metal hydride and thus also on the solvent (Scheme 5).

In THF, reaction with 1 mol NaBH$_4$ gives the methyl complex **19**[13] (20°C, 12 hours, 80%) together with minor amounts of the hydride **20** (20%). Monitoring the reaction by ^{1}H NMR indicates that the hydroxymethyl complex **21** is an intermediate (20% after 2 hours), and that its transformation to the methyl complex **19** is slow. This hydroxymethyl complex **21** (stable at 0°C) is prepared in quantitative spectroscopic yield by reaction in CH$_2$Cl$_2$ (4 hours, 20°C) and may be obtained as orange microcrystals in 80% yield after workup and crystallization from pentane. The structure of **21** is deduced from its ^{1}H and ^{13}C NMR spectra [^{1}H NMR (CD$_3$COCD$_3$) δ ppm: CH_2: 1.15 (doublet); OH: 4.72; J_{CH_2OH} = 3 Hz, ^{13}C NMR (CD$_2$Cl$_2$) δ ppm: CH_2: 67.2 (triplet in the off resonance spectrum)]. The reaction of **21** with aqueous HBr at 20°C gives Fp'Br, **22**, and 30% CH$_3$OH (from gas-liquid chromatography) according to Equation 8:

$$Fp'CH_2OH \xrightarrow{\text{aq. HBr}} Fp'Br + CH_3OH \qquad (8)$$

$$\begin{array}{ccc} & \mathbf{21} & \mathbf{22} \end{array}$$

The reduction of **2$^+$** by NaBH$_4$ in H$_2$O at 0°C gives a pale yellow suspension, which can be extracted with pentane at −90°C. The pentane solution evolves CO at $\cong$ −70°C; its infrared spectrum shows an absorption at 1,660 cm^{-1}, which disappears when CO is evolved. During CO evolution, the solution becomes purple. The ^{1}H NMR spectrum in pentane at −90°C does not display the signal usually found for a formyl ligand around $\delta \cong 15$ ppm. Instead it exhibits a sharp singlet at −16 ppm at −90°C, progressively shifting to −12 ppm with constant intensity, which does not vary above −60°C. The yellow hydride Fp'H, **20**, can be isolated in quantitative crude yield and its ^{1}H and ^{13}C NMR spectra are identical with those of the purple species (this latter color is due to trace amounts of the dimer

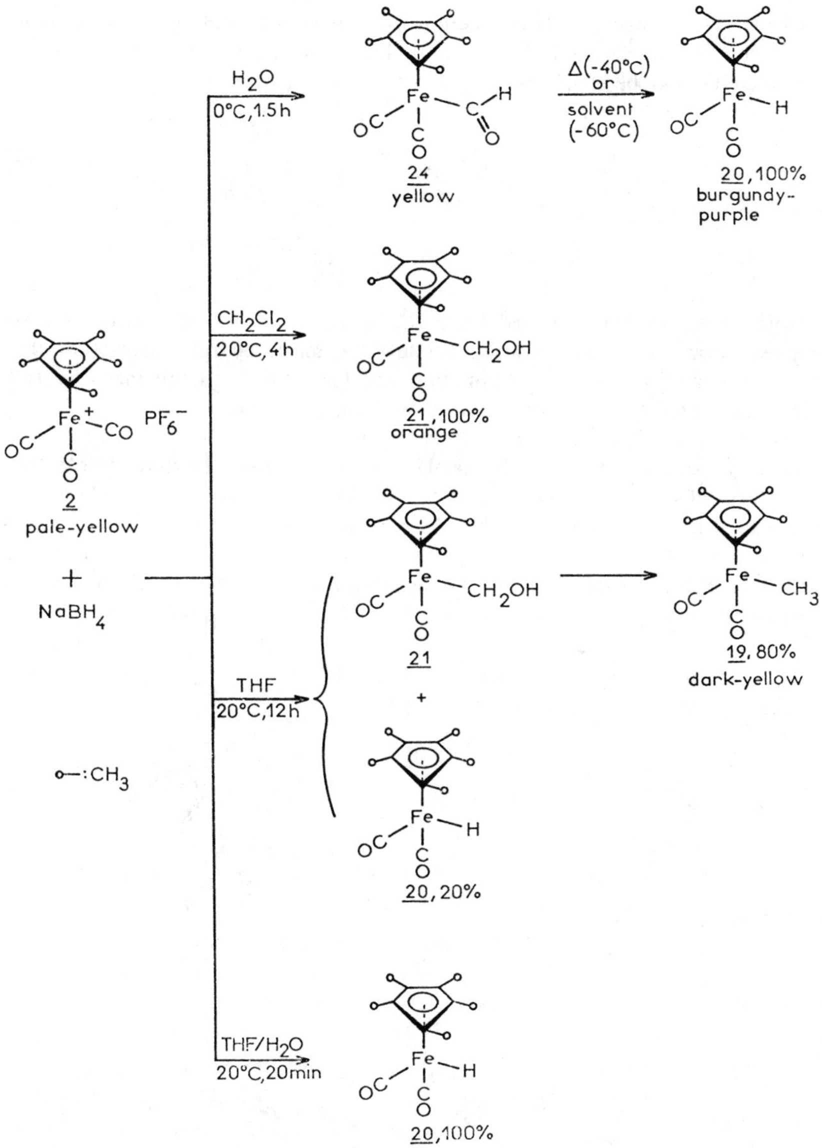
H2O
0°C,1.5h
Fe
C
H
O
OC
24
yellow
Δ(-40°C)
or
solvent
(-60°C)
20,100%
burgundy-
purple
CH2Cl2
20°C,4h
CH2OH
21,100%
orange
Fe+
CO
PF6−
2
pale-yellow
+
NaBH4
THF
20°C,12h
21
19, 80%
dark-yellow
CH3
20,20%
o—:CH3
THF/H2O
20°C,20min
20,100%

SCHEME 5

Fp$_2'$, 23). We believe these observations are best rationalized in terms of an interaction of the hydrogen atom with both a carbonyl carbon and iron giving rise to the hydride 20. In the presence of water the Fe$\cdots$H interaction may be prevented by a stronger O$\cdots$H interaction with the solvent molecules; in this latter solvent, the complex 24 might be a stabilized iron formyl. Alternatively, a bridging hydride BHFe may be invoked.

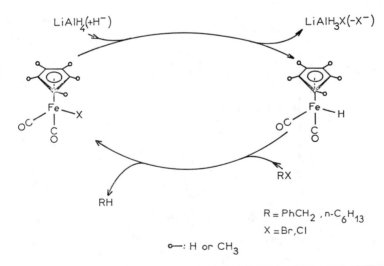

$$\text{(9)}$$

$$Cp' = C_5Me_5$$

24   **20**

Note that the extreme instability of first row transition metal formyl complexes does not hold for anionic ones; the metal hydride interaction that serves a driving force for CO elimination and formation of a full metal-hydride bond is disfavored by the anionic charge on the metal center.

The new metal hydride complex 20 is thermally stable (unlike FpH); it is also obtained by reaction of 3^+ with $NaBH_4$ in a THF/water mixture. [Note that the reaction of $C_5Me_5Re(CO)_2(NO)^+PF_6^-$ in THF/water gives the stable formyl complex $C_5Me_5Re(CO)(NO)(CHO)$].[36] Complex 20 is best synthesized by reduction of Fp'Br, 22, with $LiAlH_4$ at 20°C in THF and preliminary experiments indicate that 20 is a catalyst for the dehalogenation of $PhCH_2Cl$ by $LiAlH_4$ (SCHEME 6):

SCHEME 6. Fp and Fp' as hydride reservoir.

The iron hydride 20 and $MoH_4(dppe)_2$, 25, were also used to reduce CO in 2^+, and the results compared with those obtained with $NaBH_4$. Using 20, the reduction of 2^+ in THF gave 80% of 20 and Fp'THF, 26, together with minor amounts of **21** and **19**. Thus **20** catalyzes the decarbonylation $2^+ \rightarrow 26^+$ (Equation 10a).

$$Fp'H \quad + \quad \left[Fp'\,(THF)\right]^+ \qquad (10\,a)$$

$$\left[Fp'\,CO\right]^+ \quad + \quad \xrightarrow[\text{THF}]{\displaystyle Fp'H} \qquad Fp'\,CH_2OH \quad + \quad Fp'_2 \qquad (10b)$$

$$\xrightarrow{\displaystyle MoH_4(dppe)_2}$$

$$Fp' \quad = $$

$$\mathbf{\bullet}\!\!-\!\!:CH_3$$

In these processes (Equation 10 and SCHEME 6), **20** may be considered a "hydride reservoir:" the coordination sphere of iron in **20** remains saturated with 18 valence electrons during the catalytic cycles.

With $MoH_4(dppe)_2$, **25**, Fp'_2 and **21** are formed in equal amounts (Equation 10b), but it takes several days to reach completion of the reaction. Thus the degree of reduction of coordinated CO increases as one moves to the left of the periodic table since the strength of the coordination to oxygen of the Lewis acid formed subsequent to the hydride transfers increases in this direction. This concept, familiar to the organic chemist, has been extended by Graham in the case of borohydride reduction of CO coordinated to rhenium.[36] The results found here indicate that it applies to transition metal hydrides as well.

Rather surprisingly, we find that reaction of 3^+ with 1 mol $NaBH_4$ at 20°C in THF also gives **20** quantitatively (Equation 11). Since 19-electron species

$$Fp'PPh_3^+ \xrightarrow{\ NaBH_4\ } Fp'H + PPh_3 \qquad (11)$$
$$\mathbf{3^+} \qquad\qquad\qquad \mathbf{20}$$

bearing both CO and phosphine ligands always loose the phosphine rather than CO,[49,50] it is probable that the present odd result is the consequence of an electron-transfer path (Equation 12).

$$Fp'(PPh_3)^+ \xrightarrow[NaBH_4]{(e^-)} \text{``}Fp'(PPh_3)\text{''} \xrightarrow[-PPh_3]{} \text{``}Fp'\text{''} \xrightarrow[NaBH_4]{(H\cdot)} Fp'H \qquad (12)$$
$$\mathbf{3^+} \qquad\qquad\qquad\qquad\qquad\qquad\qquad\qquad \mathbf{20}$$

Indeed we know from section two that $NaBH_4$ is able to transfer an H atom to paramagnetic complexes (see pages 101–103).‡

‡ **Note added in proof:** we now have spectroscopic evidence for such an ET mechanism in the reduction of $[Cp'Fedppe(CO)]^+$ by $LiAlH_4$.[53]

SUMMARY AND CONCLUSION

It was possible to show that the homogeneous or heterogeneous reduction of model iron complexes proceeds by nonpairwise mechanisms, e.g., electron or hydrogen atom transfer from the main group or transition metal hydrides.

Some typical reactions are represented below in which the electron-transfer or H-atom-transfer complex is the ultimate reaction product.

$$CpFe^+(C_6Et_6) \xrightarrow[(e^-)]{NaBH_4} CpFe(C_6Et_6) \qquad (13)$$

$$CpFe(C_6Et_6) \xrightarrow[(H\cdot)]{LiAlH_4} \eta^4\text{-}CpHFe(C_6Et_6) \qquad (14)$$

$$CpFe[\eta^5\text{-}C_6Et_5(:CHMe)] \xrightarrow[(H\cdot)]{H_2/Pd/C} CpFe(C_6Et_6) \qquad (15)$$

$$CpFe(C_6Me_6) \xrightarrow[(H\cdot)]{H_2/Pd/C} CpFe(\eta^5\text{-}C_6Me_6H) \qquad (16)$$

$$(C_6Me_6)Fe(\eta^5\text{-}C_6Me_6H) \xrightarrow[(H\cdot)]{NaBH_4} (C_6Me_6)Fe(\eta^4\text{-}C_6Me_6H_2) \qquad (17)$$

These reactions indicate that "inter alia" electron-rich bulky iron sandwiches may serve as "electron reservoirs"[51] or "hydrogen-atom reservoirs"[17] in organometallic mechanisms. They are especially useful for characterizing nonpairwise reaction paths.

In several other instances, related complexes are transient intermediates in mechanisms, characterized spectroscopically in small amounts in the course of reactions.

These two categories of complexes generally have low-lying antibonding orbitals, which, together with their steric bulk, affords stabilization of the complexes and verification of nonpairwise mechanisms.

In other cases, the electron-transfer products are highly reactive species, which can be characterized only by their reaction products. Possible examples are Equations 18 and 19:

$$Fp'(CO)^+ \xrightarrow[(e^-)]{MoH_4(dppe)_2} Fp'_2 \qquad (18)$$

$$Fp'(PPh_3)^+ \xrightarrow[(e^- + H\cdot)]{NaBH_4} Fp'H \qquad (19)$$

This study has provided insight into the reduction by various metal hydrides of CO coordinated to iron. The main results are:

1. Isolation of the first iron hydroxymethyl complex $Fp'CH_2OH$ and the extreme instability of neutral iron formyl complex $Fp'CHO$, decomposing to the useful, thermally stable hydride $Fp'H$.

2. Indication that the degree of reduction of coordinated CO depends on the ability of the metal to coordinate the oxygen atom of the carbonyl as a Lewis acid, e.g., $Mo > Fe$; the dramatic role of the solvent (THF; THF/H_2O; H_2O; CH_2Cl_2) on the reduction is also displayed.

These model studies indicate that nonpairwise mechanisms occur in the reduction of both iron-carbonyl and iron-hydrocarbon complexes but that such pathways intervene in the intimate mechanism of reduction of the activated ligand only in the latter case. Finally we have designed peralkylated organo-iron complexes that may serve as "electron or H-atom reservoirs" $[CpFe^I(C_6R_6)$ and $CpFe^{II}(C_6R_5CHR')]$ and "hydride reservoirs" $[C_5R_5Fe(CO)_2H]$. We now have at hand stoichiometric and catalytic examples of e^-,[51,52] $H\cdot$, or H^- transfer using these model iron complexes.

REFERENCES

1. GAYLORD, N. G. 1956. Reduction with Metal-Hydrides. Wiley Interscience. New York, N.Y.
2. RERICK, M. N. 1968. *In* Reduction. R. L. Agustine, Ed. Marcel Dekker. New York, N.Y.
3. HOUSE, H. O. 1972. Modern Synthetic Reactions. W. A. Benjamin. Menlo Park, Calif.
4. BROWN, H. C. & S. KRISHNAMURTHY. 1979. Tetrahedron **35**: 567.
5. GLADYSZ, J. A. & W. TAM. 1978. J. Am. Chem. Soc. **100**: 2545.
6. TAM, W., W. K. WONG & J. A. GLADYSZ. 1979. J. Am. Chem. Soc. **101**: 1589.
7. CASEY, C. P. & S. M. NEUMANN. 1976. J. Am. Chem. Soc. **98**: 5395; 2544.
8. WINTER, S. R., G. W. CORNETT & E. A. THOMSON. 1977. J. Organomet. Chem. **133**: 339.
9. DAVIES, S. G., M. L. H. GREEN & D. M. P. MINGOS. 1978. Tetrahedron **34**: 3047.
10. GREVELS, F. W. & I. FISCHLER. 1981. The Organic Chemistry of Iron. E. A. Koerner von Gustorf, Ed. **2**. Academic Press. New York, N.Y.
11. RYLANDER, P. L. 1979. Catalytic Hydrogenations in Organic Synthesis (Chapter 3): 31. Academic Press. New York, N.Y.
12. KOCHI, J. K. Organometallic Mechanisms and Catalysis. 1978. Academic Press. New York, N.Y.
13. CATHELINE, D. & D. ASTRUC. 1982. J. Organomet. Chem. **226**: C52.
14. HAMON, J.-R., D. ASTRUC & P. MICHAUD. 1981. J. Am. Chem. Soc. **103**: 758.
15. CRABTREE, R. H. & G. G. HLATKY. 1981. Inorg. Chem. **21**: 1273.
16. ASTRUC, D., E. ROMAN, J.-R. HAMON & P. BATAIL. 1979. J. Am. Chem. Soc. **101**: 2240.
17. MICHAUD, P. & D. ASTRUC. 1982. Angew. Chem. Int. Ed. Engl. **12**: 918.
18. HAMON, J.-R., D. ASTRUC, E. ROMAN, P. BATAIL & J. J. MAYERLE. 1981. J. Am. Chem. Soc. **103**: 2431.
19. ASTRUC, D., J.-R. HAMON, E. ROMAN & P. MICHAUD. 1981. J. Am. Chem. Soc. **103**: 7502.
20. MICHAUD, P., D. ASTRUC & J. H. AMMETER. 1982. J. Am. Chem. Soc. **104**: 3755.
21. ASHBY, E. C., A. B. GOEL, R. N. DE PRIEST & N. S. PRASAD. 1981. J. Am. Chem. Soc. **103**: 973.
22. ASHBY, E. C., A. B. GOEL & R. N. DE PRIEST. 1980. J. Am. Chem. Soc. **102**: 7779.
23. DOHAN, J. M., S. BOUÉ & E. V. DOUCHT. 1982. J. Chem. Soc. Chem. Commun.: 917.
24. KAIM, W. 1982. Angew. Chem. Int. Ed. Engl. **21**: 141.
25. WONG, C. L., R.-J. KLINGER & J. K. KOCHI. 1980. Inorg. Chem. **19**: 423.
26. KLINGER, R.-J., K. MOCHIDA & J. K. KOCHI. 1979. J. Am. Chem. Soc. **101**: 6626.
27. JONES, W. D., J. M. HUGGINS & R. G. BERGMAN. 1981. J. Am. Chem. Soc. **103**: 4415.
28. CLARK, D. W. & L. A. P. KANE-MAGUIRE. 1979. J. Organomet. Chem. **174**: 199.
29. NESMEYANOV, N. A., N. A. VOL'KENAU, L. S. SHILOVTSEVA & V. A. PETRAKOVA. 1973. J. Organomet. Chem. **61**: 329.
30. GREEN, M. L. H., L. PRATT & G. WILKINSON. 1960. J. Chem. Soc.: 989.
31. KHAND, I. U., P. L. PAUSON & W. E. WATTS. 1969. J. Chem. Soc. C: 2024; 2261.
32. FISCHER, E. O. & F. ROHRSCHEID. 1963. Z. Naturforsch. **17b**: 483.
33. MICHAUD, P., J.-P. MARIOT, F. VARRET & D. ASTRUC. 1982. J. Chem. Soc. Chem. Commun.: 1383.
34. STEWART, R. P., N. OKAMOTO & W. A. G. GRAHAM. 1972. J. Organomet. Chem. **42**: C32.
35. SWEET, J. R. & W. A. G. GRAHAM. 1979. J. Organomet. Chem. **173**: C9.
36. SWEET, J. R. & W. A. G. GRAHAM. 1982. J. Am. Chem. Soc. **104**: 2811.
37. CASEY, C. P., M. A. ANDREWS & J. E. RINZ. 1979. J. Am. Chem. Soc. **101**: 741.
38. CASEY, C. P., M. A. ANDREWS & D. R. McALISTER. 1979. J. Am. Chem. Soc. **101**: 3371.

39. CASEY, C. P., M. A. ANDREWS & D. R. MCALISTER. 1980. J. Am. Chem. Soc. **102:** 1927.
40. GLADYSZ, J. A., J. C. SELOVER & C. E. STROUSE. 1978. J. Am. Chem. Soc. **100:** 6766.
41. VAUGHN, G. D. & J. A. GLADYSZ. 1981. J. Am. Chem. Soc. **103:** 5608.
42. TAM, W., W.-K. WONG & J. A. GLADYSZ. 1979. J. Am. Chem. Soc. **101:** 5440.
43. KIEL, W. A. & G. Y. LIN. 1980. J. Am. Chem. Soc. **102:** 3299.
44. TAM, W., G.-Y. LIN, W.-K. WONG, W. A. KIEL, V. K. WONG & J. A. GLADYSZ. 1982.
    J. Am. Chem. Soc. **104:** 141.
45. GLADYSZ, J. A. 1982. Adv. Organomet. Chem. **20:** 1.
46. COLLMAN, J. P. & S. R. WINTER. 1973. J. Am. Chem. Soc. **95:** 4089.
47. BODNAR, I., E. CORNAN, K. MENARD & A. CUTLER. 1982. Inorg. Chem.: 1272.
48. DAVISON, A., M. L. H. GREEN & G. WILKINSON. 1961. J. Chem. Soc.: 3172.
49. WINTER, S. R., G. W. CORNETT & E. A. THOMSON. 1977. J. Organomet. Chem. **133:**
    339.
50. CHEN, Y.-S. & J. E. ELLIS. 1982. J. Am. Chem. Soc. **104:** 141.
51. ASTRUC, D., J.-R. HAMON, G. ALTHOFF, E. ROMAN, P. BATAIL, P. MICHAUD, J.-P. MARIOT,
    F. VARRET & D. COZAK. 1979. J. Am. Chem. Soc. **101:** 5545.
52. BUET, A., A. DARCHEN & C. MOINET. 1979. J. Chem. Soc. Chem. Commun.: 447.
53. LAPINTE, C. & D. ASTRUC. (In preparation.)
54. HAMON, J.-R. & D. ASTRUC. 1983. J. Am. Chem. Soc. **105:** 5951.

SOME ASPECTS OF HYDROGEN ACTIVATION IN HYDROFORMYLATION

Piero Pino

Department of Industrial and Engineering Chemistry
Swiss Federal Institute of Technology (ETH)
CH-8092 Zurich, Switzerland

INTRODUCTION

Hydrogen activation is one of the most interesting aspects of homogeneous catalytic reactions involving hydrogen (SCHEME 1), some of which also have significant practical importance. Hydrogen activation has been investigated since the first homogeneous hydrogenation was described by Calvin,[1] but, in the case of hydroformylation, it is still not sufficiently understood.

Different activation paths have been postulated without substantial experimental evidence,[2–5] and some conclusions reached, particularly for the cobalt-catalyzed reactions,[6] have been unduly generalized.[7]

In view of the significance of the knowledge of hydrogen activation in the synthesis of new catalytic systems and the importance of obtaining a low-temperature activation of hydrogen, for instance in the field of asymmetric hydroformylation,[8] we are investigating this problem. In this review some recent results are summarized.

Two aspects of hydrogen activation are relevant for the understanding of the mechanism of catalytic reactions: the type of cleavage of the H—H bond (homolytic or heterolytic),[9] and the mode in which the cleavage is reached including the nature of the complex activating hydrogen and the mechanism (concerted, radical) by which activation takes place.[10, 11] The main paths for hydrogen activation recognized in hydrogenation are reported in |SCHEME 2.

In SCHEME 3 the postulated mechanisms for hydrogen activation in hydroformylation are schematically summarized. It appears that a homolytic splitting of the H—H bond is in general preferred and that oxidative addition of hydrogen either to an acylcarbonyl or to a bimetallic complex represents the preferred activation path. In the case of cobalt, activation of hydrogen by oxidative addition to a metal complex or to a π-olefin metal complex also has been proposed.[3, 4]

In order to improve the experimental evidence in the field, a series of hydroformylations has been carried out with $Co_2(CO)_8$, $Rh_4(CO)_{12}$, and $(DIOP)Pt(Cl)(SnCl_3)$ as catalyst precursors using instead of H_2 a mixture of H_2 and D_2 (1:1).[17]

As previously shown by Wilkinson in the case of olefin hydrogenation with $RhCl(PPh_3)_3$ as the catalyst precursor,[18] the use of H_2-D_2 mixtures allows us to establish whether both hydrogen atoms added to the double bonds stem from the same molecule (if H/D scramblings are not significant).

As shown in SCHEME 4, the relative amount of d_0, d_1, and d_2 reactions products allows us to distinguish among the first mechanism (4a), in which a hydride is formed as one of the first two steps in the catalytic cycle, and the other two, in which either a dihydride is formed after the olefin insertion (4b) or monohydrides are mainly

111

Hydrogenation

$$X=Y+H_2 \longrightarrow XH-\mathbf{YH}$$

$$X\equiv Y+H_2 \longrightarrow XH=YH \longrightarrow XH_2-YH_2$$

$$XY= \text{\textbackslash}C = C \text{\textbackslash} \; ; \; -C \equiv C- \; ; \; \text{\textbackslash}C=O \text{ etc.}$$

Hydroformylation

$$\text{\textbackslash}C=C\text{\textbackslash} + CO+H_2 \longrightarrow \begin{array}{c} \text{\textbackslash}C-C\text{\textbackslash} \\ | \; | \\ H \; CHO \end{array}$$

Hydrogenolysis

$$M-CH_2\text{\small\leftwave} + H_2 \longrightarrow M-H + CH_3\text{\small\leftwave}$$

Reduction of inorganic ions

$$2Fe^{III} + H_2 \longrightarrow 2Fe^{II} + 2H^+$$

H_2/D_2 Scrambling

$$H_2+D_2 \rightleftharpoons 2HD$$

Exchange between H_2 and compounds containing hydrogen atoms

$$RH + D_2 \rightleftharpoons RD + DH$$

SCHEME 1. Some catalytic homogeneous reactions involving hydrogen activation.

a) $M^{n+} + H_2 \longrightarrow MH^{(n-1)+} + H^+$ $RuCl_6^{3-} + H_2 \longrightarrow RuHCl_5^{3-} + H^+ + Cl^-$ (Ref. 9)

b) $M^a + H_2 \longrightarrow M^{(a+2)}(H_2)$ $Ir^I(Cl)(CO)(PPh_3)_2 + H_2 \longrightarrow Ir^{III}(H_2)(Cl)(CO)(PPh_3)_2$ (Ref.10)

c) $M-M + H_2 \longrightarrow 2MH$ $Co_2(CO)_8 + H_2 \longrightarrow 2HCo(CO)_4$ (Ref.12)

d) $2M\cdot + H_2 \longrightarrow 2MH$ $2[Co(CN)_5]^{3-} + H_2 \longrightarrow 2[Co(CN)_5H]^{3-}$ (Ref. 9)

SCHEME 2. Activation of hydrogen in hydrogenation.

$$[M]\text{-}CO\text{-}R + H_2 \longrightarrow [M]\overset{H}{\underset{H}{\text{-}CO\text{-}R}} \longrightarrow R\text{-}CHO + [M]\text{-}H \qquad (Rh[13], Co[5], Ir[14], Pt[7])$$

$$[M]\text{-}[M] + H_2 \longrightarrow 2[M]\text{-}H; \quad [M]\text{-}CO\text{-}R + [M]\text{-}H \longrightarrow R\text{-}CHO + [M]_2 \quad (Co[2], Rh[15])$$

$$[M] + H_2 \longrightarrow [M]H_2 \cdots\cdots \Rightarrow [M]\overset{H}{\text{-}CO\text{-}R} \longrightarrow R\text{-}CHO + [M] \qquad (Co[3,4])$$

$$[M^+X^-] + H_2 \longrightarrow [M]\text{-}H + HX; \quad [M]\text{-}CO\text{-}R + HX \longrightarrow R\text{-}CHO + [M^+X^-] \qquad (Pt[16])$$

Examples

$$[M]\text{-}CO\text{-}R = P_2(CO)_2Rh\text{-}CO\text{-}R; \quad [M]\text{-}[M] = Co_2(CO)_8; \quad [M] = [Co_2(CO)_7(Olefin)];$$

$$[M^+X^-] = P_2Pt(SnCl_3)(Cl)$$

$$P = \text{tertiary phosphine}$$

SCHEME 3. Postulated activation of H_2 in hydroformylation.

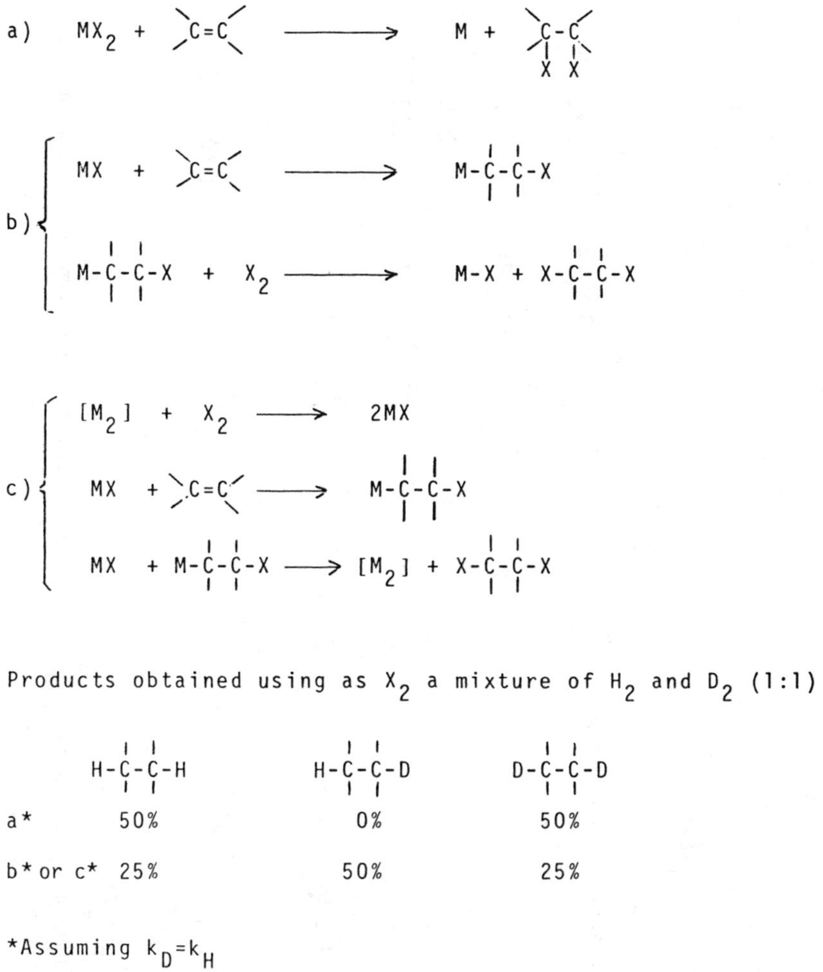

a) MX_2 + $\overset{\diagdown}{\diagup}C=C\overset{\diagup}{\diagdown}$ $\longrightarrow$ M + $\underset{X\ X}{\overset{\diagdown}{\diagup}\overset{|}{C}-\overset{|}{C}\overset{\diagup}{\diagdown}}$

b) $\left\{ \begin{array}{l} MX\ +\ \overset{\diagdown}{\diagup}C=C\overset{\diagup}{\diagdown}\ \longrightarrow\ M-\overset{|}{\underset{|}{C}}-\overset{|}{\underset{|}{C}}-X \\[2em] M-\overset{|}{\underset{|}{C}}-\overset{|}{\underset{|}{C}}-X\ +\ X_2\ \longrightarrow\ M-X\ +\ X-\overset{|}{\underset{|}{C}}-\overset{|}{\underset{|}{C}}-X \end{array} \right.$

c) $\left\{ \begin{array}{l} [M_2]\ +\ X_2\ \longrightarrow\ 2MX \\[1.5em] MX\ +\ \overset{\diagdown}{\diagup}C=C\overset{\diagup}{\diagdown}\ \longrightarrow\ M-\overset{|}{\underset{|}{C}}-\overset{|}{\underset{|}{C}}-X \\[1.5em] MX\ +\ M-\overset{|}{\underset{|}{C}}-\overset{|}{\underset{|}{C}}-X\ \longrightarrow\ [M_2]\ +\ X-\overset{|}{\underset{|}{C}}-\overset{|}{\underset{|}{C}}-X \end{array} \right.$

Products obtained using as X_2 a mixture of H_2 and D_2 (1:1)

	$H-\overset{\|}{\underset{\|}{C}}-\overset{\|}{\underset{\|}{C}}-H$	$H-\overset{\|}{\underset{\|}{C}}-\overset{\|}{\underset{\|}{C}}-D$	$D-\overset{\|}{\underset{\|}{C}}-\overset{\|}{\underset{\|}{C}}-D$
a*	50%	0%	50%
b* or c*	25%	50%	25%

*Assuming $k_D = k_H$

SCHEME 4. Products obtained in the catalytic reaction between olefins and H_2/D_2 mixtures (X = H or D).

involved in the interaction with the substrates (4c). In the following sections we summarize the results obtained up to now using this approach and some attempts in the case of cobalt catalysts to distinguish between the mechanisms schematized in 4b and 4c.

ACTIVATION OF HYDROGEN IN RHODIUM-CATALYZED HYDROFORMYLATION

In SCHEME 5 the types of hydrogen activation are shown that have been postulated when $Rh_4(CO)_{12}$ or $RhH(CO)(PPh_3)_3$ is used as catalyst precursor. In the same scheme, the expected relative amount of d_0, d_1, and d_2 products is given supposing

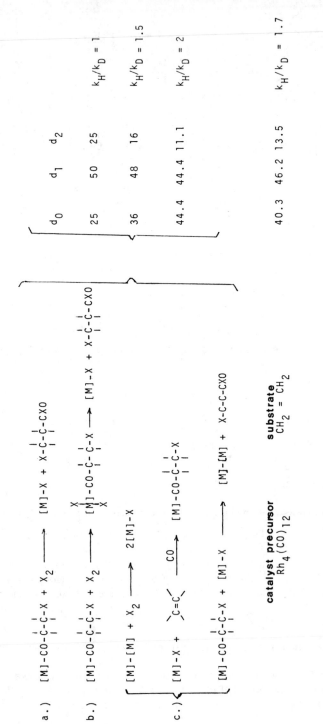

SCHEME 5. Deuterium distribution in the products of the rhodium-catalyzed hydroformylation experiments with equimolecular H_2/D_2 mixtures; comparison between expected and experimental results.

no kinetic isotope effect or a kinetic isotope effect (k_H/k_D) of 1.5 or 2 in the hydrogen activation.

The experimental results obtained in a conventional ethylene hydroformylation with an H_2/D_2 mixture[17] are shown in SCHEME 5. No significant H_2/D_2 scrambling has been observed, and a kinetic isotope effect of about 1.7 can be calculated from the change in the D_2/H_2 ratio during the reaction and from the d_0/d_2 ratio in the products (SCHEME 5).

The relative amount of the d_0, d_1, and d_2 products corresponds to that expected for the three proposed hydrogen-activation mechanisms (SCHEMES 5a–5c), showing that despite some close similarities in the catalytic steps (e.g., insertion of the olefinic substrate in an M—H bond), activation of hydrogen in some hydrogenations[18] (SCHEME 2b) and hydroformylation is largely different. The differences are probably due to the fact that CO successfully competes with hydrogen for the vacant coordination site existing in the π-olefin complex ("olefin path" in hydrogenation).[18] In other intermediates (e.g., in the acyl-metal complex) coordination of H_2 is preferred to coordination of CO, probably for steric reasons, or a very low dihydride concentration is sufficient to complete the catalytic cycle.

HYDROFORMYLATION WITH $(DIOP)Pt(Cl)(SnCl_3)$ AS THE CATALYST PRECURSOR

In platinum-catalyzed hydroformylation, activation of hydrogen via $(RCO)Pt(H_2)L_x$ (L = Cl, PR_3) has been postulated.[7] In this case the experiments carried out using D_2-H_2 mixtures (1:1) and 2-phenylpropene as olefin (SCHEME 6) are more difficult to interpret because, unlike the cobalt- or rhodium-catalyzed reaction, a rather rapid H_2/D_2 scrambling takes place. Furthermore, as already published, hydroformylation is accompanied by a remarkable hydrogenation of the substrate.[19] As shown in TABLE 1, the relative amount of d_1, d_2, and d_3 products is different for the hydrogenated substrate, 2-phenylpropene, and for the hydroformylated substrate, 3-phenylbutanal, particularly at low conversions when the amount of HD in the gas phase is about 15%. These data show that, in keeping with

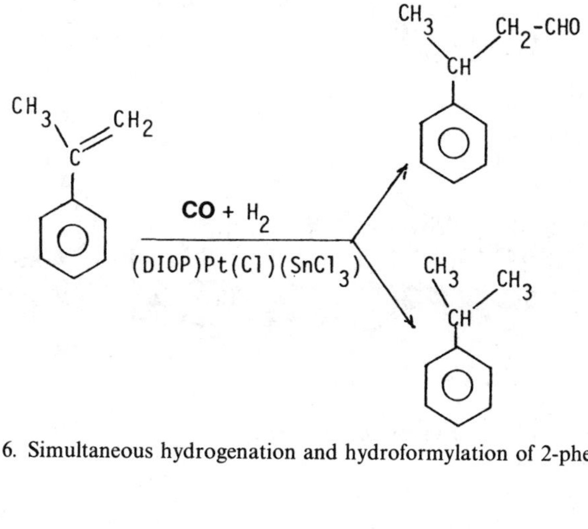

SCHEME 6. Simultaneous hydrogenation and hydroformylation of 2-phenylpropene.

TABLE 1
HYDROFORMYLATION OF 2-PHENYL-1-PROPENE WITH H_2/D_2 (1:1) AND CO MIXTURES*

Time (minutes)	H_2 (%)	HD (%)	D_2 (%)	3-Phenylbutanal			2-Phenylpropane		
				d_0 (%)	d_1 (%)	d_2 (%)	d_0 (%)	d_1 (%)	d_2 (%)
0	49	~1	49	—	—	—	—	—	—
30	42	15	43	28	46	26	54	31	15
60	26	43	31	26	44	30	37	44	19
90	23	49	28	26	46	28	32	46	23

* Catalyst precursor: Pt(Cl)(SnCl$_3$)(DIOP). Solvent: Benzene. $P_{H_2} = 45$ bar; $P_{D_2} = 45$ bar; $P_{CO} = 90$ bar.

previous results on the asymmetric hydrogenation and hydroformylation of 2-phenyl-1- butene,[19] hydrogenation and hydroformylation occur with different mechanisms (SCHEME 7). The deuterium distribution in the reaction products (TABLE 1) is in keeping with the proposed H_2-activation mechanism (SCHEME 7); however the rapid

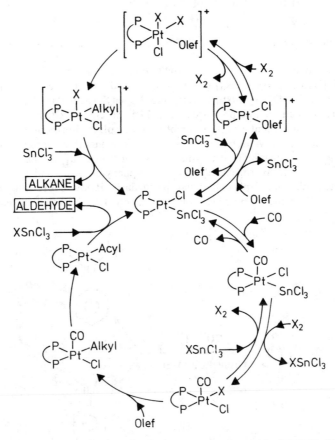

SCHEME 7. Proposed mechanism for hydrogen activation with (DIOP)Pt(Cl)(SnCl$_3$); X = hydrogen or deuterium.

	Propionaldehyde			Ref.
	$d_0\%$	$d_1\%$	$d_2\%$	
a) $Co_2(CO)_8 + H_2 \longrightarrow 2HCo(CO)_4$	25	50	25	(2)
b) $CH_3-CH_2-CO-Co(CO)_3 + H_2 \longrightarrow CH_3-CH_2-CO-\overset{H}{\underset{H}{Co}}(CO)_3$	25	50	25	(5)
c) $Co_2(CO)_8 + CH_2 = CH_2 + H_2 \longrightarrow [Co_2(CO)_7(CH_2=CH_2)(H_2)]]+CO$	50	0	50	**(4)**
d) $[Co_2(CO)_7(CH_2=CH_2)] + H_2 \longrightarrow \{Co_2(CO)_6\} + CH_3-CH_2-CHO$	50	0	50	**(3)**
Experimentally found	37.8	47.6	14.6	

$$(k_H/k_D = \sim 1.3)$$

SCHEME 8. Postulated reactions for the activation of H_2 in cobalt-catalyzed hydroformylation, and expected results in the hydroformylation of ethylene with D_2/H_2 mixtures (1:1) assuming $k_D/k_H = 1$.

H_2/D_2 scrambling as well as the presence of HD in the gas phase and of deuterium in the hydrogenation products when the reaction is carried out with H_2 in the presence of C_2H_5OD indicate that a heterolytic H_2 cleavage may play a role, not only in hydrogenation,[20] but also in hydroformylation with platinum catalysts.

HYDROFORMYLATION WITH $Co_2(CO)_8$ AS THE CATALYST PRECURSOR

Four types of hydrogen activation have been proposed for the cobalt-catalyzed hydroformylation (SCHEME 8). The results obtained in the ethylene hydroformylation using a D_2-H_2 (1:1) mixture indicate a large prevalence of d_1 products, thus excluding Mechanisms 8c and 8d. The kinetic isotopic effect (k_H/k_D) is about 1.3 ± 0.3 if calculated from the change in the gas composition and 1.6 if calculated from the d_0/d_2 ratio.

TABLE 2

REACTION BETWEEN $HCo(CO)_4$ AND 1-OCTENE IN THE PRESENCE OF H_2*

Time (minutes)	$HCo(CO)_4$ (mmol)	$Co_2(CO)_8$ (mmol × 2)	$Co_4(CO)_{12}$ (mmol × 4)	Aldehydes (mmol)	Other Metal Carbonyl Complexes† (mmol)
0	2.20	0.14	—	—	—
1.8	1.38	0.72	—	—	0.24
6	1.08	0.76	—	—	0.50
15	0.88	0.82	—	—	0.64
30	0.74	1	—	—	0.60
48	0.44	1.24	—	<0.02	0.66
1205	~0	ND‡	ND	4.06	ND

* $P_{CO} = 2.2$ atm; olefin = 83.8 mmol; $HCo(CO)_4 = 2.20$ mmol; $P_{H_2} = 100$ atm; $t = 25°$; n-hexene as the solvent.

† Calculated as total $Co_{(initial)} - [HCo(CO)_4 + Co_2(CO)_8 + Co_4(CO)_{12}]$ expressed as $RCOCo(CO)_4$.

‡ ND means not determined.

These data do not allow us to distinguish between Mechanisms 8a and 8b, or other mechanisms in which the two hydrogen atoms added to the olefin molecule in hydroformylation do not necessarily arise from the same H_2 (or D_2) molecule. In the recent literature Path 8a is preferred[21] despite the fact that activation of hydrogen via oxidative addition to $Co_2(CO)_8$ had been previously excluded because the reaction between $Co_2(CO)_8$ and H_2 appears to be much slower than hydroformylation.[4, 6] One of the grounds on which Path 8a has been favored, that is, the formation of $Co_2(CO)_8$ from $HCo(CO)_4$ in the first reaction step, does not seem very sound because in the stoichiometric reaction between $HCo(CO)_4$ and olefins, $Co_2(CO)_8$ formation parallels the $RCOCo(CO)_4$ formation and precedes the aldehyde formation[22, 23] (TABLE 2). Furthermore the IR analytical method used[21] to determine $HCo(CO)_4$ is not sufficiently accurate because of the overlapping of the band at 2.031 cm^{-1} of $Co_2(CO)_8$ with the band at 2,030 cm^{-1} of $HCo(CO)_4$. Therefore the path of H_2 activation in the cobalt-catalyzed hydroformylation, even if Paths 8c and 8d have been discarded, remains unknown.

The Reaction between HCo(CO)$_4$, Olefins, and H$_2$ or D$_2$

To obtain further information on the species activating hydrogen in cobalt-catalyzed hydroformylation, the stoichiometric reaction between olefins and HCo(CO)$_4$ has been investigated in the presence of hydrogen. In fact if hydrogen from the gas phase takes part in the reaction as expected on the basis of the reaction between Co$_2$(CO)$_8$, H$_2$, and olefins,[24, 25] the stoichiometry of the reaction should be different in the presence and in the absence of hydrogen (Scheme 9). The reaction was investigated using 3,3-dimethyl-1-butene as the substrate, which produces by hydroformylation only one aldehyde, using a low CO pressure to stabilize Co$_2$(CO)$_8$ and HCo(CO)$_4$. The reaction was monitored using a high-pressure IR cell and utilizing the band at 1,857 cm^{-1} for Co$_2$(CO)$_8$, at 2,116 cm^{-1} for HCo(CO)$_4$, at 1,867 cm^{-1} for Co$_4$(CO)$_{12}$, at 1,735 cm^{-1} for —CHO, and at 1,724 cm^{-1} for —CDO as described elsewhere.[22, 26]

As shown in Figure 1, the moles of aldehyde produced amount to more than the moles of HCo(CO)$_4$ used. The moles of Co$_2$(CO)$_8$ present show that not only H$_2$ but also CO from the gas phase has been used to produce the aldehyde or to regenerate the Co$_2$(CO)$_8$ before Co$_4$(CO)$_{12}$ is formed. In fact the last compound reacts only very slowly with CO to reform Co$_2$(CO)$_8$ under the reaction conditions used.[31]

In order to clarify the role played by the gaseous H$_2$ and considering that the exchange between HCo(CO)$_4$ and D$_2$ giving DCo(CO)$_4$ and HD, under the conditions used, is extremely slow, the reaction was repeated using D$_2$ instead of H$_2$ and the relative amount of d$_0$, d$_1$, and d$_2$ products was determined. As shown in

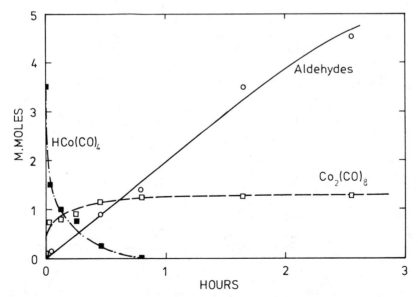

Figure 1. Hydroformylation of 1-octene with HCo(CO)$_4$ and H$_2$ ($P_{CO} = 0.65$ bar; CO = 22.6 mmol; $P_{H_2} = 100$ bar; $t = 25°$; 1-octene = 121 mmol).

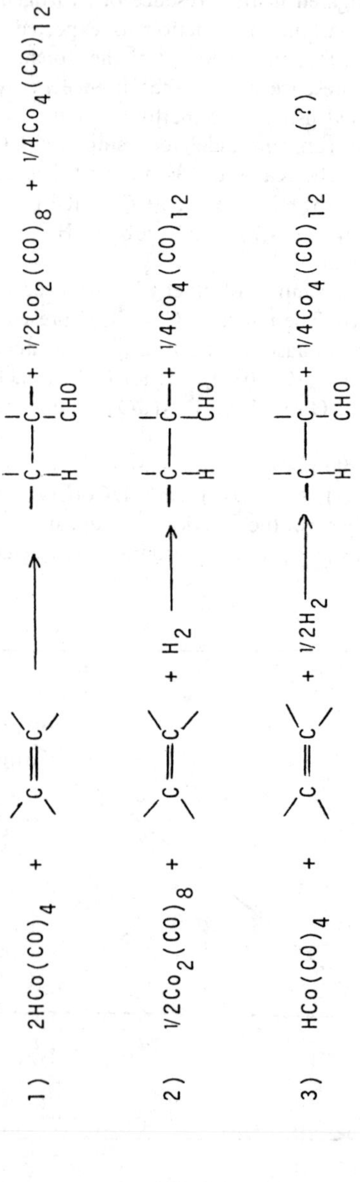

1) $2HCo(CO)_4$ + $\diagup C = C \diagdown \longrightarrow$ $-\overset{|}{\underset{|}{C}}-\overset{|}{\underset{H}{C}}-$ + $1/2Co_2(CO)_8$ + $1/4Co_4(CO)_{12}$
$\qquad\qquad\qquad\qquad\qquad\qquad\qquad CHO$

2) $1/2Co_2(CO)_8$ + $\diagup C = C \diagdown$ + $H_2 \longrightarrow$ $-\overset{|}{\underset{|}{C}}-\overset{|}{\underset{H}{C}}-$ + $1/4Co_4(CO)_{12}$
$\qquad\qquad\qquad\qquad\qquad\qquad\qquad CHO$

3) $HCo(CO)_4$ + $\diagup C = C \diagdown$ + $1/2H_2 \longrightarrow$ $-\overset{|}{\underset{|}{C}}-\overset{|}{\underset{H}{C}}-$ + $1/4Co_4(CO)_{12}$ (?)
$\qquad\qquad\qquad\qquad\qquad\qquad\qquad CHO$

SCHEME 9. Stoichiometric hydroformylation of olefins.

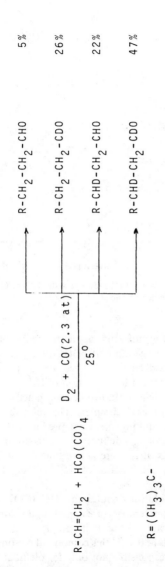

SCHEME 10. 3,3-Dimethyl-1-butene deuteroformylation in the presence of $HCo(CO)_4$.

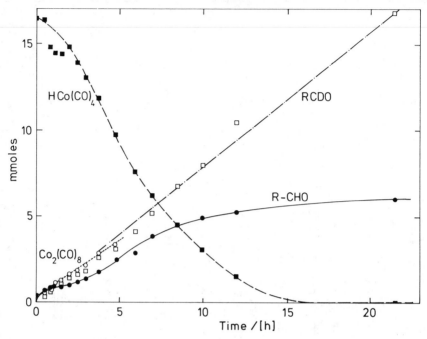

FIGURE 2. Deuterioformylation of 3,3-dimethyl-1-butene with HCo(CO)$_4$ (P_{CO} = 2.3 bar; P_{D_2} = 100 bar; t = 25°; molar ratio olefin/hydride = 33).

FIGURE 2 the formation of —CDO groups, that is, the reduction of the acyl groups by the gaseous D$_2$, starts when a large amount of HCo(CO)$_4$ is still present and continues when the HCo(CO)$_4$ has disappeared. These results clearly indicate that the deuterium-containing species formed by activation of D$_2$, the concentration of which is very small as it is not detected by IR, is much more active than HCo(CO)$_4$ in reducing the intermediate cobalt acyl complex. The relatively low activity of HCo(CO)$_4$ in comparison to other unknown species present in low concentration is also proved by the small amount of d$_0$ aldehydes found (SCHEME 10). Concerning the unknown active cobalt complex, it is more active than HCo(CO)$_4$ even in the activation of olefins, as shown by the amount of d$_1$ aldehyde formed bearing a —CHO group[22,27] (SCHEME 10).

Taking into account that in solutions containing HCo(CO)$_4$ and Co$_2$(CO)$_8$, HCo$_3$(CO)$_9$ is rapidly formed at room temperature and at subatmospheric pressure[28] and that HCo$_3$(CO)$_9$ rapidly activates olefins as it completely isomerizes 1-hexene to a hexene mixture in a few seconds at 0°C,[29] this compound or some closely related complex could be present and be responsible not only for olefin activation but also for H$_2$ (or D$_2$) activation under the conditions used. Therefore, beside SCHEMES 8a and 8b, SCHEME 11 should be considered as a possible path for hydrogen activation

$$HCo_3(CO)_9 + H_2 \longrightarrow 3HCo(CO)_3$$

SCHEME 11

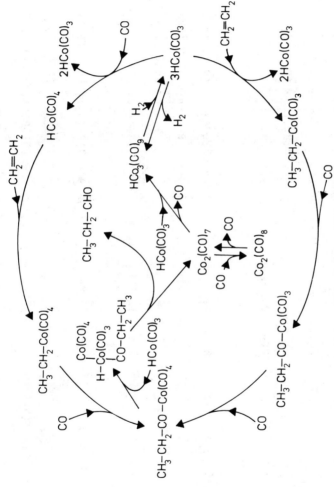

SCHEME 12. Possible paths for the hydroformylation of ethylene with $Co_2(CO)_8$ as catalyst precursor.

Catalyst precursor	Proposed Path	
	$[Co] + H_2 \longrightarrow [Co(H_2)]$	(Ref. **4**)
$Co_2(CO)_8$	$Co_2(CO)_8 + H_2 \longrightarrow 2HCo(CO)_4$	(Ref. 2)
	$(RCO)Co(CO)_3 + H_2 \longrightarrow RCOCo(H_2)(CO)_x$	(Ref. 5)
	$HCo_3(CO)_9 + H_2 \longrightarrow 3HCo(CO)_3$	
$Rh_4(CO)_{12}$	$Rh_4(CO)_{12} + 2H_2 \longrightarrow 4HRh(CO)_3$	(Ref. **14**)
$(H)Rh(CO)(PPh_3)_3$	$(RCO)Rh(CO)(PPh_3)_2 + H_2 \longrightarrow (RCO)Rh(H_2)(CO)(PPh_3)$	(Ref. 13)
	$[Rh(CO)_2(PPh_3)_2]_2 + H_2 \longrightarrow 2HRh(CO)_2(PPh_3)_2$	
$(PPh_3)_2Pt(Cl)(SnCl_3)$	$(RCO)Pt(PPh_3)_2(SnCl_3) + H_2 \longrightarrow (RCO)Pt(H_2)(PPh_3)_x(SnCl_3)_3$	(Ref. 7)
$(DIOP)Pt(Cl)(SnCl_3)$	$(DIOP)Pt(Cl)(SnCl_3) + H_2 \longrightarrow (DIOP)Pt(H)(Cl) + HSnCl_3$	

SCHEME 13. Alternative paths for the activation of H_2 in hydroformylation.

in cobalt-catalyzed hydroformylation. The resulting hypothetical mechanism for this reaction is presented in SCHEME 12, the prevalence of either path for the acyltetracarbonylcobalt formation depending on carbon monoxide pressure and temperature and leading, for monosubstituted ethylenes, to linear or branched aldehyde respectively.

CONCLUSIONS

The present state of our knowledge of hydrogen activation in hydroformylation is still very unsatisfactory. Both a homolytic and a heterolytic splitting of H_2 seem possible, but further research is necessary to clarify this aspect.

On the basis of the results of the hydroformylation with H_2/D_2 mixture, all the mechanisms involving hydrogen activation via a π-olefin dihydride complex can be excluded when $Co_2(CO)_8$, $Rh_4(CO)_{12}$, or $(DIOP)Pt(Cl)(SnCl_3)$ is used as catalyst precursor. Among the other proposed H_2-activation mechanisms (SCHEME 13), the activation of hydrogen by reaction with $Co_2(CO)_8$ with formation of $HCo(CO)_4$ seems—contrary to recent literature data—unlikely, particularly at low temperature and low carbon monoxide pressure. The activation of hydrogen via $HCo_3(CO)_9$ offers new perspectives for the formulation of a detailed hydroformylation mechanism. However, a better knowledge of the chemical properties of this interesting class of high-spin coordinatively unsaturated clusters, which includes also the recently refor-mulated complexes $HCo_3(CO)_6(PPh_3)_3$ and $HCo_3(CO)_6[P(C_4H_9)_3]_3$,[30] is needed in order to evaluate their possible role in hydroformylation. For the rhodium-catalyzed activation of hydrogen, attempts are needed to react the recently[32] identified acylrhodium carbonyls with H_2 and to isolate the corresponding dihydrides if they really form during the reaction. Finally, for the platinum-catalyzed hydroformylation using $(DIOP)Pt(Cl)(SnCl_3)$ as catalyst precursor, activation via formation of $HSnCl_3$ and a neutral platinum hydride seems more likely than the previously proposed activation via an acyl platinum dihydride.[7]

Further research should be directed to the achievement of a low-temperature hydrogen activation in the presence of carbon monoxide, a goal that should allow a rational approach to some practically important reactions like the synthesis of formaldehyde from CO and H_2 and the low-temperature asymmetric hydro-formylation.

REFERENCES

1. CALVIN, M. 1938. Trans. Faraday Soc. **34**: 1181–1191.
2. ROELEN, O. 1946. British Intelligence Objectives Sub-Committee, Final Report No. 447. London, England.
3. MARTIN, A. R. 1954. Chem. Ind. London: 1536–1537.
4. NIWA, M. & M. YAMAGUCHI. 1961. Shokubai **3**: 264–278.
5. ORCHIN, M. & W. RUPILIUS. 1972. Catal. Rev. **6**: 85–131.
6. HECK, R. F. & D. S. BRESLOV. 1961. J. Am. Chem. Soc. **83**: 4023–4027.
7. SCHWAGER, I. & J. F. KNIFTON. 1976. J. Catal. **45**: 256–267.
8. PITTMAN, C. U., JR., Y. KAWABATA & L. I. FLOWERS. 1982. J. Chem. Soc. Chem. Commun.: 473–474.
9. HALPERN, J. 1966. Chem. Eng. News **44**(45): 68–75.

10. VASKA, L. & J. W. DILUZIO. 1962. J. Am. Chem. Soc. **84**: 679–681.
11. SCHROCK, R. R. & J. A. OSBORN. 1976, J. Am. Chem. Soc. **98**: 2134–2143.
12. HALPERN, J. 1979. Pure Appl. Chem. **51**: 2171–2182.
13. EVANS, E., J. A. OSBORN & G. WILKINSON. 1968. J. Chem. Soc. A: 3133–3142.
14. MARKÓ, L. 1974. *In* Aspects of Homogeneous Catalysts. R. Ugo, Ed. **2**: 3–55. D. Reidel. Dordrecht, the Netherlands.
15. BROWN, C. K. & G. WILKINSON. 1970. J. Chem. Soc. A: 2753–2764.
16. PINO, P., F. SPINDLER & M. SCALONE. Manuscript in preparation.
17. PINO, P., F. OLDANI & G. CONSIGLIO. 1983. J. Organomet. Chem. **250**: 491–497.
18. OSBORN, J. A., F. H. JARDINE, J. F. YOUNG & G. WILKINSON. 1966. J. Chem. Soc. A: 1711–1732.
19. CONSIGLIO, G., W. ARBER & P. PINO. 1978. Chim. Ind. Milan **60**: 396–400.
20. VAN RANTWIJK, F., A. P. G. KIEBSON & H. VAN BEKKUM. 1975–76. J. Mol. Catal. **1**: 27–35.
21. ALEMDAROGLU, N. H., J. L. M. PENNINGER & E. OLTAY. 1976. Monatshefte, **107**: 1153–1165.
22. TANNENBAUM, R. 1982. Doctoral Dissertation No. 6970. ETH. Zurich, Switzerland.
23. PINO, P., G. BOR & F. SPINDLER. Unpublished results.
24. BARRIK, P. L. 1951. U.S. Patent 2,542,747.
25. PINO, P., R. ERCOLI & F. CALDERAZZO. 1955. Chim. Ind. Milan **37**: 782–786.
26. BOR, G. & R. TANNENBAUM. Manuscript in preparation.
27. PINO, P., R. TANNENBAUM, G. BOR, F. SPINDLER & A. STEFANI. Unpublished results.
28. FACHINETTI, G. Unpublished results.
29. FACHINETTI, G. & A. STEFANI. 1983. Angew. Chem. **94**: 937–938.
30. BRADAMANTE, P., G. FACHINETTI, P. PINO, A. STEFANI & P. F. ZANAZZI. 1983. J. Organomet. Chem. **251**: C47–C50.
31. BOL, G., U. DIETLER & P. PINO. 1978. J. Organomet. Chem. **154**: 301–315.
32. BROWN, J. M. & A. G. KENT. 1982. J. Chem. Soc. Chem. Commun.: 723–725.

HYDROGENATIONS AND HYDROFORMYLATIONS WITH HCo(CO)₄ AND HMn(CO)₅: RADICALS OR OTHERWISE

Milton Orchin

Department of Chemistry
University of Cincinnati
Cincinnati, Ohio 45221

Introduction

The cobalt-catalyzed hydroformylation of olefins has been known since 1938 and has been practiced commercially in the United States since 1947. Otto Roelen of Ruhrchemie A.G. who discovered the reaction thought that $HCo(CO)_4$, formed under the conditions of the reaction (3,000 psi H_2/CO, 150°C), might be the actual catalytic species although almost any form of cobalt could serve as a catalyst precursor. In 1953, Wender, Sternberg, and Orchin demonstrated that $HCo(CO)_4$ reacted stoichiometrically at room conditions with a variety of olefins to give the same products that these olefins gave on catalytic hydroformylation.[1] In 1956, Orchin, Kirch, and Goldfarb demonstrated that $HCo(CO)_4$ is formed (and indeed constitutes the principal form of cobalt) under catalytic hydroformylation conditions.[2] Thus, with the sure knowledge that $HCo(CO)_4$ was the catalytically active species in the hydroformylation reaction, many mechanistic studies were carried out using $HCo(CO)_4$ in the stoichiometric reaction conveniently carried out at 0–25°C and atmospheric pressure. Most of what follows is based on the stoichiometric reaction between olefins and $HCo(CO)_4$ or $HMn(CO)_5$.

Reactions of 1-Alkenes

The dramatic effect of CO partial pressure on the course of the stoichiometric hydroformylation with $HCo(CO)_4$ (as well as other evidence) clearly shows the importance of π complexing between the olefin and the cobalt carbonyl hydride. The π complexation is followed by olefin insertion into the H—Co bond, and this is followed by a series of reactions leading to the formation of aldehydes. The hydroformylation reaction is quite complex, involving at least 15 different, separate reactions,[3] and hence it is little wonder that no satisfactory rate expression for the reaction has been determined. The intermediate steps in the reaction are usually equilibria involving 16- and 18-electron species obtained by loss and gain of coordinated CO. There is still considerable uncertainty as to the mechanism of the final reaction step leading to aldehyde; it is frequently written as a bimolecular reaction such as

$$R-COCo(CO)_3 + HCo(CO)_4 \longrightarrow RCHO + Co_2(CO)_7$$

Little is known about the transition state involved in such a reaction; indeed it is not known with certainty whether or not these are the species actually involved.

In contrast to the very rapid reaction of $HCo(CO)_4$ with 1-alkenes, $HMn(CO)_5$ fails to react at all. Thus treatment of 1-octene with $HMn(CO)_5$ at 115°C for five hours gave no reaction.[4] The hydroformylation of 1-octene with $HCo(CO)_4$ is complete at 25°C in less than two minutes.

Reactions of Aromatic Olefins

In contrast to 1-alkenes, the reaction of $HCo(CO)_4$ with aromatic olefins such as $Ph_2C=CH_2$ gives a quantitative yield of hydrocarbon:

$$Ph_2C=CH_2 + 2\ HCo(CO)_4 \longrightarrow Ph_2CHCH_3 + CO_2(CO)_8$$
$$\mathbf{1}$$

The reaction is cleanly second order (first order in each reactant) and is unaffected by the partial pressure of CO or by solvent polarity.[5] When the reaction is carried out at −78°C in an NMR tube, the proton signal of the methyl group of Product 1 shows a doublet emission at δ 1.6, which rapidly changes to a doublet absorption.[6] Furthermore, when the reaction is carried out with $DCo(CO)_4$, the rate constant is 1.7 times that obtained with $HCo(CO)_4$. The lack of [CO] dependence, the observed chemically induced nuclear polarization (CIDNP) effect, and the inverse isotope effect are all consistent with a radical pair intermediate, and a mechanism involving such an intermediate may be written as follows:

$$Ph_2C=CH_2 + HCo(CO)_4 \rightleftharpoons \overline{Ph_2\overset{\cdot}{C}CH_3\ \overset{\cdot}{C}o(CO)_4} \qquad (1)$$
$$\mathbf{2}$$

$$\overline{Ph_2\overset{\cdot}{C}CH_3\overset{\cdot}{C}o(CO)_4} \longrightarrow Ph_2\overset{\cdot}{C}CH_3 + \overset{\cdot}{C}o(CO)_4 \qquad (2)$$

$$Ph_2\overset{\cdot}{C}CH_3 + HCo(CO)_4 \longrightarrow Ph_2CHCH_3 + \cdot Co(CO)_4 \qquad (3)$$

$$2 \cdot Co(CO)_4 \longrightarrow Co_2(CO)_8 \qquad (4)$$

Equation 1 represents the equilibrium formation of the caged (geminate) radical pair **2** (the rate-controlling step); Equation 2, the escape of the pair out of the cage; Equation 3, the rapid abstraction of an atom of hydrogen; and Equation 4, the very rapid dimerization of the tetracarbonyl radical. This mechanism is modeled after the similar mechanism proposed earlier by Sweany and Halpern for the $HMn(CO)_5$ reduction of α-methylstyrene.[7] These authors were the first to observe a CIDNP enhancement effect in a metal hydride reaction, and such an effect is considered to be rather conclusive evidence for the presence of a geminate radical pair.

The reaction of aromatic olefins with $HCo(CO)_4$ involves the transfer of a hydrogen atom to the substrate with the formation of a 17-electron cobalt carbonyl species as part of the geminate radical pair. Because of the observed inverse isotope effect, it is likely that the potential energy profile for the hydrogen atom transfer reaction involves a transition state lying close to the intermediate radical pair. However the reaction rate of the highly twisted aromatic olefin bifluorenylidene with $HCo(CO)_4$, compared to that with $DCo(CO)_4$, shows a normal isotope effect,[8] probably owing to the radical character of the ground state of this olefin.

Although this compound is a tetrasubstituted ethylene, it reacts as fast as $Ph_2C=CH_2$ (see TABLE 1).

The reaction of $HMn(CO)_5$ with aromatic olefins proceeds by exactly the same mechanism as with $HCo(CO)_4$, as TABLE 1 clearly shows, except that reactions with the latter are about 500–1,000 times faster. This rate difference is no doubt due to the fact that the dissociation energy of the H—Mn bond is about 6 kcal mol^{-1} greater than that of the H—Co bond.

REACTIONS OF CYCLOPROPENES

Although olefinic compounds having a double bond conjugated to aromatic rings ordinarily undergo hydrogenation exclusively with $HCo(CO)_4$, the reaction with 1,2-diphenylcyclopropene and some of its derivatives leads to hydroformylation as well as hydrogenation. Thus 1,2-diphenylethylene (stilbene) fails to react with $HCo(CO)_4$ (but under catalytic conditions gives 1,2-diphenylethane) while 1,2-diphenylcyclopropene reacts completely with $HCo(CO)_4$ in three minutes at 25°C to give a 34% yield of aldehyde and a 65% yield of 1,2-diphenylcyclopropanes.[9] Although $HMn(CO)_5$ fails to react at all with 1-alkenes and gives only hydrogenated products with aromatic olefins,[7] it is only recently that $HMn(CO)_5$ was found to give hydroformylation products under stoichiometric conditions.[4,9] The cyclopropenes that were used for this reaction possessed phenyl substituents, thus enhancing radical stabilization. The highly strained double bond presumably makes these substrates particularly reactive. In the reaction between 1,2-diphenyl-3,3-dimethylcyclopropene, carried out in an NMR tube, both emission and enhanced absorption were observed. The CIDNP emission was clearly due to the hydrogenated product formed by cage escape of the radical pair, while the CIDNP enhancement was ascribed to combination in the cage leading to a sigma-bonded C—Mn intermediate, which is the precursor to aldehyde. Since $HCo(CO)_4$ also leads to a mixture of hydrogenated and hydroformylated products, the reactions may be represented as shown in SCHEME 1. According to SCHEME 1 the partitioning between hydroformylation and hydrogenation might depend on the lifetime of the radical pair in the cage, since combination leads to aldehyde and escape leads to hydrogenation.

TABLE 1

REACTION RATES OF $HCo(CO)_4$ AND $HMn(CO)_5$ AND THEIR
DEUTERIO ANALOGUES WITH VARIOUS OLEFINS

	$k(L\ mol^{-1}\ s^{-1})$ at 31.5°C							
					(H)	(D)	k_H/k_D	
Compound	H—Co*	H—Mn	D—Co*	D—Mn	Co/Mn	Co/Mn	Co	Mn
$Ph_2C=CH_2$	0.44	1.4×10^{-4}	0.83	2.4×10^{-4}	31×10^2	35×10^2	0.53	0.58
$F1=CHCH_3$†	1.38	8.5×10^{-4}	2.92	1.4×10^{-3}	16×10^2	27×10^2	0.47	0.60
$F1=F1$	0.40	17×10^{-4}	0.185	7.1×10^2	2.3×10^2	2.6×10^2	2.15	2.39

* Rate values at 31.5°C were obtained from extrapolation of Arrhenius plots.

† F1= is

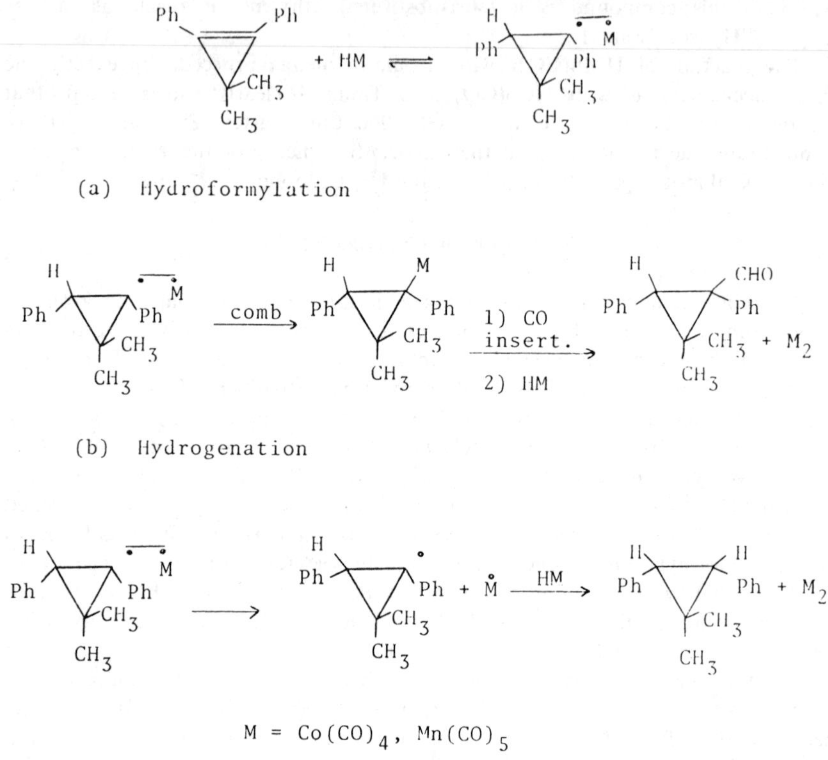

(a) Hydroformylation

(b) Hydrogenation

$$M = Co(CO)_4, \; Mn(CO)_5$$

SCHEME 1

It has been shown rather recently that sequestering the radicals $(R^1\cdot + R^2\cdot)$ generated by irradiation of $R^1C(O)R^2$ in a micelle solution leads to combination of the radicals so generated to give 100% R^1R^2 rather than a mixture containing R^1R^1 and R^2R^2 as well.[10] This arises because the radicals generated are trapped in the micelle and combine before they can escape to combine randomly. Using this technique, the reaction of several cyclopropenes in detergent solution was examined to compare the partitioning between hydrogenated and hydroformylated products with that obtained in homogeneous solution. The results are shown in FIGURE 1;[11] they indicate, as expected, that the micelle reaction enhances the formation of hydroformylation product. (For the stereochemistry of the products in SCHEME 1 and FIGURE 1 see the referenced articles.)

SYNTHESIS GAS INSERTION INTO AN ACYLMANGANESE PENTACARBONYL

A very recent report shows that the reaction of certain $RMn(CO)_5$ compounds with CO/H_2 (130 psi) in the highly polar solvent sulfolane at 75°C gives only RCHO and that similar treatment with H_2 gives only RCH_2OH.[12] When this reaction was repeated with $PhCH_2C(O)Mn(CO)_5$ and with $PhCH_2CH_2C(O)Mn(CO)_5$ but with hexane as a solvent, a completely different

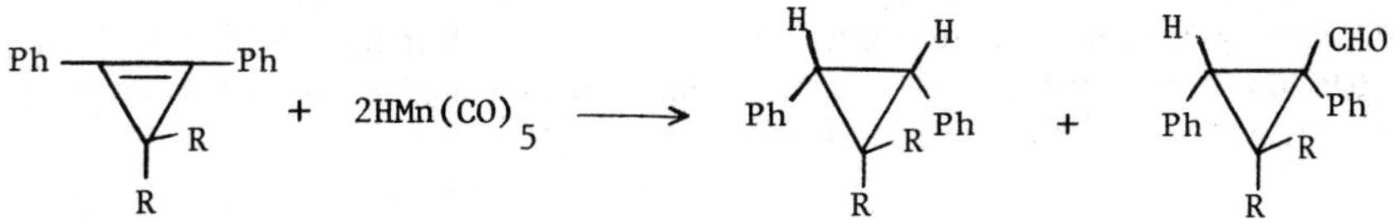

ESCAPE   COMB.

Homogeneous Solution

	ESCAPE	COMB.
R=H	59	39
R=CH$_3$	63	37

Micelle Solution

	ESCAPE	COMB.
R=H	7	93
R=CH$_3$	7	93

FIGURE 1. Comparison of homogeneous and micelle solutions.

reaction (Equation 5) was observed.[13]

$$R-\overset{\overset{\displaystyle O}{\|}}{C}-Mn(CO)_5 \longrightarrow R-\overset{\overset{\displaystyle O}{\|}}{C}-\boxed{CH_2O}-Mn(CO)_5 \qquad (5)$$

The overall reaction is equivalent to inserting CH_2O into the C—Mn bond. Although direct experimental evidence is lacking, the following sequence may account for this unusual reaction:

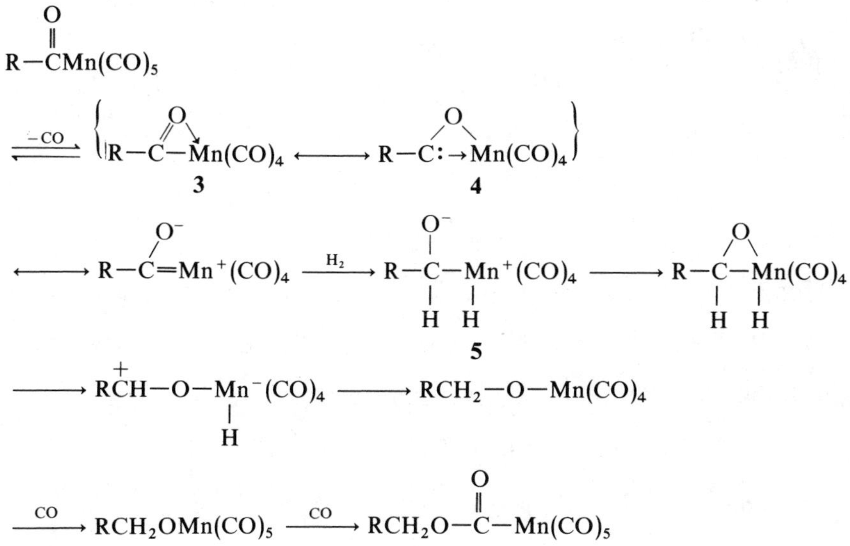

$$R-\overset{\overset{\displaystyle O}{\|}}{C}Mn(CO)_5$$

$$\underset{-CO}{\overset{\longrightarrow}{\longleftarrow}} \left\{ \underset{\mathbf{3}}{R-\overset{\overset{\displaystyle O}{\diagup\!\!\diagdown}}{C}-Mn(CO)_4} \longleftrightarrow \underset{\mathbf{4}}{R-\overset{\overset{\displaystyle O}{\diagup\diagdown}}{C}\!:\rightarrow Mn(CO)_4} \right\}$$

$$\longleftarrow R-\overset{\overset{\displaystyle O^-}{\diagup}}{C}=Mn^+(CO)_4 \xrightarrow{H_2} \underset{\mathbf{5}}{R-\overset{\overset{\displaystyle O^-}{|}}{\underset{\overset{|}{H}\ \overset{|}{H}}{C}}-Mn^+(CO)_4} \longrightarrow R-\overset{\overset{\displaystyle O}{\diagup\diagdown}}{\underset{\overset{|}{H}\ \overset{|}{H}}{C}}-Mn(CO)_4$$

$$\longrightarrow R\overset{+}{\underset{\overset{|}{H}}{C}}H-O-Mn^-(CO)_4 \longrightarrow RCH_2-O-Mn(CO)_4$$

$$\xrightarrow{CO} RCH_2OMn(CO)_5 \xrightarrow{CO} RCH_2O-\overset{\overset{\displaystyle O}{\|}}{C}-Mn(CO)_5$$

Structure 3 represents a dihapto acyl compound, and Structure 4 is an oxy carbene; such structures have been proposed for higher transition metal complexes.[14] In sulfolane, polar intermediate **5** may be prevented from forming an Mn—O bond by stabilization of the cation by the sulfolane, the solvated complex **6** collapsing instead to give the aldehyde and HMn(CO)$_4$:

$$R-\underset{\underset{H}{|}}{\overset{\overset{\bar{O}}{|}}{C}}-\overset{+}{Mn}(CO)_4 \longrightarrow RCHO + HMn(CO)_4$$

6

REFERENCES

1. WENDER, I., H. W. STERNBERG & M. ORCHIN. 1953. J. Am. Chem. Soc. **75:** 3041.
2. ORCHIN, M., L. KIRCH & I. GOLDFARB. 1956. J. Am. Chem. Soc. **78:** 5450.
3. ORCHIN, M. & W. RUPILIUS. 1972. Cat. Rev. **6:** 85.
4. NALESNIK, T. E. & M. ORCHIN. 1981. J. Organomet. Chem. **222:** C-5.
5. ROTH, J. A. & M. ORCHIN. 1979. J. Organoment. Chem. **182:** 299.
6. NALESNIK, T. E. & M. ORCHIN. 1982. Organometallics **1:** 222.
7. SWEANY, R. L. & J. HALPERN. 1977. J. Am. Chem. Soc. **99:** 8335.
8. NALESNIK, T. E., J. H. FREUDENBERGER & M. ORCHIN. 1982. J. Mol. Catal. **16:** 43.
9. NALESNIK, T. E., J. H. FREUDENBERGER & M. ORCHIN. 1982. J. Organomet. Chem. **236:** 95.
10. TURRO, N. J. & W. R. CHERRY. 1978. J. Am. Chem. Soc. **100:** 7432.
11. MATSUI, Y. & M. ORCHIN. 1983. J. Organomet. Chem. **244:** 369.
12. DOMBEK, B. D. 1979. J. Am. Chem. Soc. **101:** 6466.
13. FREUDENBERGER, J. H. & M. ORCHIN. 1982. Organometallics **1:** 1408.
14. MAATTA, E. & T. J. MARKS. 1981. J. Am. Chem. Soc. **103:** 3576.

TRANSITION METAL HYDRIDE CATALYSIS IN THE CARBON MONOXIDE/WATER SYSTEM

R. B. King and F. Ohene

Department of Chemistry
University of Georgia
Athens, Georgia 30602

INTRODUCTION

The carbon monoxide/water system is useful as a source of hydrogen through the water gas shift reaction (Equation 1). In addition, the carbon monoxide/water system can be used in place of hydrogen for the hydroformylation reaction[1] or

$$CO + H_2O \rightleftharpoons H_2 + CO_2 \qquad (1)$$

for various organic reductions, such as the reduction of nitro compounds to amines[1] or the reduction of aldehydes to the corresponding alcohols.[2]

Modern work on the homogeneous catalysis of the water gas shift reaction began with the observation by Laine, Rinker, and Ford that $Ru_3(CO)_{12}$ in alkaline solution generates an active catalyst for the water gas shift reaction at atmospheric pressure and temperatures in the range 100–120°C.[3] A variety of other metal carbonyls, such as $Rh_6(CO)_{16}$,[4,5] $Os_3(CO)_{12}$,[4] $Ir_4(CO)_{12}$,[4] $M(CO)_6$ (M = Cr, Mo, and W),[6,7] and $Fe(CO)_5$,[4,6,8] were subsequently found also to generate active catalysts for the water gas shift reaction in alkaline solution. In addition subsequent work led to the discovery of rhodium,[9,10] ruthenium,[11] and platinum/tin[12] catalysts for the water gas shift reaction that function in strong acid solution as well as platinum[13,14] and rhodium[14] complexes of bulky phosphines (e.g., triisopropylphosphine) that catalyze the water gas shift reaction in neutral solution.

During the past several years we have been engaged in detailed kinetic and mechanistic studies on the water gas shift reaction in basic solution catalyzed by various metal carbonyls. Previous papers from our laboratory have discussed the details of our work on the water gas shift catalyst systems derived from $Fe(CO)_5$ (Reference 8) and $M(CO)_6$ (M = Cr, Mo, and W)[7] in basic solution. This paper reviews these earlier results and compares them with our more recent, still unpublished, observations on the water gas shift catalyst system derived from $Ru_3(CO)_{12}$ and base.[15]

THE IRON AND GROUP VI METAL CARBONYL CATALYST SYSTEMS FOR THE WATER GAS SHIFT REACTION

A common feature of catalytic reactions of carbon monoxide (CO) at temperatures above ca. 100°C in aqueous solutions more strongly basic than formate is the rapid generation of formate according to the following reaction:

$$CO + OH^- \longrightarrow HCO_2^- \qquad (2)$$

The occurrence of this reaction is readily recognized in experiments using KOH as the base when the initial pressure of CO or $CO + H_2$ at the reaction temperature is less than that expected from the loading pressure of CO by an amount corresponding to the quantitative formation of formate by Equation 2.[7,8] Thus the strongly basic media used for the water gas shift reactions discussed in this paper contain amounts of formate equivalent to the base (generally KOH) introduced into the system but only the very small amounts of OH^- generated by the formate buffer system. Thus, during the course of the water gas shift reaction in basic medium using $Fe(CO)_5$ as the catalyst precursor, the pH starts at 8.6 and falls gradually to 7.4,[8,16] indicative of the buffering effects of the formate produced by Equation 2.

Our experimental observations on the water gas shift reaction in basic solution using $Fe(CO)_5$ as the catalyst precursor are in agreement with the following catalytic cycle (M = Fe):[8,16]

$$M(CO)_5 + OH^- \xrightarrow{k_1} M(CO)_4CO_2H^- \tag{3a}$$

$$M(CO)_4CO_2H^- \xrightarrow{k_2} HM(CO)_4^- + CO_2 \tag{3b}$$

$$HM(CO)_4^- + H_2O \xrightarrow{k_3} H_2M(CO)_4 + OH^- \tag{3c}$$

$$H_2M(CO)_4 \xrightarrow{k_4} M(CO)_4 + H_2 \tag{3d}$$

$$M(CO)_4 + CO \xrightarrow{k_5} M(CO)_5 \tag{3e}$$

Such a mechanism has been called an "associative mechanism" since the catalyst precursor enters the catalytic cycle through a bimolecular reaction with a nucleophile (Equation 3a).[17]

Application of a conventional steady-state approximation to the Mechanism 3a–3e (M = Fe) leads to the following rate law for the production of hydrogen:[8,16]

$$\frac{d[H_2]}{dt} = k_1[Fe(CO)_5][OH^-] \tag{4}$$

The following experimental observations support this rate law:[6,16] (1) The rate of H_2 production is first order in $Fe(CO)_5$ concentration. (2) The rate of H_2 production is independent of CO pressure (provided enough CO is present to prevent total decomposition of the iron carbonyl system through complete decarbonylation).[8] (3) The rate of H_2 production is essentially independent of the base concentration because the formate buffer system arising through Equation 2 keeps the OH^- concentration essentially independent of the amount of base introduced into the system.

Our experimental observations on the water gas shift reaction in basic medium using one of the group VI metal carbonyls $M(CO)_6$ (M = Cr, Mo, and W) as the catalyst precursor are in agreement with the following rather different type of catalytic cycle:

$$M(CO)_5 + HCO_2^- \xrightarrow{k_1} M(CO)_5OCOH^- \tag{5a}$$

$$M(CO)_5OCOH^- \xrightarrow{k_2} HM(CO)_5^- + CO_2 \tag{5b}$$

$$HM(CO)_5^- + H_2O \xrightarrow{k_3} H_2M(CO)_5 + OH^- \tag{5c}$$

$$H_2M(CO)_5 \xrightarrow{k_4} H_2 + M(CO)_5 \tag{5d}$$

This catalytic cycle is entered through the reversible dissociation of the metal hexacarbonyl to give a *coordinately unsaturated* M(CO)$_5$ fragment by the following reaction:

$$M(CO)_6 \underset{k_a}{\overset{k_d}{\rightleftharpoons}} M(CO)_5 + CO \tag{6}$$

Because of the entry of the catalyst precursor [i.e., M(CO)$_6$] into the catalytic cycle (5a–5d) through dissociation of CO to give a coordinately unsaturated M(CO)$_5$ fragment (Equation 6) a mechanism such as (5a–5d + 6) has been called a "dissociative mechanism."[17]

Application of a conventional steady-state approximation to the Mechanism 5a–5d + 6 leads to the following rate law for the production of hydrogen:[7,16]

$$\frac{d[H_2]}{dt} = (k_d/k_a)k_1 \frac{[M(CO)_6][HCO_2^-]}{[CO]} \tag{7}$$

The following experimental observations support this rate law:[7,16] (1) The rate of H$_2$ production is first order in M(CO)$_6$ concentration. (2) The rate of H$_2$ production is inversely proportional to the CO pressure. (3) The rate of H$_2$ production is approximately first order in the base concentration, which is equivalent to the formate concentration from Equation 2.

Our previously reported kinetic studies[7,8,16] including the kinetic analyses of the Mechanisms 3a–3e and 5a–5d + 6 by steady-state methods suggest the following roles of experimental kinetic data in elucidating possible mechanisms for the water gas shift reaction in basic solution catalyzed by different homogeneous transition metal systems:

1. The dependencies of H$_2$ production on CO pressure and the catalyst precursor concentration suggest reactions that may be required for the catalyst precursor to enter the catalytic cycle. For example, the inverse CO pressure dependence of the water gas shift reaction catalyzed by the group VI metal carbonyls M(CO)$_6$ (M = Cr, Mo, and W) relates to the dissociative step (Equation 6) necessary for these carbonyls to enter the Catalytic Cycle 5a–5d.

2. The dependence of H$_2$ production on base concentration distinguishes between associative mechanisms where OH$^-$ is the active species arising from the base and dissociative mechanisms where formate is the active species arising from the base. This relates to the properties of the formate buffer system in holding the OH$^-$ concentration nearly constant over wide differences in base concentration.

The remainder of this paper explores the mechanistic aspects of Ru$_3$(CO)$_{12}$ as a catalyst precursor in basic solution for the water gas shift reaction based on preliminary kinetic data of types similar to those used in our earlier work with Fe(CO)$_5$ and M(CO)$_6$ (M = Cr, Mo, and W) catalyst precursors as outlined above.[16] This work with Ru$_3$(CO)$_{12}$ is instructive since it suggests novel processes for metal clusters to enter water gas shift reaction catalytic cycles. This work with Ru$_3$(CO)$_{12}$ is also significant since the experiments by Laine, Rinker, and Ford reported in 1977 using Ru$_3$(CO)$_{12}$ as the catalyst precursor[3] initiated modern interest in the homogeneous water gas shift reaction.

THE RUTHENIUM CARBONYL CATALYST SYSTEM FOR THE
WATER GAS SHIFT REACTION

Experiments to assess the operational characteristics and kinetics of $Ru_3(CO)_{12}$ as a catalyst for the water gas shift reaction were conducted under the following conditions:[15] (1) 0.0001 to 0.0004 M $Ru_3(CO)_{12}$ as the catalyst precursor; (2) 0.1 to 0.5 M base [KOH or $(C_2H_5)_3N$]; (3) 3:1 methanol/water as the solvent; and (4) 1.5 to 40 atmospheres CO pressure. Details of the autoclaves used to conduct these experiments and the methods for gas analyses (CO, CO_2, H_2, Ar) are the same as in previous work from this laboratory.[7,8]

The reaction rates on the ordinates of FIGURES 1–5 are expressed in turnover frequencies (TF) defined as follows:

$$TF = \frac{\text{mmol } H_2}{\text{mmol } Ru_3(CO)_{12}} \text{ per unit time} \qquad (8)$$

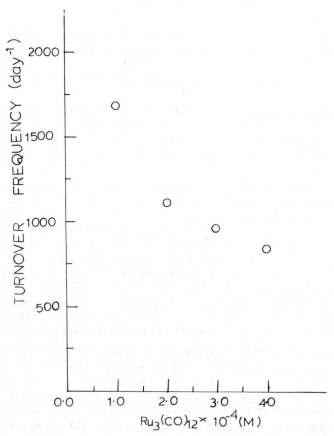

FIGURE 1. Turnover frequency of H_2 production plotted as a function of the $Ru_3(CO)_{12}$ concentration. Reaction conditions: 140°C and 400 psia CO pressure; KOH (50 mmol); 1:3 H_2O–CH_3OH (200 ml).

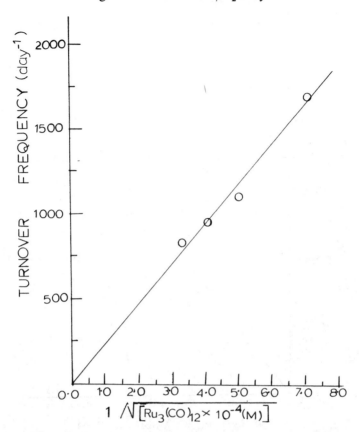

FIGURE 2. The data from FIGURE 1 replotted as a function of the reciprocal of the square root of the Ru$_3$(CO)$_{12}$ concentration.

For the data from our laboratory (FIGURES 1, 2, 4, 5, and 6) the time unit is taken to be one day (24 hours).

Presently available data indicate the following effects of the indicated variables on the turnover frequencies for H$_2$ production as defined in Equation 8:

1. *Ru$_3$(CO)$_{12}$ concentration:* the definition of turnover frequency in Equation 8 indicates that if water gas shift reaction is first order in Ru$_3$(CO)$_{12}$ concentration, a plot of turnover frequency versus Ru$_3$(CO)$_{12}$ concentration should give a horizontal line. However, FIGURE 1 indicates that the turnover frequency decreases with increasing Ru$_3$(CO)$_{12}$ concentration in a nonlinear manner. Thus the water gas shift reaction in basic solution using Ru$_3$(CO)$_{12}$ as the catalyst precursor is clearly not first order in Ru$_3$(CO)$_{12}$ in contrast to the analogous water gas shift reactions using M(CO)$_6$ (M = Cr, Mo, and W) and Fe(CO)$_5$, which are all first order in the metal carbonyl catalyst precursors.[7,8,16]

FIGURE 2 replots the same data of FIGURE 1 as turnover frequency versus the reciprocal of the square root of the Ru$_3$(CO)$_{12}$ concentration. In this case a

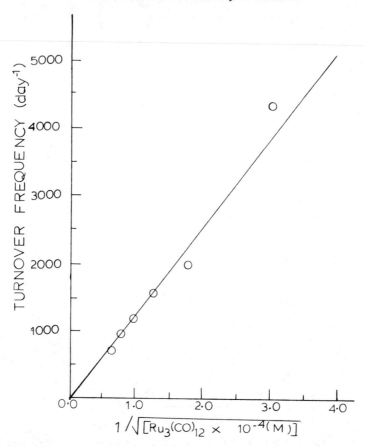

FIGURE 3. Data from Figure 2 of the paper by Slegeir, Sapienza, and Easterling[17] replotted as turnover frequency (after five hours) versus the square root of the $Ru_3(CO)_{12}$ concentration. Reaction conditions: 100°C and 415 psia CO pressure; 5 g $(CH_3)_3N$; 15 g H_2O diluted to 100 ml with tetrahydrofuran.

straight line passing through the origin is obtained. Similarly, replotting turnover data obtained by Slegeir, Sapienza, and Easterling (see Figure 2 in Reference 17) versus the reciprocal of the square roots of their $Ru_3(CO)_{12}$ concentrations also leads to a straight line passing through the origin (FIGURE 3). Combining the observations in FIGURES 2 and 3 with the definition of turnover frequency in Equation 8 indicates that the water gas shift reaction catalyzed by $Ru_3(CO)_{12}$ in basic solution under the conditions investigated (see above) is half order in $Ru_3(CO)_{12}$.

2. *CO Pressure*: a plot of turnover frequency versus CO pressure over a fairly wide pressure range (25 to 600 psi) at a constant $Ru_3(CO)_{12}$ concentration (0.0001 M) leads to a curved line (FIGURE 4). Again replotting the data of FIGURE 4 as turnover frequency versus the square root of CO pressure gives a straight line passing through the origin (FIGURE 5). This suggests that the water gas shift

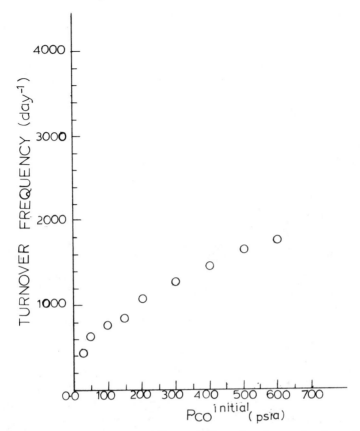

FIGURE 4. Turnover frequency of H_2 production plotted as a function of the CO pressure. Reaction conditions: 140°C; $Ru_3(CO)_{12}$ (0.02 mmol); KOH (50 mmol); 1:3 $H_2O–CH_3OH$ (200 ml).

reaction catalyzed by $Ru_3(CO)_{12}$ in basic solution under the conditions investigated is half order in CO. The apparent discrepancies between this observation and earlier observations[3,18] may relate to the fact that over a more limited CO pressure range, the dependence of rate or turnover frequency on CO pressure could appear to be linear.

3. *Base concentration*: plots of turnover frequency versus base concentration at constant $Ru_3(CO)_{12}$ concentration (0.0001 M) and CO pressure yield straight lines passing through the origin for either KOH or $(C_2H_5)_3N$ as the base (FIGURE 6). The higher reaction rate using tertiary amines as the base relative to hydroxide indicated in FIGURE 6 was previously noted by Slegeir, Sapienza, and Easterling.[17]

Our experimental observations thus suggest that the water gas shift reaction catalyzed by $Ru_3(CO)_{12}$ in basic medium is half order in $Ru_3(CO)_{12}$ and CO and first order in base. These observations can be explained by a mechanism that is

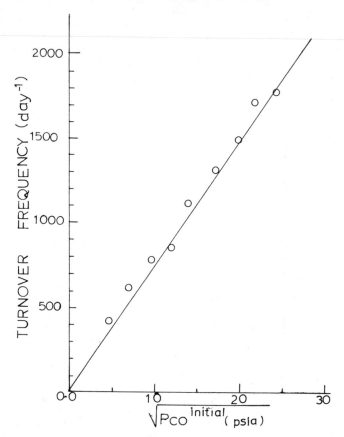

FIGURE 5. The data from FIGURE 4 replotted as a function of the square root of the CO pressure (in psia).

divided into the following three stages: (1) conversion of $Ru_3(CO)_{12}$ to the potentially catalytically active species $Ru(CO)_5$ and $Ru_2(CO)_8$; (2) the $Ru_2(CO)_8$ catalytic cycle; and (3) the $Ru(CO)_5$ catalytic cycle. These stages will be considered separately.

Conversion of $Ru_3(CO)_{12}$ to the Catalytically Active Species

The conversion of $Ru_3(CO)_{12}$ to the catalytically active species can occur through the following equilibria:

$$Ru_3(CO)_{12} + CO \rightleftharpoons Ru_3(CO)_{13} \qquad (9a)$$

$$Ru_3(CO)_{13} \rightleftharpoons Ru(CO)_5 + Ru_2(CO)_8 \qquad (9b)$$

In Step 9a, the entering CO breaks one of the three Ru—Ru bonds of the

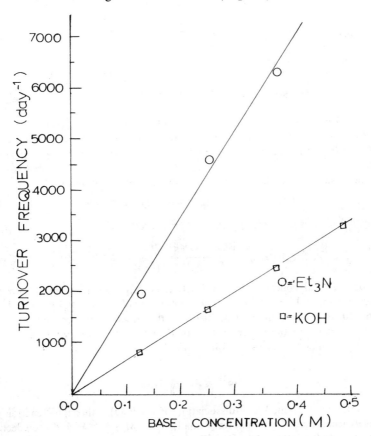

FIGURE 6. Turnover frequency of H_2 production plotted as a function of base concentration. Reaction conditions: 140°C and 400 psia CO pressure; $Ru_3(CO)_{12}$ (0.02 mmol); 1:3 H_2O–CH_3OH (200 ml). The upper line refers to experiments using $(C_2H_5)_3N$ as the base. The lower line refers to experiments using KOH as the base.

$Ru_3(CO)_{12}$ triangle (I) to form an $Ru_3(CO)_{13}$ with a trimetallacyclobutanone structure (II) and only two Ru—Ru bonds.

In Step 9b, the $Ru_3(CO)_{13}$ intermediate (II) eliminates $Ru(CO)_5$ to leave $Ru_2(CO)_8$, which may have a structure such as III with a Ru=Ru double bond analogous

III   IV

to the $Rh=Rh$ double bond found in the related complex $[(CH_3)_5C_5RhCO]_2$ (IV).[19] Also, evidence has been obtained for the formation of an osmium analogue $Os_2(CO)_8$ during the decomposition of $Os_2(CO)_9$ in heptane solution at room temperature.[20]

The net result of Equilibria 9a and 9b is the following equilibrium:

$$Ru_3(CO)_{12} + CO \rightleftharpoons Ru_2(CO)_8 + Ru(CO)_5 \tag{10}$$

The constant for this equilibrium can be expressed as follows:

$$K_{eq} = \frac{[Ru_2(CO)_8][Ru(CO)_5]}{[Ru_3(CO)_{12}][CO]} \tag{11}$$

Let M be the total concentration of $Ru_3(CO)_{12}$ originally introduced into the catalytic system and let x be the concentration of $Ru_2(CO)_8$ produced by Equation 10. Because of the stoichiometry of this reaction, the concentration of $Ru_2(CO)_8$ is equal to the concentration of $Ru(CO)_5$. Substituting M and x so defined into Equation 11 gives the following:

$$K_{eq} = \frac{x^2}{(M - x)[CO]} \tag{12}$$

Now assume that the amount of dissociation of $Ru_3(CO)_{12}$ through Reaction 10 is relatively small (i.e., K_{eq} is relatively small). Then $M - x \approx M$ and the effective concentration of $Ru_3(CO)_{12}$ {i.e., $[Ru_3(CO)_{12}]$} can be considered to be equivalent to the original concentration of $Ru_3(CO)_{12}$. Under these conditions, solving Equation 12 for x gives

$$x = [Ru_2(CO)_8] = [Ru(CO)_5] = k_{eq}\sqrt{[Ru_3(CO)_{12}][CO]} \tag{13}$$

Equation 13 already suggests how half-order (i.e., square root) dependencies on $Ru_3(CO)_{12}$ concentration and CO pressure can arise from an entry of $Ru_3(CO)_{12}$ into catalytic cycles through a process represented by Equation 10 in the following situation: (1) the amount of conversion of $Ru_3(CO)_{12}$ into $Ru_2(CO)_8$ and $Ru(CO)_5$ is small so that the concentrations of the latter two species remain low under the actual reaction conditions; and (2) the reaction of $Ru_2(CO)_8$ with CO to give the known[20] $Ru_2(CO)_9$ does not occur under the reaction conditions, possibly because the polar solvent mixture (aqueous methanol) favors reactions of $Ru_2(CO)_8$ with species more polar than CO.

The $Ru_2(CO)_8$ Catalytic Cycle

The $Ru=Ru$ derivative $Ru_2(CO)_8$ (III) can function as a coordinately unsaturated species like $M(CO)_5$ (M = Cr, Mo, and W) in the dissociative mechanism

(Equations 5a–5d + 6) for the water gas shift reaction in basic medium with $M(CO)_6$ as the catalyst precursor. An analogous catalytic cycle involving formate addition to $Ru_2(CO)_8$ as the initial step involves the following steps:

$$Ru_2(CO)_8 + HCO_2^- \xrightarrow{k_1} HCO_2Ru_2(CO)_8^- \quad (14a)$$

$$HCO_2Ru_2(CO)_8^- \xrightarrow{k_2} HRu_2(CO)_8^- + CO_2 \quad (14b)$$

$$HRu_2(CO)_8^- + H_2O \xrightarrow{k_3} H_2Ru_2(CO)_8 + OH^- \quad (14c)$$

$$H_2Ru_2(CO)_8 \xrightarrow{k_4} Ru_2(CO)_8 + H_2 \quad (14d)$$

Application of a conventional steady-state approximation to the Mechanism 14a–14d leads to the following rate law for the production of hydrogen through the $Ru_2(CO)_8$ catalytic cycle (the "dissociative" pathway):

$$\left(\frac{d[H_2]}{dt}\right)_{diss} = k_1[Ru_2(CO)_8][HCO_2^-] \quad (15)$$

The Ru(CO)$_5$ Catalytic Cycle

A catalytic cycle of the type (**3a–3e**) where M = Ru is possible for $Ru(CO)_5$. The corresponding rate law

$$\left(\frac{d[H_2]}{dt}\right)_{ass} = k_1[Ru(CO)_5][OH^-] \quad (16)$$

is analogous to Equation 4 for the corresponding reaction with $Fe(CO)_5$.

The total rate of H_2 production from the water gas shift reaction catalyzed by $Ru_3(CO)_{12}$ in basic solution is the sum of the H_2 production from the $Ru_2(CO)_8$ catalytic cycle (**14a–14d**) and the $Ru(CO)_5$ catalytic cycle (**3a–3e**, M = Ru), i.e.

$$\frac{d[H_2]}{dt} = \left(\frac{d[H_2]}{dt}\right)_{diss} + \left(\frac{d[H_2]}{dt}\right)_{ass} \quad (17)$$

Combination of Equations 13, 15, 16, and 17 gives the total rate equation for the water gas shift reaction catalyzed by $Ru_3(CO)_{12}$ in basic solution as follows:

$$\frac{d[H_2]}{dt} = k_{eq}\sqrt{[Ru_3(CO)_{12}][CO]}(k_1[HCO_2^-] + k_1'[OH^-]) \quad (18)$$

The following points are apparent from Equation 18: (1) Equation 18 agrees with the observed half-order dependencies on $Ru_3(CO)_{12}$ concentration and CO pressure. (2) Because of the properties of the formate buffer system, the value of $k_1'[OH^-]$ relative to $k_1[HCO_2^-]$ can be estimated by extrapolating the turnover frequency in the straight lines in FIGURE 6 to zero base concentration. The intersection of these straight lines with the origin in FIGURE 6 indicates that $k_1'[OH^-] \ll k_1[HCO_2^-]$ and that essentially all of the catalysis of the water gas shift reaction by $Ru_3(CO)_{12}$ in basic solution occurs through the dissociative mechanism involving the $Ru_2(CO)_8$

intermediate (III) (Equations 14a–14d) rather than the associative mechanism involving $Ru(CO)_5$ (Equations 3a–3e with M = Ru). This can be reconciled with the known catalytic activity of $Fe(CO)_5$ for the water gas shift reaction[8,16] and the enhanced reactivity[21] of $Ru(CO)_5$ relative to $Fe(CO)_5$ toward OH^- as indicated in Equation 3a by noting that the concentrations of $Ru(CO)_5$ produced by Equilibrium 10 are several orders of magnitude below the concentrations of $Fe(CO)_5$ introduced into the catalytic system in the reported studies. Thus the product $k'_1[Ru(CO)_5][OH^-]$ from Equation 16 in a water gas shift catalyst system derived from basic $Ru_3(CO)_{12}$ is much smaller than the product $k_1[Fe(CO)_5][OH^-]$ from Equation 4 in the water gas shift catalyst system derived from basic $Fe(CO)_5$ because the $[Ru(CO)_5]$ factor in Equation 16 is several orders of magnitude less than the $[Fe(CO)_5]$ factor in Equation 4.

SUMMARY

Kinetic studies on the water gas shift reaction in basic solution using $Fe(CO)_5$ and $M(CO)_6$ (M = Cr, Mo, and W) as catalyst precursors establish two fundamentally different mechanisms for this reaction. Thus the reaction using $Fe(CO)_5$ as a catalyst precursor proceeds by an associative mechanism (Equations 3a–3e, M = Fe). However, water gas shift reactions using $M(CO)_6$ (M = Cr, Mo, and W) as a catalyst precursor proceed through dissociative mechanisms (Equations 4a–5d + 6). Experimentally, these two rather different mechanisms can be distinguished by the dependence of the reaction rate on the base concentration.

The metal cluster $Ru_3(CO)_{12}$ is an interesting catalyst precursor for the water gas shift reaction in basic solution since its reaction with CO can lead to entries into either associative catalytic cycles [i.e., that involving $Ru(CO)_5$ analogous to $Fe(CO)_5$] or dissociative catalytic cycles [i.e., that involving the unsaturated binuclear species $Ru_2(CO)_8$ (III)]. Thus the kinetic studies on the water gas shift reaction in basic solution using $Ru_3(CO)_{12}$ as the catalyst precursor provide an unprecedented opportunity to compare associative and dissociative catalytic cycles in the same catalyst system. An important indication from this work is that associative catalytic cycles require much larger metal concentrations within the catalytic cycle than do dissociative catalytic cycles. Thus if a catalyst precursor such as $Ru_3(CO)_{12}$ reacts to give similar concentrations of intermediates for both associative and dissociative catalytic cycles, reaction through the dissociative catalytic cycle will dominate. In other words, an associative reaction pathway for the water gas shift reaction in basic solution requires a much higher concentration of the active intermediate than does a dissociative reactive pathway in order to achieve the same reaction rate. In practice, this difference is frequently balanced by the fact that an associative water gas shift catalytic cycle is entered through coordinately saturated stable metal carbonyl derivatives [e.g., $Fe(CO)_5$] obtainable in high concentrations whereas a dissociative water gas shift catalytic cycle is entered through coordinately unsaturated metal carbonyl derivatives [e.g., $M(CO)_5$, where M = Cr, Mo, or W, or $Ru_2(CO)_8$], which can only be obtained in low concentrations. This point is significant in planning and interpreting attempts to detect intermediates in catalytic reactions by physical and spectroscopic techniques such as infrared spectroscopy.[22]

REFERENCES

1. PETTIT, R., K. CANN, T. COLE, C. H. MAULDIN & W. SLEGEIR. 1979. Adv. Chem. Ser. **173:** 121.
2. THOMSON, W. J. & R. M. LAINE. 1981. Am. Chem. Soc. Symp. Ser. **152:** 133.
3. LAINE, R. M., R. G. RINKER & P. C. FORD. 1977. J. Am. Chem. Soc. **99:** 252.
4. KANG, H. C., C. H. MAULDIN, T. COLE, W. SLEGEIR, K. CANN & R. PETTIT. 1977. J. Am. Chem. Soc. **99:** 8323.
5. LAINE, R. M. 1978. J. Am. Chem. Soc. **100:** 6451.
6. KING, R. B., C. C. FRAZIER, R. M. HANES & A. D. KING, JR. 1978. J. Am. Chem. Soc. **100:** 2925.
7. KING, A. D., JR., R. B. KING & D. B. YANG. 1981. J. Am. Chem. Soc. **103:** 2699.
8. KING, A. D., JR., R. B. KING & D. B. YANG. 1980. J. Am. Chem. Soc. **102:** 1028.
9. CHENG, C.-H., D. E. HENDRIKSEN & R. EISENBERG. 1977. J. Am. Chem. Soc. **99:** 2791.
10. BAKER, E. C., D. E. HENDRIKSEN & R. EISENBERG. 1980. J. Am. Chem. Soc. **102:** 1020.
11. FORD, P. C., P. YARROW & H. COHEN. 1981. Am. Chem. Soc. Symp. Ser. **152:** 95.
12. CHENG, C.-H. & R. EISENBERG. 1978. J. Am. Chem. Soc. **100:** 5968.
13. YOSHIDA, T., Y. UEDA & S. OTSUKA. 1978. J. Am. Chem. Soc. **100:** 3941.
14. YOSHIDA, T., T. OKANO & S. OTSUKA. 1981. Am. Chem. Soc. Symp. Ser. **152:** 79.
15. KING, A. D., JR., R. B. KING & F. OHENE. 1982. Unpublished results.
16. KING, R. B., A. D. KING, JR. & D. B. YANG. 1981. Am. Chem. Soc. Symp. Ser. **152:** 123.
17. SLEGEIR, W. A. R., R. S. SAPIENZA & B. EASTERLING. 1981. Am. Chem. Soc. Symp. Ser. **152:** 325.
18. UNGERMANN, C., V. LANDIS, S. A. MOYA, H. COHEN, H. WALKER, R. G. PEARSON, R. G. RINKER & P. C. FORD. 1979. J. Am. Chem. Soc. **101:** 5922.
19. NUTTON, A. & P. M. MAITLIS. 1979. J. Organomet. Chem. **166:** C21.
20. MOSS, J. R. & W. A. G. GRAHAM. 1977. J. Chem. Soc. Dalton Trans.: 95
21. GROSS, D. C. & P. C. FORD. 1982. Inorg. Chem. **21:** 1702.
22. KING, R. B., A. D. KING, JR. & M. Z. IQBAL. 1979. J. Am. Chem. Soc. **101:** 4893.

THE REMARKABLY MILD, REGIOSPECIFIC, AND CATALYTIC HOMOGENEOUS HYDROESTERIFICATION OF OLEFINS*

Bertrand Despeyroux and Howard Alper†

*Ottawa-Carleton Institute for Research
and Graduate Studies in Chemistry
Department of Chemistry
University of Ottawa
Ottawa, Ontario, Canada K1N 9B4*

One of the most important classes of catalytic homogeneous processes involving metal hydrides as intermediates is the hydroesterification (and hydrocarboxylation) of olefins.[1] Linear and branched chain esters can be produced in these reactions, with the linear isomer usually being formed as the major isomer. Useful straight-chain esters include fatty acid esters, while the branched chain isomers are of importance in the generation of monomers for polymerization processes (e.g., methyl methacrylate), and for the synthesis of certain pharmacologically active compounds.

Numerous patents and primary literature publications have appeared on the hydroesterification reaction. For example, cobalt carbonyl or cobalt salts can, in the presence of pyridine, catalyze the formation of methyl esters of fatty acids from suitable olefins, carbon monoxide, and methanol, at 160°C and 160 atm.[2] Palladium(II) complexes have been widely employed for the hydroesterification of terminal olefins [e.g., 1-heptene/CO/CH_3OH/$PdCl_2(PPh_3)_2$/$SnCl_2$/70°C/141 bar gives methyl *n*-octanoate in 93% yield].[3,4] A recent patent has claimed that hydrocarboxymethylation of propene catalyzed by bis(triarylarsine)palladium dichloride/triarylarsine/$AlCl_3$/HCl (nitrogen atmosphere) gives the branched chain ester methyl 2-methylpropionate as the major product along with methyl *n*-butyrate (4–11:1 ratio of branched to straight-chain esters).[5] Finally, diesters were formed by the reaction of olefins with carbon monoxide, oxygen, alcohol, palladium, and cupric chlorides in the presence of triethylamine.[6] All of these reactions require drastic conditions, are not regiospecific, and work well only for α-olefins. We now describe a homogeneous palladium-based catalytic system that enables one to effect hydroesterification under remarkably mild conditions, is applicable to internal as well as terminal olefins, and, with one exception, is completely regiospecific for the production of branched chain esters.

EXPERIMENTAL SECTION

Olefins were commercial materials. Palladium chloride (Strem Chemicals) and cupric chloride (J. T. Baker) were used as received. Solvents were dried by standard methods prior to use.

* We are grateful to British Petroleum for support of this Research.
† To whom correspondence should be addressed.

General Procedure for the Hydroesterification of Olefins

Palladium chloride is added to methanol and benzene or toluene through which is bubbling carbon monoxide. After one minute, concentrated hydrochloric acid is added. Once the solution turns yellow ($PdCl_2$ is dissolved), copper(II)chloride (5–10:1 ratio of $CuCl_2/PdCl_2$) is added and oxygen is bubbled through the solution

TABLE 1

PRODUCTS OBTAINED FROM THE CARBONYLATION OF OLEFINS WITH
$CO/O_2/PdCl_2/CuCl_2/HCl/CH_3OH$

Substrate	Product	Yield (%)
1-Decene	$C_8H_{17}\underset{\underset{\displaystyle CH_3}{\mid}}{C}HCOOCH_3$	100
Vinylcyclohexane	$C_6H_{11}\underset{\underset{\displaystyle CH_3}{\mid}}{C}HCOOCH_3$	98
1,7-Octadiene	$CH_3OCOCH(CH_2)_4\overset{\overset{\displaystyle CH_3}{\mid}}{C}HCOOCH_3$ with CH_3 below	100
1,9-Decadiene	$CH_3OCOCH(CH_2)_6\overset{\overset{\displaystyle CH_3}{\mid}}{C}HCOOCH_3$ with CH_3 below	70
Propene	$(CH_3)_2CHCOOCH_3$	90–100*
2-Methyl-1-undecene	$C_9H_{19}C(CH_3)_2COOCH_3$	63
	$C_9H_{19}C(CH_3)_2OCH_3$	18
	internal olefin	8
4-Methylstyrene	$4\text{-}CH_3C_6H_4\underset{\underset{\displaystyle CH_3}{\mid}}{C}HCOOCH_3$	90
	$4\text{-}CH_3C_6H_4\underset{\underset{\displaystyle CH_3}{\mid}}{C}HOCH_3$	8
	internal olefin	2
cis-2-Decene	$CH_3(CH_2)_7\underset{\underset{\displaystyle CH_3}{\mid}}{C}HCOOCH_3$	100
trans-2-Decene	$CH_3(CH_2)_7\underset{\underset{\displaystyle CH_3}{\mid}}{C}HCOOCH_3$	60
cis-4-Methyl-2-pentene	$(CH_3)_2CHCH_2\underset{\underset{\displaystyle CH_3}{\mid}}{C}HCOOCH_3$	92
trans-4-Methyl-2-pentene	$(CH_3)_2CHCH_2\underset{\underset{\displaystyle CH_3}{\mid}}{C}HCOOCH_3$	70
cis-3-Heptene	$C_2H_5\underset{\underset{\displaystyle COOCH_3}{\mid}}{C}HC_4H_9$ and $C_3H_7\underset{\underset{\displaystyle COOCH_3}{\mid}}{C}HC_3H_7$	98
Cyclododecene	methyl cyclodedecanecarboxylate	90
Dicyclopentadiene	monoester of undetermined structure	>90

* The yield is approximate, as the exact weight of propene was not determined.

(as well as carbon monoxide). The olefin (10–25:1 ratio of olefin/$PdCl_2$) is now added, and the reaction mixture is stirred for one to four hours at 25–35°C. During this time, additional concentrated HCl is added to the reaction mixture. The ester is then extracted with hexane and subsequent evaporation of the hexane usually gives an exceedingly pure product (i.e., no distillation). Distillation is used when product mixtures are formed.

RESULTS AND DISCUSSION

Treatment of a terminal olefin with carbon monoxide, oxygen, copper chloride, hydrochloric acid, and palladium chloride as the catalyst, in methanol or methanol-toluene (or benzene) at 25–35° for one to four hours at atmospheric pressure, affords branched chain methyl esters in high yields. This reaction is completely regiospecific when applied to monosubstituted terminal olefins (e.g., vinyl-cyclohexane) and diolefins (e.g., 1,7-octadiene) (TABLE 1). In the case of 4-methyl-styrene, a small amount of ether by-product is formed but no straight-chain ester. The 1,1-disubstituted olefin 2-methyl-1-undecene also reacts in a regiospecific manner, as far as ester formation is concerned.

The hydroesterification of cyclic olefins such as cyclododecene and dicyclopenta-diene is also very facile. Olefins having a double bond at the 2-position also react regiospecifically with cis-olefins affording esters in higher yields than with the trans-isomer (e.g., 2-decene). Only the 2-substituted ester was obtained. The regiospecificity breaks down for cis-3-heptene, which gives a mixture of esters.

Several other alcohols were employed in these reactions, and gave esters in acceptable yields (TABLE 2). However, water cannot be used, as the reaction is not catalytic, and a black precipitate is formed.

The above-described process is superior to all other known methodology in this area, not only in terms of the mildness of the reaction conditions, but also because of the observed regiospecificity. Two species, oxygen and hydrochloric acid, are

TABLE 2

PRODUCTS OBTAINED FROM THE CARBONYLATION OF 1-DECENE WITH CO, O_2, $PdCl_2$, $CuCl_2$, HCl, AND ROH

ROH where R =	Product	Yield (%)
H CH_3	$C_8H_{17}\underset{\underset{CH_3}{\mid}}{C}HCOOCH_3$	100
C_2H_5	$C_8H_{17}\underset{\underset{CH_3}{\mid}}{C}HCOOC_2H_5$	90
n-C_3H_7	$C_8H_{17}\underset{\underset{CH_3}{\mid}}{C}HCOOC_3H_7$	90
i-C_3H_7	$C_8H_{17}\underset{\underset{CH_3}{\mid}}{C}HCOO i\text{-}C_3H_7$	30
n-C_4H_9	$C_8H_{17}\underset{\underset{CH_3}{\mid}}{C}HCOO n\text{-}C_4H_9$	70

TABLE 3

PRODUCTS OBTAINED FROM THE CARBONYLATION OF OLEFINS WITH CO, $PdCl_2$, $CuCl_2$, HCl, AND CH_3OH (No O_2)

Substrate	Product	Yield (%)
1-Decene	$C_8H_{17}CHCOOCH_3$ $\quad\quad\mid$ $\quad\quad CH_3$	100
1-Pentene	$C_3H_7CHCOOCH_3$ $\quad\quad\mid$ $\quad\quad CH_3$	90
2-Methyl-1-undecene	$C_9H_{19}C(CH_3)_2COOCH_3$	6
	$C_9H_{19}C(CH_3)_2OCH_3$	14
4-Methylstyrene	$4\text{-}CH_3C_6H_4CHCOOCH_3$ $\quad\quad\quad\quad\mid$ $\quad\quad\quad\quad CH_3$	50
	$4\text{-}CH_3C_6H_4CHOCH_3$ $\quad\quad\quad\quad\mid$ $\quad\quad\quad\quad CH_3$	50

critical to the success of this process. One can do the reaction without oxygen but, in most instances, at a significant reduction in reaction rate and product yield (TABLE 3—e.g., 4-methylstyrene and 2-methyl-1-undecene). Another disadvantage of the latter method is that copper chloride has to be added repeatedly at intervals during the reaction, making this a tedious process. Hydrochloric acid must be present, otherwise the conversion of olefin to ester is low, and palladium black precipitates out of solution. Essentially no reaction occurs when $CuCl_2$ is absent. Another interesting point is that added phosphine does *not* have a beneficial effect on the hydroesterification reaction.

One possible mechanism for the reaction involves the palladium hydride $HPdL_nCl$, proposed by Tsuji[7] and others. However, the formation of esters by formal Markownikoff addition to the olefin can also be rationalized on the basis of a carbonium ion intermediate.[8] Recently, a new mechanism has been proposed for the cobalt carbonyl–catalyzed carbalkoxylation of olefins for which a carbo-alkoxycobalt complex is a key intermediate.[9] The analogous palladium species may also participate in the hydroesterification reaction described herein.

REFERENCES

1. MULLEN, A. 1980. *In* New Syntheses with Carbon Monoxide. J. Falbe, Ed.: 243–308. Springer-Verlag. Berlin, Federal Republic of Germany.
2. HOFMANN, P., K. KOSSWIG & W. SCHAEFER. 1980. Ind. Eng. Chem. Prod. Res. Dev. **19**: 330.
3. KNIFTON, J. F. 1972. DE Offen. 2, 303, 118 (2/2/72).
4. KNIFTON, J. F. 1978. J. Am. Oil Chem. Soc. **55**: 496.
5. SQUIRE, E. N. & F. J. WALLER. 1980. DE Offen. 2739096 (du Pont).
6. Ube Industries. 1976. Japan Pat. 53050 709 (9/25/76).
7. TSUJI, J. 1969. Acc. Chem. Res. **2**: 144.
8. FENTON, D. M. 1973. J. Org. Chem. **38**: 3192.
9. MILSTEIN, D. & J. L. HUCKABY. 1982. J. Am. Chem. Soc. **104**: 6150.

SELECTIVE HOMOGENEOUS CATALYSTS FOR THE CONVERSION OF METHANOL TO ETHANOL*

Michael J. Chen, Harold M. Feder, and Jerome W. Rathke†

Chemical Technology Division
Argonne National Laboratory
Argonne, Illinois 60439

Stanley A. Roth and Galen D. Stucky

School of Chemical Sciences
University of Illinois
Urbana, Illinois 61801

INTRODUCTION

Ethanol is an important chemical feedstock and fuel extender, which is currently produced from petroleum-based ethylene or from feed grains by fermentation. Among the most extensively studied techniques for diverting the raw material base for such chemicals from oil to coal is through the catalytic reactions of synthesis gas, a mixture of H_2 and CO, which can be made via coal gasification. In this context, the direct conversion of synthesis gas to ethanol[1] and the indirect reaction of synthesis gas with methanol to yield ethanol (homologation) (for reviews see References 2 and 3) have been studied as possible alternative processes for the production of ethanol from coal. By far, the largest amount of effort on the homologation of methanol has been focused on modifications of the $HCo(CO)_4$-catalyzed reaction. The first report of alcohol homologation using cobalt carbonyls was that of Wender,[4] although an earlier (1941) patent using a CoO/NiO catalyst exists.[5] Recent reports by workers at Union Carbide,[6] Exxon,[7] Gulf,[8] British Petroleum,[9] and Celanese[10] attest to the current industrial interest in the cobalt carbonyl–catalyzed reaction. Our own efforts, however, have focused on a new and unusually general catalytic method occurring in methanol-amine solutions of metal carbonyls at synthesis gas pressures near 300 atm and temperatures near 200°C.[11,12] Ethanol is formed according to the following stoichiometry:

$$CH_3OH + 2 CO + H_2 = CH_3CH_2OH + CO_2 \qquad (1)$$

In contrast to the $HCo(CO)_4$-based catalysts, the new catalysts show essentially no tendency to further homologate the product ethanol. They produce anhydrous ethanol while consuming less H_2, because CO_2 rather than H_2O is the oxygenated by-product. One additional feature of the new method, which is particularly helpful in its necessary further development, is that the mechanism uses component reactions that are widespread for metal carbonyls. In contrast, the cobalt systems all appear to require a feature unique to $HCo(CO)_4$, namely,

* This research was supported by the office of Chemical Sciences, Division of Basic Energy Sciences, U.S. Department of Energy.
† To whom correspondence should be addressed.

152

its unusually high acidity for a metal complex that can be formed by activation of hydrogen. This article summarizes some of our recent efforts in developing the new method.

RESULTS AND DISCUSSION

Mechanism

In general, the new catalysts use methylammonium ions as methyl carriers, transition metal complex anions as nucleophilic methyl acceptors, and catalytic decomposition of formic acid to remove protons generated in hydrogen activation steps.

In contrast to the $HCo(CO)_4$-catalyzed reaction, in which methyl groups are activated by protonation of methanol, the new systems activate methyl groups by the following equilibria, which are established rapidly in methanol solutions of amine (B).

$$CH_3OH + CO \xrightarrow{\quad B \quad} HCO_2CH_3 \qquad (2)$$

$$HCO_2CH_3 + B \rightleftharpoons CH_3B^+ + HCO_2^- \qquad (3)$$

Reaction 2 is catalyzed by amines and Reaction 3 generates the active methyl transfer reagent, a methylammonium ion, CH_3B^+. When methanol solutions containing trimethylamine, 1-methylpiperidine, or 1,3-bis(1-methyl-4-piperidyl) propane (3.3 M) are heated to 200°C at 300 atm 3 CO/H_2 pressure, Reactions 2 and 3 both reach equilibrium within 20 minutes and greater than 80% of each amine is converted to its methylammonium ion. The method's characteristic selectivity for methanol vs. ethanol homologation arises in Step 3. Attack (S_N2) on methyl formate by trimethylamine is preferred for steric reasons over attack on ethyl formate. In S_N2 reactions involving poor leaving groups, methyl transfers can be as much as 150 times faster than ethyl transfers (see Reference 13 and references therein).

Under the reaction conditions, $Fe(CO)_5$ reacts as follows:

$$Fe(CO)_5 + H_2 + B \rightleftharpoons HFe(CO)_4^- + CO + BH^+ \qquad (4)$$

This reaction has been studied previously by Wada and Matsuda.[14] Formate ion generated in Reaction 3 is not inert. We have observed that methanol solutions of $[N(CH_3)_3H][HCO_2]$ rapidly react according to Equation 5 in the presence of $Fe(CO)_5$.

$$BH^+ + HCO_2^- \longrightarrow H_2 + CO_2 + B \qquad (5)$$

The effect of Reaction 5 is to drive the Equilibrium 4 to the right by removal of protons. Reaction 5 amounts to a catalytic decomposition of formic acid in basic solution. Without formic acid decomposition the net overall reaction would be stoichiometric in amine, all of which would soon become protonated via Reaction 4. Reaction 5 is the source of CO_2 produced in the net catalytic reaction in Equation 1.

The result of Reactions 2-5 is that CH_3B^+ and $HFe(CO)_4^-$ are both present when methanol solutions of $Fe(CO)_5$ and amine are heated under CO and H_2

TABLE 1

REACTIVITY AND SELECTIVITY PATTERNS FOR VARIOUS CATALYSTS*

Experiment No.	Complex (mmol)	Reaction Time (hours)	Products		Turnover Frequency (per hour)
			C_2H_5OH† (mmol)	CH_4 (mmol)	
1	$Fe(CO)_5$ (16.0)‡	6.0	34	73	1.1
2	$RhCl_3 \cdot 3H_2O$ (16.0)	6.0	160 (160)§	30 (21)§	2.0
3	RhI_3 (5.0)‡¶	2.0	42	19	6.1
4	$Ru_3(CO)_{12}$ (5.3)‡§	3.4	26	26	0.96
5	$Mn_2(CO)_{10}$ (11.5)	6.0	102	22	0.90
6	$Cr(CO)_6$ (16.0)	6.0	0	0	0
7	$Fe(CO)_5$ (16.0) $P(n\text{-}Bu)_3$ (32.0)	6.0	17	19	0.38
8	$Mn_2(CO)_{10}$ (11.5) MeI (80.0)	6.0	300	50	2.5

* 200°C and 300 atm (3:1) CO/H_2 with continuous gas purge of 600 ml/minute. Catalyst solutions (160 ml) contain 2.0 M N-methylpiperidine.
† Includes HCO_2Et.
‡ Gas purging was not used.
§ Trimethylamine replaces N-methylpiperidine.
¶ Volume of solution = 50 ml.

pressure. The ions appear to react as in Equations 6–10. It is necessary to

$$CH_3B^+ + HFe(CO)_4^- \longrightarrow CH_3HFe(CO)_4 + B \tag{6}$$

$$CH_3HFe(CO)_4 + CO \longrightarrow CH_3C(O)HFe(CO)_4 \tag{7}$$

$$CH_3HFe(CO)_4 + CO \longrightarrow CH_4 + Fe(CO)_5 \tag{8}$$

$$CH_3C(O)HFe(CO)_4 + CO \longrightarrow CH_3C(O)H + Fe(CO)_5 \tag{9}$$

$$CH_3C(O)H + H_2 \xrightarrow{\ HFe(CO)_4^-\ } CH_3CH_2OH \tag{10}$$

note, at this point, that Equations 2–7 and 9–10 add up to give the measured overall stoichiometry, Equation 1. Under the most typical reaction conditions, Equation 6 is the rate-limiting step and, as will be discussed later, has been the subject of a kinetic study. Reactions 7–10 or their analogues are well precedented.[15–17]

Catalysts

TABLE 1 lists complexes that we have examined for activity for catalytic methanol homologation in amine-methanol solution. These tests were made with the reasonable expectation that if the transition metal carbonyl anions frequently observed in similar solutions (most studied are aqueous or alcoholic solutions containing OH^-) can be generated reversibly in amine-methanol solution, reactions similar to those discussed for $Fe(CO)_5$ will occur. There are several routes for generation of nucleophilic transition metal carbonyl anions in these solutions. For example, we have examined the equilibrium reaction between H_2 and

$Mn_2(CO)_{10}$ (ca. 0.1 M) in methanol solutions at 200°C and at 300 atm of 3:1 CO/H_2. Under these conditions, about 25% of the $Mn_2(CO)_{10}$ is converted to $HMn(CO)_5$. However, in methanol solutions containing $N(CH_3)_3$ (other conditions approximately the same), the $HMn(CO)_5$ ($pK_a \sim 7$) is deprotonated and the equilibrium reaction (11) is established.

$$Mn_2(CO)_{10} + H_2 + 2B \longrightarrow 2\ Mn(CO)_5^- + 2BH^+ \qquad (11)$$

With the amine present (2.0 M), at least prior to onset of additional reactions, essentially all of the $Mn_2(CO)_{10}$ is converted to the nucleophile $Mn(CO)_5^-$.

Another pathway for the formation of transition metal anions exists in amine-methanol solutions even for those complexes that do not react with H_2. For example, King has suggested Reaction 12 as a key step in the mechanism for the catalytic water gas shift reaction by group VI transition metal carbonyls.[18] Darensbourg has subsequently shown that the $M(CO)_5(O_2CH)^-$ (M = Cr, Mo, or W) ion does indeed decompose smoothly to give $HM(CO)_5^-$ and CO_2.[19]

$$M(CO)_6 + O_2CH^- \xrightarrow{-CO} [M(CO)_5(O_2CH)^-] \longrightarrow HM(CO)_5^- + CO_2 \qquad (12)$$

Since one mole of O_2CH^- is produced for each mole of CH_3OH that is reacted in Equations 2 and 3, an abundant supply of this ion is present in amine-methanol solutions for reactions similar to (12).

Under the reaction conditions given in TABLE 1, at least early in the reactions, the major species present in the RhI_3, $Mn_2(CO)_{10}$, and $Ru_3(CO)_{12}$ systems were $Rh(CO)_2I_2^-$, $Mn(CO)_5^-$, and $HRu_3(CO)_{11}^-$ respectively, identified by their infrared (IR) spectra. Although observation of these ions, alone, does not establish that they are the reactive species in these systems, their presence, together with the observed catalysis of the homologation reaction, fits the pattern discussed for $Fe(CO)_5$. In the case of the $Mn_2(CO)_{10}$ system, additional evidence for a mechanism similar to that proposed for $Fe(CO)_5$ was obtained by observation of the reaction of $N(CH_3)_4^+$ with $Mn(CO)_5^-$, in the presence of H_2 and CO, to produce CH_3CH_2OH and CH_4. This reaction is the subject of a kinetic study, to be discussed later.

It is conceivable that certain metal complexes may not possess all of the properties necessary to complete the catalytic homologation cycle, but may possess properties particularly advantageous for certain of the steps involved in it. In order to explore this possibility, metal complexes were tested for activity in the presence of a complex capable of completing the entire catalytic cycle, namely, $Fe(CO)_5$. These experiments are summarized in TABLE 2. In some cases, the rate and selectivity observed in the mixed systems appear to be a composite of what might be expected to occur if the complexes were tested individually. For example, Experiment 5 (TABLE 2) shows that the mixed $Fe(CO)_5/Mo(CO)_6$ system gives essentially the same rate and selectivity observed in Experiment 1 (TABLE 1) for $Fe(CO)_5$ alone. Presumably, $Mo(CO)_6$, by itself, is inactive. On the other hand, this is clearly not the case in the $Fe(CO)_5/Re_2(CO)_{10}$ system in TABLE 2, Experiment 7. In this case, the presence of $Re_2(CO)_{10}$ inhibits completely the activity of $Fe(CO)_5$. The most interesting observation to evolve from these experiments, however, is the promotional effect observed in the $Fe(CO)_5/Mn_2(CO)_{10}$ system (TABLE 2, Experiment 1). In this mixed system, the selectivity is essentially

TABLE 2

REACTIVITY AND SELECTIVITY FOR MIXED CATALYSTS*

Experiment No.	"Cocatalysts"† (mmol)	Reaction Time (hours)	Products	
			C_2H_5OH (mmol)	CH_4 (mmol)
1	$Mn_2(CO)_{10}$ (11.5)	6.0	330	67
2	$RhCl_3 \cdot 3H_2O$ (16.0)	6.0	160	
3	$Mn_2(CO)_{10}$ (11.5) $RhCl_3 \cdot 3H_2O$ (16.0)	6.0	309	42
4	$Ru_3(CO)_{12}$ (5.3)	6.0	60	150
5	$Mo(CO)_6$ (16.0)	6.0	34	89
6	$NiBr_2$ (16.0)	2.0	7	27
7	$Re_2(CO)_{10}$ (8.0)	6.0	0	0
8	$Mn_2(CO)_{10}$ (11.5) $P(n\text{-Bu})_3$ (48)	6.0	330	38

* Reaction at 200°C and 300 atm 3:1 CO/H_2. The reactor was purged with the same CO/H_2 mixture at 600 ml/minute.
† All the catalyst solutions (160 ml) contain 16.0 mmol $Fe(CO)_5$ and 320 mmol $MeNC_5H_{10}$ except for Experiment 3, in which no $Fe(CO)_5$ was added.

the same as that observed for the $Mn_2(CO)_{10}$ system without added $Fe(CO)_5$ (TABLE 1, Experiment 5), but the rate is more than three times as large. Our interpretation of these observations is as follows. Under the reaction conditions, the pure manganese system is quickly driven to a lower pH than observed in the pure iron system. At the lower pH, formation of methyl formate and methylammonium ion is inhibited, and the catalytic reaction is slower. The lower pH in the manganese system suggests that the system is less effective for formic acid decomposition in Reaction 5. Consistent with this view, addition of $Fe(CO)_5$, a known catalyst for Reaction 5, accelerates the methanol homologation reaction. In the mixed system, essentially all of the iron is in the form of nonnucleophilic $Fe(CO)_5$, and the products, therefore, retain the high selectivity toward ethanol formation characteristic of the manganese system.

Kinetic Measurements

Because so many reactions occur simultaneously when metal carbonyls are heated in amine-methanol solution under synthesis gas pressure, we have found it most worthwhile to study the kinetics of certain steps individually in nonreactive solvents. For example, reactions of $N(CH_3)_4^+$ and $(CH_3)_2NC_5H_{10}^+$ with $HFe(CO)_4^-$ under CO and H_2 pressure in 1-methyl-2-pyrrolidinone solution exhibit all of the chemistry in Equations 6–10 and are easily amenable to kinetic studies. Reaction of the tetramethylammonium salt occurs according to Equation 13,

$$N(CH_3)_4^+ + HFe(CO)_4^- + a\,H_2 + (1 + a)CO$$

$$= (1 - a)CH_4 + a\,C_2H_5OH + Fe(CO)_5 + N(CH_3)_3 \quad (13)$$

where the coefficient a is dependent upon the reaction conditions. Kinetics for the reaction in 1-methyl-2-pyrrolidinone solution were measured by monitoring the disappearance of the ions, which decreased in a precise 1:1 ratio. Rates at 200°C were first order with respect to $N(CH_3)_4^+$ and $HFe(CO)_4^-$, but were independent of H_2 pressures between 61 and 184 atm. They were also independent of CO pressures varied over the same range. The sum of the methane and ethanol produced in the reaction corresponded well with the amount of $N(CH_3)_4^+$ reacted. The rate law is therefore:

$$\frac{-d[N(CH_3)_4^+]}{dt} = \frac{d([CH_4] + [C_2H_5OH])}{dt} = k^{(2)}[N(CH_3)_4^+][HFe(CO)_4^-] \quad (14)$$

An Eyring plot using $k^{(2)}$ values measured over the range 180–210°C is linear, and the activation parameters $\Delta H^\ddagger$, 44.1 kcal, and $\Delta S^\ddagger$, 16.8 entropy units, were obtained. The large positive entropy of activation is consistent with a reaction in which ions react to form neutral species and release solvating molecules. This observation and the lack of CO and H_2 dependence and the first-order dependence on $[N(CH_3)_4^+]$ and $[HFe(CO)_4^-]$ all clearly point to Equation 6 as the rate-limiting step for the reaction of $N(CH_3)_4^+$ with $HFe(CO)_4^-$. Further, in the case where $Fe(CO)_5$ (0.10 M) and $N(CH_3)_3$ (2.0 M) catalyze the homologation reaction in methanol solution at 200°C in the presence of 300 atm 3 CO/H_2, formation rates of these ions are high and Reactions 2 and 3 are at or near equilibrium. In this situation, the methyl transfer step in Equation 6 is also apparently rate limiting.

We have also measured the rates of reaction of $N(CH_3)_4^+$ with $Mn(CO)_5^-$ to yield C_2H_5OH and CH_4 in 1-methyl-2-pyrrolidinone solution. A rate law analogous to that for $HFe(CO)_4^-$ was found. At 200°C, the rate constant, $k^{(2)}$, for $Mn(CO)_5^-$ is 5.3×10^{-4} M^{-1} s^{-1}. Under the same conditions, the rate constant for $HFe(CO)_4^-$ is 2.0×10^{-4} M^{-1} s^{-1}. As one might expect from their relative basicities [pK_a, $HMn(CO)_5 = 7.1$; pK_a, $H_2Fe(CO)_4 = 4.4$], $Mn(CO)_5^-$ is a better nucleophile for this reaction than is $HFe(CO)_4^-$.

CONCLUSION

This article has demonstrated that the new catalytic method discussed here allows unusual flexibility in terms of catalyst design. Choices can be made from an array of nucleophiles, catalysts for formic acid decomposition, and even (in logical extensions of the principles involved) in the selection of the methyl carrier and base to complete the catalytic cycles. Further development of the method will be in the exploration of these possibilities.

ACKNOWLEDGMENT

We thank Professor Jack Halpern for useful discussions.

REFERENCES

1. ICHIKAWA, M. 1978. J. Chem. Soc. Chem. Commun.: 566.
2. PIACENTI, F. & M. BIANCHI. 1977. *In* Organic Syntheses via Metal Carbonyls. I. Wender & P. Pino, Eds. 2: 1–42. John Wiley & Sons, Inc. New York, N.Y.

3. SLOCUM, D. 1980. *In* Catalysis in Organic Syntheses. W. Jones, Ed.: 245–276. Academic Press, Inc. New York, N.Y.
4. WENDER, I., R. LEVINE & M. ORCHIN. 1949. J. Am. Chem. Soc. **71:** 4160.
5. WEITZEL, G., K. EDER & A. SCHEUERMANN. 1941. (BASF); DE-DS 867 849.
6. WALKER, W. E. 1981. U.S. Patent 4277634.
7. DOYLE, G. 1981. J. Mol. Catal. **13:** 237.
8. PRETZER, W. R., T. KOBYLINSKI & J. E. BOZIK. 1979. U.S. Patent 4133966.
9. BARLOW, M. T. 1981. European Patent 29723.
10. KOERMER, G. S. & W. E. SLINKARD. 1978. Ind. Eng. Chem. Prod. Res. Dev. **17:** 231.
11. FEDER, H. M. & M. J. CHEN. 1981. U.S. Patent 4301312.
12. CHEN, M. J., H. M. FEDER & J. W. RATHKE. 1982. J. Am. Chem. Soc. **104:** 7346.
13. DE LA MARE, P. B. D., L. FOWDEN, E. D. HUGHES, C. K. INGOLD & J. D. H. MACKIE. 1955. J. Chem. Soc.: 3200.
14. WADA, F. & T. MATSUDA. 1973. J. Organomet. Chem. **61:** 365.
15. COOKE, M. P., JR. 1970. J. Am. Chem. Soc. **92:** 6080.
16. STERNBERG, H. W., R. MARKBY & I. WENDER. 1957. J. Am. Chem. Soc. **79:** 6116.
17. KANG, H. C., C. H. MAULDIN, T. COLE, W. SLEGEIR, K. CANN & R. PETTIT. 1977. J. Am. Chem. Soc. **99:** 8323.
18. KING, A. D., JR., R. B. KING & D. B. YANG. 1981. J. Am. Chem. Soc. **103:** 2699.
19. DARENSBOURG, D. J., A. ROKICKI & M. Y. DARENSBOURG. 1981. J. Am. Chem. Soc. **103:** 3223.

REGIOSELECTIVITY IN HYDROFORMYLATION OF LINEAR AND BRANCHED OCTENES USING HCo(CO)$_4$

Barry L. Haymore, Allen van Asselt, and Gary R. Beck

Corporate Research Laboratories
Monsanto Company
St. Louis, Missouri 63167

INTRODUCTION

The catalytic conversion of olefins to aldehydes using H$_2$ and CO in the presence of HCo(CO)$_4$ is one of the most studied catalytic reactions.[1-3]

$$C-C-C-C=C + H_2 + CO \xrightarrow{\text{Co}} C-C-C-C-\overset{\overset{\text{CHO}}{|}}{C} + C-C-C-\overset{\overset{\text{CHO}}{|}}{C}-C$$

$$+ C-C-\overset{\overset{\text{CHO}}{|}}{C}-C-C$$

One characteristic of this reaction is that the unmodified cobalt catalyst, HCo(CO)$_4$, isomerizes the starting olefin *and* hydroformylates it at the same time. If the isomerization were slow relative to hydroformylation, only two aldehydes would be produced, those with the formyl groups located at the two original trigonal carbon atoms. If isomerization were fast relative to hydroformylation, all possible aldehydes would be produced with the formyl groups ending up on all possible carbon atoms regardless of the initial location of the olefinic double bond. In this case, the product distribution of the resulting aldehydes would be more or less the same for all olefins possessing the same carbon skeleton. In point of fact, neither extreme case correctly describes the relative isomerization/hydroformylation rates under commonly used hydroformylation conditions. The relative rates are not too different; indeed, all possible aldehydes are generated from a simple monoolefin, and the relative amounts of the various aldehyde products are strongly dependent on the initial location of the olefinic double bond.

For branched monoolefins of intermediate size (C$_6$–C$_{10}$), the aldehyde product distributions have not been studied in detail. Owing to our interest in butene dimers (branched and linear octenes), we have carried out a series of catalytic hydroformylation reactions using pure octenes under the same set of reaction conditions, thus allowing us to determine the effects of branching and positional isomerism on product distributions of octenes that possess five different carbon skeletons: linear octenes, 2-methylheptenes, 3-methylheptenes, 2,4-dimethylhexenes, and 3,4-dimethylhexenes.

EXPERIMENTAL SECTION

The hydroformylation reactions were conducted in a 300 mL Magnedrive Autoclave (Autoclave Engineers, Inc.), which was stirred at 1,200 rpm. Unless otherwise specified, a standard set of experimental conditions was used: constant pressure

159

of 200 atm of 1:1 H_2/CO, 120°C, 3.0 g olefin in 75 mL C_6H_6, and 115 mg freshly sublimed $Co_2(CO)_8$ (Strem). The reaction mixture was sampled periodically to monitor the course of the reaction.

The entire group of linear and branched octenes was also hydroformylated using $RhH(CO)(PPh_3)_3$ in the presence of excess triphenylphosphine.[4] Typical experimental conditions were as follows: 75 mg $RhH(CO)(PPh_3)_3$, 150 mg PPh_3, 100 mg olefin, 1.5 mL C_6H_6, 100°C, and 8 atm 1:1 H_2/CO. Because this hydroformylation catalyst isomerized the free and coordinated olefins slowly in comparison to the rate of hydroformylation, the Rh catalyst principally formylated the original trigonal carbon atoms and could be used as a method of aldehyde synthesis and identification. After all positional isomers of an olefin with a particular carbon skeleton were hydroformylated, each aldehyde was then uniquely identified and characterized by gas chromatography/mass spectrometry (G€/MS). Each positional isomer produced two principal aldehyde products, except for trisubstituted and tetrasubstituted olefins. With trisubstituted olefins, the Rh catalyst formylated only the monosubstituted trigonal carbon atom (ignoring a small amount of isomerization). The hydroformylation of 3,4-dimethylhexene-3 with Rh proceeded too slowly to be useful. Twenty-four nine-carbon aldehydes were prepared by octene hydroformylation using rhodium.*

The tertiary aldehydes (2,2-disubstituted aldehydes) were prepared by alkylation at the 2-position of smaller aldehydes using KH in tetrahydrofuran (THF) at 25°C.[5] The hydroformylation of 2,4-dimethylhexene-2 and 2,4-dimethylhexene-3 using Rh did not produce 2-isopropyl-3-methylpentanal in significant yield, so the alkylation procedure was used. The following gives a synopsis of our aldehyde synthesis via the alkylation route: 2,2-dimethylheptanal (44% yield, 95% C-alkylated) from heptanal (Aldrich) and methyl iodide; 2-methyl-2-ethylhexanal (54% yield, 94% C-alkylated) from 2-ethylhexanal (Eastman) and methyl iodide; 2,3-dimethyl-2-ethylpentanal (41% yield, 54% C-alkylated) from 2,3-dimethylpentanal (Wiley) and ethyl iodide; 2,2,4-trimethylhexanal (33% yield, 95% C-alkylated) from 4-methylhexanal and methyl iodide; 2,4-dimethyl-2-ethylpentanal (61% yield, 93% C-alkylated) from 4-methyl-2-ethylpentanal and methyl iodide; 2-isopropyl-3-methylpentanal (30% yield, 28% C-alkylated) from 3-methylpentanal (Aldrich) and isopropyl iodide. 4-Methylhexanal was prepared by the hydroformylation [$RhH(CO)(PPh_3)_3$ + PPh_3]

---

* The following aldehydes were prepared by Rh-catalyzed hydroformylation: nonanal from octene-1; 2-methyloctanal from octene-1 and *trans*-octene-2; 2-ethylheptanal from *trans*-octene-2 and *trans*-octene-3; 2-propylhexanal from *trans*-octene-4; 3-methyloctanal from 2-methylheptene-1; 2-isopropylhexanal from 2-methylheptene-2; 2-propyl-4-methylpentanal from *trans*-2-methylheptene-3 and *trans*-6-methylheptene-3; 2-ethyl-5-methylhexanal from *trans*-6-methylheptene-3 and *trans*-6-methylheptene-2; 2,6-dimethylheptanal from 6-methylheptene-1 and *trans*-6-methylheptene-2; 7-methyloctanal from 6-methylheptene-1; 4-methyloctanal from 3-methylheptene-1; 2,3-dimethylheptanal from 3-methylheptene-1 and 3-methylheptene-2; 3-ethylheptanal from 2-ethylhexene-1; 2-propyl-3-methylpentanal from *E*-3-methylheptene-3 and *trans*-5-methylheptene-3; 2-ethyl-4-methylhexanal from *trans*-5-methylheptene-3 and *trans*-5-methylheptene-2; 2,5-dimethylheptanal from *trans*-5-methylheptene-2 and 5-methylheptene-1; 6-methyloctanal from 5-methylheptene-1; 4,5-dimethylheptanal from 3,4-dimethylhexene-1; 2,3,4-trimethylhexanal from 3,4-dimethylhexene-1 and *E*-3,4-dimethylhexene-2; 3-ethyl-4-methylhexanal from 2-ethyl-3-methylpentene-1; 3,5-dimethylheptanal from 2,4-dimethylhexene-1; 3-ethyl-5-methylhexanal from 2-ethyl-4-methylpentene-1; 2,3,5-trimethylhexanal from 3,5-dimethylhexene-2 and 3,5-dimethylhexene-1; 4,6-dimethylheptanal from 3,5-dimethylhexene-1.

TABLE 1
WITTIG SYNTHESES OF BRANCHED OCTENES

Olefin	Aldehyde/Ketone	Isolated Yield, Second Distillation	Boiling Point (uncorrected)	Alkyl Halide
3,4-Dimethyl-1-hexene (erythro + threo)	2,3-dimethylpentanal (Wiley)	52%	111°C	methyl bromide
2,4-Dimethyl-3-hexene (cis + trans)	isobutyraldehyde (Aldrich)	38%	108°C	2-iodobutane
3,5-Dimethyl-1-hexene	2,4-dimethylpentanal (Reference 17)	76%	103°C	methyl bromide
2,4-Dimethyl-2-hexene	2-methylbutanal (Aldrich)	51%	110°C	isopropyl iodide
5-Methyl-3-heptene (cis + trans)	2-methylbutanal (Aldrich)	49%	113°C	n-propyl bromide
5-Methyl-2-heptene (cis + trans)	3-methylpentanal (K & K)	59%	117°C	ethyl bromide
6-Methyl-3-heptene (cis + trans)	isovaleraldehyde (Aldrich)	65%	114°C	n-propyl bromide
3-Methyl-2-heptene (cis + trans)	2-hexanone (Aldrich)	42%	122°C	ethyl bromide
3,5-Dimethyl-2-hexene (cis + trans)	4-methyl-2-pentanone (Aldrich)	22%	112°C	ethyl bromide
2-Ethyl-4-methyl-1-pentene	5-methyl-3-hexanone (K & K)	53%	108°C	methyl bromide
3-Methyl-3-heptene (cis + trans)	methyl ethyl ketone (Aldrich)	31%	121°C	n-butyl bromide

of 3-methylpentene-1 (Aldrich); this product contained 9% 2,3-dimethylpentanal, which was identified by comparison to a commercial sample. 4-Methyl-2-ethyl-pentanal was prepared by oxidation of 4-methyl-2-ethyl-1-pentanol (K & K).[6]

Eleven branched octenes (*cis/trans* mixtures) were synthesized by the usual Wittig coupling[7] of aldehydes or ketones with phosphorous ylides, which were prepared from phenyl lithium and the respective phosphonium salts. Labeled (99 + % ^{13}C) octene-1 was similarly prepared from heptanal and $^{13}CH_3I$ (Stohler Isotope Chemicals). The phosphonium salts were prepared in good yields from PPh_3 and the respective alkyl halide in refluxing benzene or toluene; the reaction with CH_3Br was carried out in benzene at room temperature. The product olefins were carefully distilled twice to get high-purity material (>98%) (see TABLE 1 for details). 5-Methylheptene-1 was prepared by allowing allyl bromide to react with the Grignard reagent formed from 1-bromo-2-methylbutane (Pfaltz and Bauer).[8,9] The remaining olefins were purchased from Wiley Organics, Columbus, Ohio. All olefins were carefully analyzed for purity and the results are given in TABLE 2. All olefins (42 isomers in all) were also subjected to analysis by ^{13}C NMR spectroscopy to verify their exact structures.[10] High-field (360 MHz) ^{1}H NMR spectra allowed us to distinguish between the diastereomers of 3,4-dimethylhexene-1 by comparison with literature data.[11,12]

Quantitative analyses of the eight-carbon olefins and of the nine-carbon aldehydes were carried out with the aid of gas chromatography. Two capillary columns were

TABLE 2
ANALYSES OF STARTING OLEFINS

Olefin	Cis (Z)	Trans (E)	C$_8$ Olefin, Same Skeleton	C$_8$ Olefin, Different Skeleton	All Other
Octene-1	99.9		—	0.1	—
trans-Octene-2	0.1	99.6	0.2	0.1	—
cis-Octene-3	97.7*	1.9	0.3	0.1	—
trans-Octene-3	0.3*	99.4*	0.2	0.1	—
cis-Octene-4	98.3*	1.5	0.1	0.1	—
trans-Octene-4	0.6*	99.2	0.1	0.1	—
2-Methylheptene-1	99.6		0.1	0.1	0.2
2-Methylheptene-2	99.4		0.2	0.3	0.1
trans-2-Methylheptene-3	0.1	99.4	0.2	0.1	0.2
cis-2-Methylheptene-3	84.9	13.9	0.5	0.5	0.2
trans-6-Methylheptene-3	0.4	99.2	0.3	—	0.1
trans-6-Methylheptene-2	24.2	75.1	0.5	0.1	0.1
6-Methylheptene-1	98.6		0.5	0.8	0.1
3-Methylheptene-1	98.6		—	1.1	0.1
E + Z-3-Methylheptene-2	52.7	45.7	0.3	1.2	0.1
Z-3-Methylheptene-3	68.8	27.4	3.3	0.4	0.1
E-3-Methylheptene-3	29.7	68.4	1.7	0.2	—
trans-5-Methylheptene-3	8.9	89.9	0.6	0.2	0.4
trans-5-Methylheptene-2	31.1	68.1	0.2	0.4	0.2
5-Methylheptene-1	98.9		0.6	0.4	0.1
2-Ethylhexene-1	99.0		0.8	0.1	0.1
2,4-Dimethylhexene-1	98.5		1.0	0.3	0.2
2,4-Dimethylhexene-2	98.4		1.2	0.3	0.1
E + Z-2,4-Dimethylhexene-3	56.3	41.5	0.2	1.7	0.3
E + Z-3,5-Dimethylhexene-2	57.0*	42.0*	0.7	0.2	0.1
3,5-Dimethylhexene-1	98.4		0.3	1.1	0.2
2-Ethyl-4-methylpentene-1	98.4		1.4	0.1	0.1
er + th-3,4-Dimethylhexene-1	54.2 (erythro)	45.6 (threo)	0.1	0.1	—
Z-3,4-Dimethylhexene-2	96.2	3.0	0.7	—	0.1
E-3,4-Dimethylhexene-2	0.3	99.6	—	0.1	—
E-3,4-Dimethylhexene-3	13.2	85.8	0.3	0.6	0.1
2-Ethyl-3-methylpentene-1	99.0		0.7	0.1	0.2

* Determined by bromination of olefin and analysis of vicinal dibromides.

used with flame ionization detectors (FID). Column A was a 60-meter (0.27 mm inner diameter) SP2100 WCOT glass capillary column and was operated isothermally at 30°C (olefins) or at 60°C (aldehydes). Column B was a 30-meter (0.50 mm inner diameter) squalane SCOT stainless steel capillary column and was operated isothermally at 50°C (olefins) or at 90°C (aldehydes). The data from both chromatographic columns taken together allowed us to completely analyze the olefins and aldehydes corresponding to each carbon skeleton.

All syntheses and hydroformylation experiments were carried out under atmospheres of CO, N$_2$, or Ar. All solvents (ethers, aliphatic hydrocarbons, aromatic hydrocarbons) were carefully purified by distillation from benzophenone/potassium under nitrogen.

Mass spectral data were acquired using a Finnigan TSQ gas chromatograph/mass spectrometer operating in the electron impact mode at 70 eV. Longer chain aliphatic aldehydes are known to undergo ordinary hydrocarbon fragmentation (HF) (m/e = 29, 43, 57, 71, 85, 99) and McLafferty rearrangements (MR) (m/e = 44, 58, 72, 86).[13] 2-Propylhexanal and 2-ethylheptanal yielded HF peaks that were small enough that they did not interfere with MR peaks that were used for the analyses. 2-Methyloctanal had an HF peak at m/e 57 which was large enough to interfere with the MR peak at m/e 58. Mass spectra of unlabeled and labeled 2-methyloctanal (prepared from $^{13}CH_3I$ and octanal) were used to calculate correction factors, which were applied to the MR peaks at m/e 58 and 59 (^{13}C labeled). The fragmentation pattern for nonanal was so complicated in the region of m/e 44 (MR peak) that it could not be used. Thus, our nonanal hydroformylation product (Co) was dimethylated with $^{13}CH_3I$ forming 2,2-dimethylnonanal with MR peaks at m/e 74 (doubly labeled) and 75 (triply labeled). Ordinary nonanal and labeled nonanal (from Rh hydroformylation of octene-1 labeled at C-1) were both dimethylated with $^{12}CH_3I$ and $^{13}CH_3I$. The mass spectra of these four products (unlabeled, singly labeled, doubly labeled, and triply labeled) allowed us to determine that the peaks at m/e 74 and 75 were essentially pure MR peaks.

RESULTS AND DISCUSSION

A series of linear and branched octene isomers have been subjected to catalytic hydroformylation conditions in order to determine the product distributions of the resulting aldehydes. The same conditions were used in all cases so that the results could be compared: 200 atm 1:1 synthesis gas (H_2/CO), 4–5% olefin in benzene solvent, 120°C, octene:Co = 40:1. We attempted to obtain samples when the reactions were 40–60% complete (16–24 turnovers) so that the product distributions could be compared at similar olefin conversions.

n-*Octenes*

The linear octenes were hydroformylated so that we could determine the aldehyde product distributions in olefins without branches. The results are given in TABLE 3.

TABLE 3
ALDEHYDES FROM LINEAR OCTENES†

$\overset{*}{C}$=C—C—C—C—C—C	C—C=C—C—C—C—C—C
65.5 13.8 3.6 2.3 1.6 1.3 3.0 8.9	62.0 21.4 9.9 6.7
74.4 16.8 4.9 3.9 → sums	
66% conversion	55% conversion
C—C—C=C—C—C—C—C	C—C—C—C=C—C—C—C
56.7 17.8 13.0 12.5: *trans*	56.2 17.0 11.0 15.8: *trans*
57.2 17.9 12.7 12.2: *cis*	54.4 17.5 11.6 16.5: *cis*
52% conversion—*trans*	52% conversion—*trans*
58% conversion—*cis*	60% conversion—*cis*

† Number indicates amount and location of formyl group in product.

A difficulty in the interpretation of the results is readily apparent. Although the starting olefin may not have symmetry, the octene carbon skeleton does. Thus, only four aldehydes are produced from an eight-carbon olefin; formylation at C-8 or C-1 yields the same product aldehyde, etc. In one case, octene-1, we prepared the olefin which was isotopically labeled (99 + % ^{13}C) at C-1. Using GC/MS techniques, the four product aldehydes were separated and then subjected to mass spectroscopy. The McLafferty rearrangement products were monitored in order to obtain the relative ratios of labeled and unlabeled products (see Experimental Section).

$$\overset{*}{C}-C-C-C-C-\overset{\overset{\displaystyle CHO}{|}}{C}-C-C \longrightarrow CH_3CH_2CH=CH-\overset{+}{O}H \quad m/e\ 72\ (27\%)$$
2-ethylheptanal

$$\overset{*}{C}-\overset{\overset{\displaystyle CHO}{|}}{C}-C-C-C-C-C-C \longrightarrow \overset{*}{C}H_3CH_2CH=CH-\overset{+}{O}H \quad m/e\ 73\ (73\%)$$
2-ethylheptanal

We assumed that any kinetic isotope effects could be ignored. This information coupled with the distribution of the four aldehyde products allowed us to calculate

Aldehyde	Percent Unlabeled McLafferty Rearrangement Product
$C-C-C-C-C-C-\overset{\overset{\displaystyle CHO}{|}}{C}-C$	12%
$C-C-C-C-C-C-\overset{\overset{\displaystyle CHO}{|}}{C}-C$	18%
$C-C-C-C-C-\overset{\overset{\displaystyle CHO}{|}}{C}-C-C$	27%
$C-C-C-C-\overset{\overset{\displaystyle CHO}{|}}{C}-C-C-C$	41%

the entire distribution for the formylation of all eight carbon atoms in the initial octene-1 (see TABLE 3). Our results on octene-1 seem to parallel published work on pentene-1 and hexene-1 (100°C, 160 atm);[14] however, we see more isomerization at 120°C.

The product distribution for octene-1 is interesting; 65.5% of the product is nonanal resulting from formylation of C-1, while 8.9% of the product is nonanal resulting from formylation of C-8. This clearly illustrates both the isomerization capability and preference of terminal formylation by HCo(CO)$_4$. A comparison with trans-octene-4 shows that formylation of C-1 amounts to 28.1% of the product, much less than the 65.5% in octene-1; however, the amount of formylation at C-8 increases from 8.9% (octene-1) to 28.1% (trans-octene-4), and the resulting total nonanal production only decreases from 74.4% to 56.2%. These data support what may be an obvious conclusion: the more carbon atoms between a terminal carbon atom (or any other carbon atom for that matter) and the trigonal carbon atoms of the olefinic double bond, the less frequently the terminal carbon atom will be formylated. We now have some quantitative notion about this effect for linear octenes and can compare this to the branched octenes. An interesting observation for octene-1 is that formylation percentages decrease going from C-1 to C-6 and then start increasing again with C-7 and C-8. Once the cobalt gets near the other end of the linear eight-carbon skeleton, it seems to "sense" the terminal carbon atom and preferentially formylates C-8 over C-5, C-6, or C-7. A similar effect is seen in *trans*-octene-4, for which formylation at C-3 is less than at C-2 or C-4.

The product distribution data for labeled octene-1 illustrate another general observation, which is applicable to many of the olefins we studied. When other factors, such as nearby branching, do not distort the distributions too much, the ratio of formylation at a terminal carbon atom (C-8 in octene-1) to formylation at the adjacent carbon atom (C-7 in octene-1) is 3.0 ± 0.1 when the double bond *is not* located between these two carbon atoms. This simply gives us a quantitative measure of the preference for terminal formylation. In contrast, that same ratio is 4.7 ± 0.1 when the double bond *is* located between the two carbon atoms (C-1, C-2 in octene-1). The increase in this ratio gives us a quantitative measure of the degree to which the cobalt attaches itself to the terminal carbon in an α-olefin and hydroformylates directly without isomerization.

2-Methylheptenes

The hydroformylation results for the positional isomers of 2-methylheptene are given in TABLE 4. We note that there are seven unique aldehydes obtainable from the eight-carbon skeleton because formylation at the methyl branch (C-8) or at C-1 yields the same aldehyde, 3-methyloctanal. Several features of the aldehyde distribu-

TABLE 4
ALDEHYDES FROM 2-METHYLHEPTENES*

```
        C                                      C
        |                                      |
  C=C—C—C—C—C—C                          C—C=C—C—C—C—C
78.7  3.8  0.1  1.6  2.2  3.5  10.1       63.9  1.2  0.2  3.3  4.3  6.9  20.2

     79% conversion                           68% conversion


        C                                      C
        |                                      |
  C—C—C=C—C—C—C                          C—C—C—C=C—C—C
27.6  0.4  0.7  7.2  8.7  13.9  41.5      18.3  0.3  0.2  5.3  10.1  15.5  50.3: trans
                                          18.2  0.3  0.3  5.6  10.6  15.2  49.8: cis
     67% conversion
                                            78% conversion—trans
                                            70% conversion—cis


        C                                      C
        |                                      |
  C—C—C—C—C=C—C                          C—C—C—C—C—C=C
 9.1  0.1  0.1  3.1  8.9  19.4  59.3       5.4  0.1   0   1.6  4.9  15.3  72.7

     78% conversion                           85% conversion
```

* Number indicates amount and location of formyl group in product.

tions stand out. The cobalt catalyst will formylate the sterically hindered tertiary carbon atom (C-2); the amount of 2,2-dimethylheptanal produced varies from 0.1 % to 3.8 % depending on the location of the double bond. Even more noteworthy is the near absence of hydroformylation at C-3, the position adjacent to the tertiary carbon. In all cases except for 2-methylheptene-3, formylation at C-2 was preferred over formylation at C-3.

The presence of branching enhances terminal formylation, apparently by two mechanisms: (1) the branch inhibits isomerization through the tertiary carbon; and (2) it inhibits formylation of the tertiary carbon and its attached neighboring carbon atoms (except when they are terminal carbons) even when the cobalt is attached to these carbon atoms during the catalytic cycle. From 2-methylheptene-1 and 2-methyl-heptene-2, we can obtain the relative formylation distributions for the three branches; each distribution is normalized to 100 % (below, left). It is instructive to compare this distribution along the five-carbon branch in 2-methylheptenes to the normalized distribution of a similar fragment (C-4 through C-8) in octene-1 (below, right).

$$
\begin{array}{cc}
\begin{array}{c}
100 \\
\text{C} \\
|
\end{array} &
\begin{array}{c}
\text{C*} \\
\|
\end{array} \\
\text{C}-\text{C}-\text{C}-\text{C}-\text{C}-\text{C}-\text{C} &
\text{C}-\text{C}-\text{C}-\text{C}-\text{C}-\text{C}-\text{C} \\
100 \quad \frac{1}{2} \quad 9 \quad 12\frac{1}{2} \quad 20 \quad 58 &
13\frac{1}{2} \quad 9\frac{1}{2} \quad 7\frac{1}{2} \quad 17\frac{1}{2} \quad 52
\end{array}
$$

The methyl branch in 2-methylheptenes shifts formylation away from C-3 (-13%) to C-5 $(+5\%)$, C-6 $(+2\%)$, and C-7 $(+6\%)$. A comparison of the product distribution along this five-carbon chain (above, left) with analogous normalized distributions of 2-methylheptenes (below) that have double bonds in positions 3, 4, 5, and 6 shows

$$
\begin{array}{cc}
\begin{array}{c}
\text{C} \\
|
\end{array} &
\begin{array}{c}
\text{C} \\
|
\end{array} \\
\text{C}-\text{C}-\text{C}=\text{C}-\text{C}-\text{C}-\text{C} &
\text{C}-\text{C}-\text{C}-\text{C}-\text{C}-\text{C}=\text{C} \\
1 \quad 10 \quad 12 \quad 19\frac{1}{2} \quad 57\frac{1}{2} &
0 \quad 2 \quad 5 \quad 16 \quad 77
\end{array}
$$

$$
\begin{array}{cc}
\begin{array}{c}
\text{C} \\
|
\end{array} &
\begin{array}{c}
\text{C} \\
|
\end{array} \\
\text{C}-\text{C}-\text{C}=\text{C}-\text{C}-\text{C} &
\text{C}-\text{C}-\text{C}-\text{C}-\text{C}=\overset{*}{\text{C}} \\
0 \quad 6\frac{1}{2} \quad 12\frac{1}{2} \quad 19 \quad 62 &
2 \quad 2\frac{1}{2} \quad 4 \quad 16 \quad 75\frac{1}{2}
\end{array}
$$

$$
\begin{array}{c}
\text{C} \\
| \\
\text{C}-\text{C}-\text{C}-\text{C}-\text{C}=\text{C}-\text{C} \\
0 \quad 3\frac{1}{2} \quad 10 \quad 21\frac{1}{2} \quad 65
\end{array}
$$

a "memory effect"; that is, the amounts of aldehydes produced from formylation of the original trigonal carbon atoms are too large in a relative sense. These data show that 2-methylheptene-3 gives too much 2-propyl-4-methylpentanal; 2-methyl-heptene-4 gives too much 2-ethyl-5-methylhexanal; 2-methylheptene-5 gives too much 2,6-dimethylheptanal. There are other trends and correlations in these data,

Formylated Carbon	2-Methyl-heptene-1, -2		2-Methyl-heptene-3		2-Methyl-heptene-4		2-Methyl-heptene-5		2-Methyl-heptene-6
C-4	9	→	10†	→	6$\frac{1}{2}$	→	3$\frac{1}{2}$	→	2
C-5	12$\frac{1}{2}$	→	12	→	12$\frac{1}{2}$†	→	10	→	5
C-6	20	→	19$\frac{1}{2}$	→	19	→	21$\frac{1}{2}$†	→	16

† Percentages are too high.

which will be left for the reader to explore. Finally, we note that the normalized product distribution along the five-carbon chain in 2-methylheptene-3 is similar to those in 2-methylheptene-1 and 2-methylheptene-2. There is a gradual shift in distribution along the series to 2-methylheptene-6, which can then be compared to the same distribution in octene-1 (carbons C-1 through C-5). Except for lower amounts of formylation at C-3 (-2%) in 2-methylheptene-6, these two normalized distributions are quite similar.

3-Methylheptenes

The hydroformylation results for the isomers of 3-methylheptene are given in TABLE 5. From these eight-carbon olefins, eight different aldehydes are produced in varying amounts; however, four of the eight have two diastereomeric forms and 12 distinct products are formed. Although there are some differences in diastereomeric product ratios depending on the geometric isomerism (*cis/trans*) and on the positional isomerism of the starting olefin, we have ignored these effects and have presented the product distributions as sums of diastereomers where appropriate. The 3-methylheptenes show that moving the methyl branch closer to the center of the molecule

TABLE 5
ALDEHYDES FROM 3-METHYLHEPTENES*

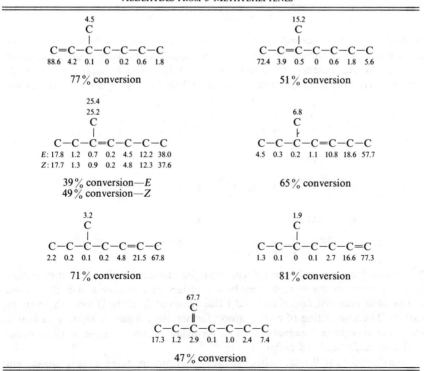

* Number indicates amount and location of formyl group in product.

increases terminal hydroformylation compared to linear octenes and 2-methyl-heptenes. The methyl branch effectively drives the formylation to the ends of the two-carbon and four-carbon chains. We observe small but significant amounts of formylation at the tertiary carbon (C-3) and significantly reduced amounts of formylation at C-4. In general, formylation is preferred in the order C-2 > C-3 > C-4, but the differing locations of the double bond can change this ordering for a specific positional isomer.

The normalized product distribution for the two-carbon chain (C-1 + C-2) can be calculated from the hydroformylation results of 5-methylheptene-1, 5-methyl-heptene-2, and 5-methylheptene-3. The normalized product distribution for the four-carbon chain (C-4 + C-5 + C-6 + C-7) can be calculated from the hydroformylation results of 3-methylheptene-1 and 3-methylheptene-2. The results given below (upper left) can be compared to those for the 2-methylheptenes and octene-1. The movement of the methyl branch from the 2-position to the 3-position shifts formylation away from C-4 and C-5 to C-6 and C-7.

$$
\begin{array}{cccccccc}
& & 100 & & & & & \\
& & \text{C} & & & & & \\
& & | & & & & & \\
\text{C} & \text{C} & \text{C} & \text{C} & \text{C} & \text{C} & \text{C} & \\
93 & 7 & & \tfrac{1}{2} & 8 & 22 & 69\tfrac{1}{2} &
\end{array}
\qquad
\begin{array}{cccccccc}
& & \text{C*} & & & & & \\
& & \| & & & & & \\
\text{C} & \text{C} & \text{C} & \text{C} & \text{C} & \text{C} & \text{C} & \\
& & & & 11 & 9 & 20 & 60
\end{array}
$$

$$
\begin{array}{ccccccc}
& \text{C} & & & & & \\
& | & & & & & \\
\text{C} & \text{C} & \text{C} & \text{C} & \text{C} & \text{C} & \text{C} \\
& & 9 & 12\tfrac{1}{2} & 20 & 58\tfrac{1}{2} &
\end{array}
\quad
\text{from 2-methylheptene-1 and 2-methylheptene-2}
$$

A comparison of the normalized product distributions along the four-carbon chains in the positional isomers of 3-methylheptenes (below) shows the memory effect noted before. 5-Methylheptene-3 gives too much 2-ethyl-4-methylhexanal, and 5-methyl-heptene-2 gives too much 2,5-dimethylheptanal. The composite picture obtained from

$$
\begin{array}{ccccccc}
& \text{C} & & & & & \\
& | & & & & & \\
\text{C} & \text{C} & \text{C}=\text{C} & \text{C} & \text{C} & \text{C} \\
& & \tfrac{1}{2} & 8 & 22 & 69\tfrac{1}{2} &
\end{array}
\qquad
\begin{array}{ccccccc}
& \text{C} & & & & & \\
& | & & & & & \\
\text{C} & \text{C} & \text{C} & \text{C} & \text{C} & \text{C}=\text{C} \\
& & 0 & 3 & 17 & 80 &
\end{array}
$$

$$
\begin{array}{ccccccc}
& \text{C} & & & & & \\
& | & & & & & \\
\text{C} & \text{C} & \text{C} & \text{C}=\text{C} & \text{C} & \text{C} \\
& & 1 & 12\tfrac{1}{2} & 21 & 65\tfrac{1}{2} &
\end{array}
\qquad
\begin{array}{ccccccc}
& \text{C} & & & & & \\
& | & & & & & \\
\text{C} & \text{C} & \text{C} & \text{C} & \text{C} & \text{C}=\text{C} \\
& & 2 & 5 & 16 & 77 &
\end{array}
$$

$$
\begin{array}{ccccccc}
& \text{C} & & & & & \\
& | & & & & & \\
\text{C} & \text{C} & \text{C} & \text{C}=\text{C} & \text{C} & \text{C} \\
& & 0 & 5 & 22\tfrac{1}{2} & 72\tfrac{1}{2} &
\end{array}
\qquad
\begin{array}{ccccccc}
& \text{C} & & & & & \\
& | & & & & & \\
\text{C} & \text{C} & \text{C} & \text{C} & \text{C} & \text{C}=\overset{*}{\text{C}} \\
& & 3 & 4 & 16 & 77 &
\end{array}
$$

the 2-methylheptenes and 3-methylheptenes is that the steric inhibition of formylation resulting from a methyl branch is felt by the tertiary carbon atom and by the carbon atoms once removed (α-carbon) and twice removed (β-carbon) from the tertiary carbon. The formylation of carbon atoms farther away from the tertiary carbon is indirectly affected in the reverse fashion; formylation here increases at the expense of the sterically hindered carbon atoms.

What happens to the normalized product distributions along a short two-carbon chain that is attached to a tertiary carbon atom? Because it is difficult to realign product distributions along these short chains, a larger fraction of the aldehyde product arises from formylation of carbon atoms on the other chains after isomerization through the tertiary carbon atom. We also note that the ratio of C-1 to C-2 formylation in monobranched α-olefins is unaffected by methyl branching at the 6- and 5- positions but is strongly affected by methyl branching at the 3-position. Although we didn't study such olefins with 4-methyl branching, it is likely that this formylation ratio would be altered by a methyl branch in this position (see below).

```
        C                              C
        |                              |
  C—C—C=C—C—C—C              C=C—C—C—C—C—C
  93½  6½                    82½  17½

        C                              C
        |                              |
  C—C=C—C—C—C—C              C=C—C—C—C—C—C
  95   5                     82½  17½

        C                              C
        |                              |
  C=C—C—C—C—C—C              *C=C—C—C—C—C—C
  95½  4½                    82½  17½
```

One piece of mechanistic information that would be useful to have is the fraction of aldehyde product that is produced by the cobalt catalyst after the metal goes through the tertiary carbon atom. The experimental aldehyde product distribution for an olefin can be considered as a weighted average of two product distributions. The first product distribution should be the same for all 3-methylheptenes and is the distribution one would obtain from hydroformylating the tertiary alkyl (1-methyl-1-ethylpentyl) complex of cobalt. The second product distribution is unique for each isomer of 3-methylheptene and is the statistical distribution obtained from hydroformylation of the chain (one of three) in which the double bond was originally located. Simply adding up the amount of hydroformylated product on the two chains not originally containing the double bond would not give the desired information; the reason is that the cobalt can isomerize out to the tertiary carbon atom and beyond and then return to the original chain before aldehyde elimination. The needed piece of information is the first product distribution, that from the hydroformylation of the tertiary alkyl of cobalt. Although this information might have been obtained experimentally by adding $Co(CO)_4^-$ to 2-ethyl-2-methyhexanoyl chloride under hydroformylation conditions, the experiment was not carried out. In lieu of these experimental data, we devised a method for estimating it. We have the relative product distributions for each of the three branches (C_1, C_2, C_4). All we need to know are the relative amounts of formylation on the three branches (A:B:C). The experimental results are given below. We see that the three ratios are

```
            B
           ‾‾
            C
            |
    C—C—C—C—C—C—C
    ‿‿‿‿    ‿‿‿‿‿‿
      A        C
```

almost but not quite mutually compatible, and some of our assumptions must not be entirely correct. Notwithstanding, we simply used the best "average" values of

Ratio	Obtained from
A/C = 1.70	2-ethylhexene-1
B/C = 1.80, 1.90 (1.85 average)	3-methylheptene-1, 3-methylheptene-2
B/A = 1.33, 1.42, 1.34, 1.36	3-methylheptene-3, 5-methylheptene-3,
(1.36 average)	5-methylheptene-2, 5-methylheptene-1

A (0.34), B (0.44) and C (0.22) realizing the limitations of the data. When these numbers are applied to the distribution data for the three chains (C_1, C_2, C_4), we obtain the composite result below.

$$
\begin{array}{c}
43.5 \\
C \\
| \\
C-C-C-C-C-C-C \\
31.2 \quad 2.3 \quad 1.3 \quad 0.1 \quad 1.7 \quad 4.8 \quad 15.1
\end{array}
$$

The fact that B > A > C is mechanistically plausible. Formylation will be preferred on the branch whose terminal carbon is nearest the tertiary carbon atom (C-3) to which the cobalt was originally attached. Now we can calculate what fraction of the formylation proceeded through the tertiary carbon atom (coumn B). The three olefins in which C-3 is a trigonal carbon all have high percentages; all others are low

Olefin	A Amount of Product with Formyl Not on Branch with Double Bond	B Amount of Product Formed after Co Attached to Tertiary Carbon
3-methylheptene-1	7.2%	10%
3-methylheptene-2	23.7%	37%
2-ethylhexene-1	32.3%	54%
3-methylheptene-3	45.1%	57%
5-methylheptene-3	11.8%	15%
5-methylheptene-2	5.7%	7%
5-methylheptene-1	3.3%	4%

($\leq 15\%$). That 3-methylheptene-2 is less than 3-methylheptene-3 and that 3-methyl-heptene-1 is less than 5-methylheptene-3 simply reflects the greater facility of the cobalt at finding a terminal carbon atom (C-1) when the initial double bond is located on the shorter two-carbon chain. That 6-methylheptene-1 is less than 3-methylheptene-1 indicates that in 6-methylheptene-1 the cobalt is less likely to find its way to C-3 because it is farther away from the double bond. Although 2-ethyl-hexene-1 is an α-olefin, the 2,2-disubstitution retards formylation of the branching methyl group (C-8). Sixty-eight percent of the hydroformylation product from 2-ethylhexene-1 is 3-ethylheptanal, but only two-thirds of this amount comes from direct formylation of C-8; the remaining one-third isomerizes through C-3 and eventually back to C-8 to give the additional 3-ethylheptanal!

3,4-Dimethylhexenes

The hydroformylation results for the isomers of 3,4-dimethylhexenes are given in

TABLE 6
ALDEHYDES FROM 3,4-DIMETHYLHEXENES*

$\quad$ 1.2 $\qquad$ C $\quad$ C $\qquad$ $\|\quad\|$ $\qquad$ C=C—C—C—C—C $\qquad$ 96.3 2.5 0 $\qquad\qquad$ 71% conversion	$\qquad\qquad$ 8.8 $\qquad\qquad$ 9.0 $\qquad$ C $\quad$ C $\qquad$ $\|\quad\|$ $\qquad$ C—C=C—C—C—C $\qquad$ 87.6 3.4 0.2: E $\qquad$ 87.2 3.6 0.2: Z $\qquad$ 44% conversion—E $\qquad$ 43% conversion—Z
$\qquad$ 67.7 $\qquad$ C $\quad$ C $\qquad$ $\|\quad\|$ $\qquad$ C—C—C=C—C—C $\qquad$ 29.8 1.3 1.2 $\qquad\qquad$ 24% conversion	$\qquad$ 85.7 $\qquad$ C $\quad$ C $\qquad$ $\|\|\quad\|$ $\qquad$ C—C—C—C—C—C $\qquad$ 11.8 0.4 2.1 $\qquad\qquad$ 58% conversion

* Number indicates amount and location of formyl group in product.

TABLE 6. Because the carbon skeleton is symmetric there are only four possible aldehydes; however, three of them exist in two diastereomeric forms and one, 2,3,4-trimethylhexanal, exists in *four* diastereomeric forms. The two branches significantly reduce the rates of hydroformylation especially in the case of the tetrasubstituted olefin 3,4-dimethylhexene-3. Terminal hydroformylation occurs in 96–97% of the product for all isomers. This result is not unexpected because four of the carbon atoms are terminal, and the other four central atoms (C-2 through C-5) are sterically hindered. How easily can the cobalt move from one tertiary carbon (C-3) to the other (C-4)? Based on results for other olefins, we believe that this *does not* readily occur and that nearly all of the formylation occurs in the same half of the molecule as does the original double bond; however, we have not carried out the isotope labeling experiments in order to prove this point. We will defer further discussion of the 3,4-dimethylhexenes.

2,4-Dimethylhexenes

The hydroformylation results for the isomers of 2,4-dimethylhexenes are given in TABLE 7. Seven different aldehydes can be produced from these olefins; three of them occur as pairs of diastereomers. Carbon atom C-3 is located α to two tertiary carbon atoms, and this effectively prevents formylation of this carbon atom at all, even though formylation at C-2 and C-4 seems largely unaffected. Another noteworthy feature is the small amount of aldehyde product that results from isomerization through C-3. In 2,4-dimethylhexene-1, 2,4-dimethylhexene-4, and 2-ethyl-4-methyl-pentene-1, only 0.3–1.5% of the product comes from this isomerization. Once the cobalt moves through C-3 the first time, the likelihood of going back through C-3 again is negligible. Thus α,γ-dibranching in an olefin is far more effective than mono-branching in preventing isomerization through the tertiary carbon atoms. What happens when the initial location of the double bond is between the branches? Our qualitative observations indicate that 2,4-dimethylhexene-2 and 2,4-dimethyl-hexene-3 hydroformylate at rates similar to other trisubstituted olefins and faster

TABLE 7
ALDEHYDES FROM 2,4-DIMETHYLHEXENES*

```
        0.3                                    2.1
    C       C                              C       C
    |       |                              |       |
  C=C—C—C—C—C                            C—C=C—C—C—C
 96.4 3.1  0  0  0 0.2                   95.0 1.4  0  0.1 0.1 1.3

      64% conversion                          58% conversion


          59.0                                   8.1
    C       C                              C       C
    |       |                              |       |
  C—C—C=C—C—C                            C—C—C—C=C—C
 9.6 0.1 0.2 1.5 1.6 28.0               0.3  0  0  0.3 4.7 86.6

      54% conversion                          41% conversion


          1.6                                    77.5
    C       C                              C       C
    |       |                              |       ‖
  C—C—C—C—C=C                            C—C—C—C—C—C
 0.1  0  0  0  4.3 94.0                 1.5  0  0  2.1 1.0 17.9

      75% conversion                          67% conversion
```

* Number indicates amount and location of formyl group in product.

than 3,4-dimethylhexene-3. In 2,4-dimethylhexene-3, the more distant branching carbon attached to C-2 is more effective at directing the isomerization than is the closer branching carbon on C-4. The methyl branch is significantly more effective at inhibiting formylation at an α-carbon atom than at its own tertiary carbon atom. The product distribution data for 2,4-dimethylhexene-2 lead to similar conclusions, but the distribution is shifted because the initial location of the double bond is no longer equidistant from C-1 and C-6. The results for 2,4-dimethylhexene-3 show that a small but significant amount ($\sim 10\%$) of the aldehyde product can result from isomerization through C-3 without formylating it to an appreciable extent.

General Comparisons

The dimethylhexenes allow us to observe the effect of methyl branching in the 4-position on product distributions in α-olefins. The normalized distributions below are largely unaffected by the second methyl branch until it occupies the 4-position. Then, formylation at C-1 is enhanced at the expense of formylation at C-2. The actual percentages (rounded to the nearest $\frac{1}{2}\%$) when compared to the normalized ones show that the principal effect of the 4-methyl group in 3,4-dimethylhexene-1 is to reduce formylation at C-2 which is two carbon atoms removed from the tertiary carbon atom in question.

```
 C8          C7                C    C              C   C
 |           |                 |    |              |   |
 C6—C5—C4—C3—C2=C1           C—C—C—C—C=C        C—C—C—C—C=C
normalized:  0   5   95     normalized: 0  4½ 95½  normalized: 0  2½ 97½
actual:      0   4½  88½     actual:     0  4½ 94   actual:     0  2½ 96½
```

The following comparisons illustrate the effect of a second methyl branch on the hydroformylation of a hexene-3 backbone. The 5-methyl group in 3,5-dimethylhexene-3 (2,4-dimethylhexene-3) is able to direct formylation to carbon atoms that lie on the other side of the tertiary carbon (C-3). The normalized distributions show that as one branching methyl group moves from C-6 to C-5 to C-4, formylation at the other methyl group (C-7) is increased at the expense of formylation at C-1, apparently because of the increased tendency of cobalt to formylate the nearest terminal carbon atom.

$$
\begin{array}{ccc}
& \overset{56\frac{1}{2}}{C_7} & \\
C_8 & | & \\
| & | & \\
C_6-C_5-C_4=C_3-C_2-C_1 & \\
\text{normalized:} \quad 1\frac{1}{2} \quad\; 2\frac{1}{2} \quad 39\frac{1}{2}
\end{array}
$$

$$
\begin{array}{ccc}
C & \overset{65\frac{1}{2}}{C} \\
| & | \\
C-C-C=C-C-C \\
\text{normalized:} \quad 1\frac{1}{2} \quad 2 \quad 31
\end{array}
$$

$$
\begin{array}{ccc}
C & \overset{67\frac{1}{2}}{C} \\
| & | \\
C-C-C=C-C-C \\
\text{normalized:} \quad 1 \quad 1\frac{1}{2} \quad 30
\end{array}
$$

45 % of product formylated on C_1, C_2, C_3, C_7

90 % of product formylated on C_1, C_2, C_3, C_7

50 % of product formylated on C_1, C_2, C_3, C_7

The following data underscore the effects of placing a second methyl branch nearer to the center of the principal carbon chain: more terminal hydroformylation and less isomerization through the two branches, especially when the two tertiary carbon atoms are separated by one other carbon atom.

$$
\begin{array}{l}
\quad\;\; C \quad\;\; \overset{67\frac{1}{2}}{C} \\
\quad\;\; | \quad\;\; \| \\
C-C-C-C-C-C \\
\qquad\qquad\quad 17\frac{1}{2}
\end{array}
$$

$$
\begin{array}{l}
\quad\;\; C \qquad\;\; C \\
\quad\;\; | \qquad\;\; | \\
C-C-C-C-C=C \\
\qquad\qquad\qquad\quad 78\frac{1}{2}
\end{array}
$$

$$
\begin{array}{l}
\quad\;\; C \qquad\;\; C \\
\quad\;\; | \qquad\;\; | \\
C-C-C-C=C-C \\
\qquad\qquad\qquad\quad 64
\end{array}
$$

$$
\begin{array}{l}
\quad\;\; \overset{77\frac{1}{2}}{C} \\
\quad\;\; | \quad\;\; \| \\
C-C-C-C-C-C \\
\qquad\qquad\quad 18
\end{array}
$$

$$
\begin{array}{l}
\quad\;\; C \qquad\;\; C \\
\quad\;\; | \qquad\;\; | \\
C-C-C-C-C=C \\
\qquad\qquad\qquad\quad 96\frac{1}{2}
\end{array}
$$

$$
\begin{array}{l}
\quad\;\; C \qquad\;\; C \\
\quad\;\; | \qquad\;\; | \\
C-C-C-C=C-C \\
\qquad\qquad\qquad\quad 95
\end{array}
$$

The final comparison that we want to make is the amount of formylation at C-1 in monosubstituted α-olefins. Terminal hydroformylation increases as the methyl branch gets closer to C-1, reaches a maximum when the branch is at the 3-position, and then decreases. Of course, with a methyl branch at the 2-position, the α-olefin is 1,1-disubstituted, and with a methyl branch in the 1-position, the olefin is no longer an α-olefin. Although we did not hydroformylate 4-methylheptene-1 (note skeletal symmetry), we estimate terminal hydroformylation at C-1 to be about 82 %.

$$
\begin{array}{l}
\quad\;\; C \\
\quad\;\; | \\
C-C-C-C-C-C=\overset{*}{C} \\
\qquad\qquad\qquad\qquad 65\frac{1}{2}
\end{array}
$$

$$
\begin{array}{l}
\qquad\qquad\; C \\
\qquad\qquad\; | \\
C-C-C-C-C-C=C \\
\qquad\qquad\qquad\qquad 88\frac{1}{2}
\end{array}
$$

$$
\begin{array}{l}
\quad\;\; C \\
\quad\;\; | \\
C-C-C-C-C-C=C \\
\qquad\qquad\qquad 72\frac{1}{2}
\end{array}
$$

$$
\begin{array}{l}
\qquad\qquad\; C \\
\qquad\qquad\; | \\
C-C-C-C-C-C=C \\
\qquad\qquad\qquad\qquad <78\frac{1}{2}
\end{array}
$$

$$
\begin{array}{l}
\quad\;\; C \\
\quad\;\; | \\
C-C-C-C-C-C=C \\
\qquad\qquad\qquad 77\frac{1}{2}
\end{array}
$$

$$
\begin{array}{l}
\qquad\; C-C \\
\qquad\quad\; | \\
C-C-C-C-C=C \\
\qquad\qquad\qquad\; 67\frac{1}{2}
\end{array}
$$

$$
\begin{array}{l}
\qquad\qquad\qquad C \\
\qquad\qquad\qquad | \\
C-C-C-C-C-C=C \\
\qquad\qquad\qquad\qquad <21\frac{1}{2}
\end{array}
$$

Other Factors Governing Product Distributions

Although a detailed discussion is beyond the scope of the present paper, we will briefly discuss other factors that govern the aldehyde product distributions from the hydroformylation of linear and branched octenes. Several experimental variables had *little* influence on product distributions: (1) pressure, (2) olefin concentration, (3) olefin-to-cobalt ratio, (4) geometric (*cis/trans*) isomerism, and (5) extent of reaction (conversion). The pressure of the synthesis gas (1:1 H_2/CO) was varied from 130 atm to 270 atm at 120°C for 2-ethylhexene-1, 3-methylheptene-3, and octene-2. We noted insignificant changes in product distributions with conversions in the 40% to 60% range. This is in agreement with previous work.[15] We did not vary the ratio of H_2 and CO in the synthesis gas. The olefin concentration for 2-ethylhexene-1 was varied from 0.5% to 5% to 45% in benzene with the olefin:Co ratio = 40. No changes in product distribution were observed. The olefin-to-cobalt ratio was varied from 10 to 100 for 5% 2-ethylhexene-1 in benzene at 120°C and 200 atm. Again no changes in product distributions were observed. Five pairs of *cis/trans* isomers were hydroformylated (see TABLES 3–7). We found that geometric isomerism had little effect upon the product distributions except for octene-4, and here the differences were only a few percent (TABLE 3). During most of our hydroformylation experiments, product distributions were monitored as a function of time. Only very small changes in product distribution were observed through 60% conversion and in many cases through 80% conversion. When the unreacted olefin was also analyzed at the same time intervals, it was found that the unreacted olefin was isomerized by the cobalt catalyst relatively slowly. Yet the product aldehyde resulted from extensive and rapid isomerization at even low conversions. Thus, extensive isomerization of unreacted olefin only occurs at higher conversions; even in these cases (75–95% conversion), the smaller amounts of isomerized olefin did not make large changes in overall product distributions. Consequently, our data, which were measured at varying conversions, can be compared with reasonable confidence. At the molecular level, this means that the cobalt catalyst rapidly isomerizes the coordinated olefin, but the catalyst preferentially hydroformylates the isomerized olefin rather than releasing it back into solution.[14–16]

Several experimental variables had significant influence on product distributions: (1) positional isomerism, (2) branching, (3) temperature, and (4) stirring rate (gas-liquid mixing). The effects of positional isomerism and branching have already been discussed in detail. In one case, we varied the temperature from 100°C to 160°C. The data below show that product distributions at 100°C and 120°C were very similar. At 140°C some changes became evident, and at 160°C large changes were observed. The higher temperatures produced more aldehyde product with the formyl groups on internal carbon atoms. Finally, we note that with very slow stirring

```
            25.9
            25.4
            26.7
            31.9
             C
             |
     C—C—C=C—C—C—C
```

							Temp.	Conversion	Selectivity
18.0	1.2	0.9	0.1	4.9	12.3	36.7	100	41%	99%
17.8	1.2	0.7	0.2	4.5	12.2	38.0	120	39%	99%
17.2	1.4	1.0	0.3	5.8	13.3	34.3	140	52%	96%
14.3	2.5	2.0	0.8	11.1	17.8	19.6	160	45%	87%

rates (poor gas-liquid mixing), the product distributions changed; there was more internal formylation and less terminal formylation. The effects of temperature and gas-liquid mixing on product distributions have been studied before[15] and are likely to be more pronounced for α-olefins.

SUMMARY

The aldehyde product distribution from the hydroformylation of an olefin is dependent upon the location of the olefinic double bond in the carbon skeleton. Using one common set of reaction conditions, we have determined the relative product distributions of all nine-carbon aldehydes produced from the hydroformylation of all positional isomers of *n*-octenes, 2-methylheptenes, 3-methylheptenes, 2,4-dimethylhexenes, and 3,4-dimethylhexenes. The presence of methyl groups in the branched olefins inhibits formylation of the tertiary carbon atom and carbon atoms near the tertiary carbon atom, thus promoting formylation of the terminal carbon atoms. The variety of aldehyde products derived from a single olefin results from extensive isomerization of the coordinated olefin by cobalt; however, the unreacted olefin is isomerized relatively slowly, and the aldehyde product distributions do not vary greatly with degree of reaction (conversion).

ACKNOWLEDGMENTS

We wish to acknowledge R. G. Kaley for obtaining mass spectral data, D. E. Willis for assistance in developing chromatographic analyses, and J. M. Stern for assistance with high-pressure equipment. We also appreciate helpful discussions with D. Forster and G. F. Schaefer.

REFERENCES

1. CORNILS, B. 1980. Hydroformylation. Oxo syntheses, Roelen reaction. *In* New Syntheses with Carbon Monoxide. J. Falbe, Ed: 1–225. Springer-Verlag. New York, N.Y.
2. ORCHIN, M. 1981. Acc. Chem. Res. **14**: 259.
3. PINO P., F. PIACENTI & M. BIANCHI. 1976. Reactions of carbon monoxide and hydrogen with olefinic substrates: the hydroformylation reaction. *In* Organic Syntheses via Metal Carbonyls. I. Wender & P. Pino, Eds. **2**: 43–232. Wiley-Interscience Publishers. New York, N.Y.
4. LEVISON, J. J. & S. D. ROBINSON. 1970. J. Chem. Soc. A: 2947.
5. GROENWEGEN, P., H. KALLENBERG & A. VAN DER GEN. 1978. Tetrahedron Lett.: 491.
6. COREY, E. J. & J. W. SUGGS. 1975. Tetrahedron Lett.: 2647.
7. WITTIG, G. & U. SCHOELLKOPF. 1954. Chem. Ber. **87**: 1318.
8. BARBIER, P. & V. GRIGNARD. 1904. Bull. Soc. Chim. **31**: 841.
9. MULLIKEN, S. P., R. L. WAKEMAN & H. T. GERRY. 1935. J. Am. Chem. Soc. **57**: 1605.
10. DORMAN, D. E., M. JAUTELAT & J. D. ROBERTS. 1971. J. Org. Chem. **36**: 2757.
11. STEPHANI, A. 1973. Helv. Chim. Acta **56**: 1192.
12. RITTER, W., W. HULL & H. CANTOW. 1978. Tetrahedron Lett.: 3093.
13. GILPIN, J. A. & F. W. MCLAFFERTY. 1957. Anal. Chem. **29**: 990.
14. BIANCHI, M., F. PIACENTI, P. FREDIANI & U. MATTEOLI. 1977. J. Organomet. Chem. **135**: 387.
15. PIACENTI, F., P. PINO, R. LASSARONI & M. BIANCHI. 1966. J. Chem. Soc. C: 488.
16. CASEY, C. P. & C. R. CYR. 1973. J. Am. Chem. Soc. **95**: 2240.
17. VANDENBROUCKE, W. & M. ANTEUNIS. 1972. J. Chem. Soc. Perkin Trans. 2: 123.

METAL HYDRIDES IN HOMOGENEOUS CATALYTIC REDUCTION OF CARBON MONOXIDE

B. D. Dombek

Union Carbide Corporation
South Charleston, West Virginia 25303

INTRODUCTION

In transition metal–catalyzed hydrogenation reactions, metal hydrides are certain to be key intermediates, although in some cases they may be too elusive or reactive to be actually observed. It must be kept in mind, however, that catalytically important species are, by nature, unstable and are therefore difficult to study directly. Information about their reactions is often obtainable only through kinetic investigations and by model studies of closely related systems. I would like to describe here three aspects of our research on the involvement of metal hydrides in the catalytic reduction of carbon monoxide. These studies have shown that the properties of metal hydrides have profound effects on the activity, selectivity, and reaction pathways of the catalytic process.

MANGANESE HYDRIDE, ALKYL, AND ACYL COMPLEXES AS MODELS

The first system I will describe is not a catalytic one, but it has provided us with a great deal of information about the possible mechanism of related catalytic processes. This system was designed to model intermediates possibly involved in the ruthenium-catalyzed reduction of CO in carboxylic acid solvents, which will be described in the next section. Reduction of CO has often been postulated to proceed through a hydroxymethyl intermediate (Equation 1).[1-5] Most α-hydroxyalkyl complexes are quite unstable under normal conditions.[6,7]

$$M-CO \xrightarrow{3/2H_2} M-CH_2OH \xrightarrow{H_2, CO} \text{organic products} \qquad (1)$$

This does not diminish the probability that they are important catalytic intermediates, but it renders the study of their reactions more difficult. We reasoned that such a catalytic intermediate in a carboxylic acid solvent might be converted to the carboxylate ester (Equation 2). It was therefore of interest to study the

$$M-CH_2OH \xrightarrow[-H_2O]{RCO_2H} M-CH_2O\overset{\overset{\displaystyle O}{\|}}{C}R \qquad (2)$$

behavior of such acyloxymethyl complexes, both as relatively stable models for hydroxymethyl compounds and as even closer models for intermediates in CO reductions carried out in carboxylic acid solvents.

Two major questions were to be studied through this model system. First, does the electronic character of the ester group substantially affect the ease of alkyl group

176

migration (Equation 3)? Second, does the steric bulk or possible chelating ability of this ligand perturb its reactivity in alkyl migration?

$$M-CH_2OCR \xrightarrow{\ CO\ } M-CCH_2OCR \qquad (3)$$

These questions were investigated in complexes of the formula $(CO)_5MnCH_2OCR$ ($R = Bu^t$, CH_3), which are stable and easily prepared from $NaMn(CO)_5$ and the chloromethyl esters. It was found that the complexes do indeed react under a CO atmosphere to provide the acyl derivatives (Equation 3). Although these reactions required somewhat higher temperatures than those used for methyl migration in $(CO)_5MnCH_3$ (50–60°C vs. 25°C),[8] it did not appear that the electronic character of the ligand provided more than a minor retarding influence on this reaction. Likewise, there was no evidence that steric or coordinating properties of the ester group were significant in determining the rate of the reaction, since very similar results were obtained in complexes of alkoxymethyl ligands, such as $(CO)_5MnCH_2OCH_2CH_3$. One implication of these results is that the analogous hydroxymethyl complex, were it stable under these conditions, would also be expected to undergo the alkyl migration process.

A second part of our study involved the reaction of these acyloxymethyl complexes with H_2. King had reported that hydrogenolysis of the $(CO)_5MnCCH_3$ acyl complex produced acetaldehyde,[9] and we were interested in studying our complexes under similar conditions. It was found that exposure of the *alkyl* complexes to H_2 under relatively mild conditions (100 psig H_2, 75°C) gave good yields of *alcohol* products.[10] It appears that these reactions proceed through initial alkyl migration to give an acyl intermediate (Equation 4). (Only under very low pressures of H_2

$$(CO)_5MnCH_2OCR \longrightarrow (CO)_4MnCCH_2OCR \xrightarrow{[H]} HOCH_2CH_2OCR \quad (4)$$

could the CH_3OCR product be detected; presumably the slowness of the reaction under those conditions allows a competitive hydrogenolysis pathway involving CO loss to become observable.) Various alkyl groups were found to be hydroxymethylated in this manner (Equations 5 and 6).

$$(CO)_5MnCH_3 \xrightarrow{\ H_2\ } HOCH_2CH_3 \qquad (5)$$

$$(CO)_5MnCH_2OCH_2CH_3 \longrightarrow HOCH_2CH_2OCH_2CH_3 \qquad (6)$$

Since we were interested in the migratory aptitude of the hydride ligand, it seemed that the reaction of $HMn(CO)_5$ in the above manner could lead to a formyl intermediate, which could be hydrogenated to methanol. (It had been postulated that such a formyl was involved as an intermediate in CO substitution.)[11] Several experiments were therefore carried out under increasingly severe conditions, but

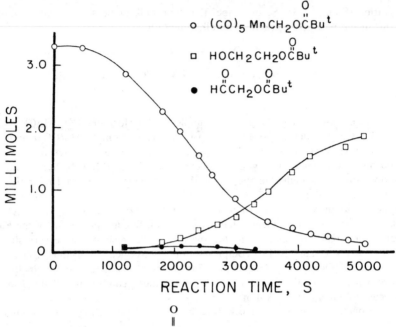

FIGURE 1. Reaction of $(CO)_5MnCH_2O\overset{O}{\overset{\|}{C}}Bu^t$ with H_2 at 65 psi, 75°C in sulfolane solvent.

only traces of methanol were ever observed (Equation 7). It appears that the migratory

$$(CO)_5MnH \xrightarrow[\times]{H_2} HOCH_3 \qquad (7)$$

ability of the hydride ligand in this complex is much lower than that of alkyl ligands (or perhaps the subsequent hydrogenolysis of a formyl is much more difficult than for acyl complexes).

The mechanism of hydrogenolysis in these reactions was uncertain, so some simple kinetic experiments were carried out. The $(CO)_5MnCH_2O\overset{O}{\overset{\|}{C}}Bu^t$ complex was allowed to react with H_2 under controlled conditions, and the concentrations of starting complex and organic products were monitored by IR spectroscopy and vapor phase chromatography (VPC), respectively. It was found (FIGURE 1) that an induction period preceded any observable reaction, and that when the reaction began it exhibited autocatalytic behavior. The initial organic product observed was the aldehyde $HC\overset{O}{\overset{\|}{C}}CH_2O\overset{O}{\overset{\|}{C}}Bu^t$, but this product was rapidly hydrogenated to the alcohol as the reaction proceeded (Equation 8). Carbon monoxide, when introduced

$$(CO)_5MnCH_2O\overset{O}{\overset{\|}{C}}Bu^t \xrightarrow{[H]} HC\overset{O}{\overset{\|}{C}}CH_2O\overset{O}{\overset{\|}{C}}Bu^t \xrightarrow{[H]} HOCH_2CH_2O\overset{O}{\overset{\|}{C}}Bu^t \qquad (8)$$

into the system, greatly retarded the hydrogenolysis and was apparently even more effective at inhibiting the *hydrogenation* process of Equation 8, since the aldehyde could then be obtained as the major product.

The autocatalytic behavior of the hydrogenolysis reaction suggested that this process did not involve the simple oxidative addition of H_2 to a coordinatively unsaturated metal acyl intermediate. Instead, it appeared more probable that H_2 activation was occurring on highly coordinatively unsaturated species generated by the reaction in the absence of CO. Studies of H_2 activation by manganese and rhenium carbonyls have been reported by Byers and Brown,[12] and application of their results to our reaction suggested the following pathway for the process:

$$(CO)_5Mn-R \;\rightleftharpoons\; (CO)_4Mn-\overset{\displaystyle O}{\overset{\displaystyle \|}{C}}R \qquad (9)$$

$$(CO)_4Mn-\overset{\displaystyle O}{\overset{\displaystyle \|}{C}}R + HMn(CO)_4 \;\longrightarrow\; (CO)_4Mn^\cdot + (CO)_4Mn\overset{\displaystyle H}{\overset{\displaystyle |}{}}\overset{\displaystyle O}{\overset{\displaystyle \|}{C}}R \qquad (10)$$

$$(CO)_4Mn\overset{\displaystyle H}{\overset{\displaystyle |}{}}\overset{\displaystyle O}{\overset{\displaystyle \|}{C}}R \;\longrightarrow\; H\overset{\displaystyle O}{\overset{\displaystyle \|}{C}}R + (CO)_4Mn^\cdot \qquad (11)$$

$$(CO)_4Mn^\cdot + H_2 \;\longrightarrow\; H_2Mn(CO)_4^{\,\cdot} \qquad (12)$$

$$H_2Mn(CO)_4^{\,\cdot} + (CO)_4Mn^\cdot \;\longrightarrow\; 2\,HMn(CO)_4 \qquad (13)$$

The observed induction period might be expected based on this scheme, since the reaction must somehow be initiated. Autocatalytic behavior would also be predicted because an additional active $(CO)_4Mn^\cdot$ species is generated as each molecule of aldehyde is produced. Presumably the hydrides involved in this process are hydrogenation catalysts as well, and are responsible for converting the aldehyde to alcohol product. As described above, the presence of CO retards the process, and it can be seen that it would have several opportunities to do so in this scheme. The observed Mn product in the presence of CO is $Mn_2(CO)_{10}$; the metal-containing products from reactions in the absence of CO have not been identified with any certainty.

To demonstrate that a hydrogen atom transfer of the type postulated in Equation 10 is feasible, we carried out reactions between manganese alkyl complexes and the relatively stable hydride $HMn(CO)_5$, with no H_2 present. These reactions were found to proceed similarly (Equation 14), producing the same

$$(CO)_5Mn-R + \text{excess } HMn(CO)_5 \;\longrightarrow\; HOCH_2R + H\overset{\displaystyle O}{\overset{\displaystyle \|}{C}}R \qquad (14)$$

aldehyde and alcohol products observed previously from reactions with H_2. Similar studies involving reactions of transition metal alkyls with hydrides have been reported by Bergman,[13,14] and Halpern has reported a more general study of reactions involving manganese hydrides with alkyl and acyl complexes.[15]

Our studies on this model system showed that acyloxymethyl complexes could be converted quite easily to two-carbon products even under mild conditions. We next hoped to produce the analogous manganese hydroxymethyl complex *in situ*, and show that it could react similarly. Reaction of $HMn(CO)_5$ under H_2

or H_2/CO with paraformaldehyde (at depolymerizing temperatures) was found to produce CH_3OH, but no two-carbon products could be detected. Under the same conditions, $(CO)_5MnCH_2O\overset{\overset{\displaystyle O}{\|}}{C}R$ complexes were found to give two-carbon glycol esters with high selectivity (Equation 8). These observations made it seem unlikely that a hydroxymethyl complex was actually produced in the reactions of formaldehyde, but suggested that the hydrogenation to methanol was proceeding through another intermediate (Equation 15).

$$(CO)_5MnH + CH_2O \begin{cases} (CO)_5Mn-OCH_3 \xrightarrow{[H]} HOCH_3 \\ \\ \cancel{(CO)_5Mn-CH_2OH} \longrightarrow HOCH_2CH_2OH \end{cases}$$

$$(15)$$

Mononuclear metal carbonyl complexes containing coordinated alkoxides are extremely rare and apparently quite reactive. We were unsuccessful in attempts to prepare $(CO)_5MnOCH_3$, but this certainly does not rule it out as an intermediate in catalysis. It was found that a phosphine-substituted analogue, $(CO)_3(dpe)MnOCH_2CH_3$ [dpe = bis(diphenylphosphino)ethane], could be prepared and isolated. Although this complex was also quite unstable, hydrogenolysis under H_2 did give the expected alcohol (ethanol) product.

Roth and Orchin have reported that $HCo(CO)_4$ reacts with CH_2O under CO to give a two-carbon product, glycolaldehyde, which is apparently produced through a hydroxymethyl intermediate.[16] Thus, reactions of transition metal hydrides with formaldehyde appear capable of proceeding in two different directions, depending on the character of the metal hydride (Equation 16). The most obvious difference between the manganese and cobalt carbonyl hydrides is in their acidities;

$$CH_2O + (CO)_nMH \begin{cases} \underset{n=4}{\overset{M=Co}{}} (CO)_nM-CH_2OH \longrightarrow HC\overset{\overset{\displaystyle O}{\|}}{}CH_2OH \\ \\ \underset{n=5}{\overset{M=Mn}{}} (CO)_nM-OCH_3 \longrightarrow HOCH_3 \end{cases}$$

$$(16)$$

$HCo(CO)_4$ is a strong acid in water,[17] while $HMn(CO)_5$ is much less acidic (and presumably more oxophilic), with a pKa of about 7.[18] Thus the direction of addition of the hydride ligand to CH_2O may depend on whether it assumes the characteristics of a proton or a hydride ion in the addition process.

In the ruthenium-catalyzed reduction of CO to be described below, it was found that carboxylic acids had a significant effect on product selectivity. We therefore attempted the reaction of $HMn(CO)_5$ with formaldehyde and H_2 in acetic acid solvent. This solvent was indeed found to influence the selectivity of the reaction; although only methanol product was previously observed [in tetrahydrofuran (THF) or sulfolane solvent], now ethylene glycol diacetate was found, in addition to methyl acetate (Equation 17). This two-carbon product is presumed to be formed through an acyloxymethyl intermediate, which could be carbonylated as described above. The possible mechanism of formation of this presumed acyloxy-methyl complex will be discussed below.

$$HMn(CO)_5 + CH_2O \Big< \begin{array}{l} \xrightarrow{\text{sulfolane}} (CO)_5Mn-OCH_3 \xrightarrow{H_2} HOCH_3 \\ \\ \xrightarrow{\text{HOAc}} (CO)_5Mn-CH_2OAc \xrightarrow[\text{HOAc}]{H_2} \overset{(AcO)}{HOCH_2CH_2OAc} \end{array} \quad (17)$$

Studies on this model system have provided information on the facility of migration of the acyloxymethyl group, the hydrogenolysis and hydrogenation of the resulting acyl ligand, and the reactions of formaldehyde with metal hydrides. Further information on the difficulty of metal hydride conversion to metal formyls (or organic products) was also obtained. The usefulness of these studies in under-standing a catalytic system should become apparent in the next section.

RUTHENIUM-CATALYZED CO REDUCTION IN CARBOXYLIC ACID SOLVENTS

Catalytic hydrogenation of CO by ruthenium catalysts to methanol and methyl formate has been described by Bradley[19] and also by King.[20] It was reported that no two-carbon products could be observed from this reaction. We had also obtained similar results, although traces of ethylene glycol could be detected in some reactions.[21] We also found that when the catalytic process was carried out in a carboxylic acid solvent, a significant amount of ethylene glycol ester could be observed in addition to the major methyl ester product (Equation 18), even when

$$CO + H_2 \xrightarrow{Ru_3(CO)_{12},\ HOAc} CH_3OAc + AcOCH_2CH_2OAc \quad (18)$$

the reaction was performed under relatively low pressures (3,000–5,000 psi).[22,23] Questions about the mechanism of this process and the involvement of the carboxylic acid stimulated some of the model studies already described. (Similar studies of catalysis in carboxylic acid solvents have been reported by Knifton,[24] although most of these reactions are complicated by the presence of added salts, which greatly modify the behavior of the system—see the following section.)

Other hydroxylic solvents and additional protonic acids were investigated, but carboxylic acids appeared to be unique in their ability to effectively promote formation of the two-carbon product. (Rates of CO hydrogenation in carboxylic acids were similar to those obtainable in other solvents.) The rate of formation of the glycol product was found to be approximately second order in acid concentration, but other responses (to H_2 and CO partial pressures and ruthenium catalyst concentration, for example) are essentially the same for one- and two-carbon product formation. It thus appeared that the same catalyst was involved in the generation of these two products, and that the acid might be involved in determining product selectivity at an intermediate stage of the reaction.

Studies of solutions during catalysis by high-pressure infrared spectroscopy showed that the major ruthenium complex observed was $Ru(CO)_5$. Together with the observed first-order dependence on Ru concentration,[19,22] this indicates that the catalyst remains mononuclear, and that the processes are intramolecular in the metal catalyst.

Reaction of $Ru(CO)_5$ with H_2 under pressure has been observed to produce the unstable $H_2Ru(CO)_4$.[25] Although this hydride was not observed in our high-pressure IR studies, it is presumed to be the initial metal hydride produced in this system. Metal formyl complexes are postulated to be intermediates in CO reduction,[1,2-6,22,26-28] even though (as pointed out above) their formation from metal hydrides appears to be a difficult process in many systems. (There are now two examples of metal formyls being produced by reactions of metal hydrides with CO.)[27,28] If a metal formyl is indeed an intermediate in this catalytic system, it must be produced by the intramolecular migration of a hydride ligand to coordinated CO, presumably according to Equation 19.

$$H_2Ru(CO)_4 \xrightarrow{\text{CO}} (CO)_4HRu-\overset{\overset{\text{O}}{\|}}{C}H \qquad (19)$$

Since acyl hydride complexes are known to reductively eliminate aldehydes,[29] it is not unreasonable to propose that the formyl hydride complex of Equation 19 could eliminate formaldehyde. (Although formaldehyde is a thermodynamically unfavorable product under these conditions,[3] a kinetically significant amount could perhaps be involved as an intermediate.[4] It should also be noted that *coordinated* formaldehyde should be significantly stabilized.) The possibility of a formaldehyde (perhaps coordinated) intermediate then presented a reasonable explanation for the selectivity of this catalytic system for methanol production (in the absence of carboxylic acid solvents). The formaldehyde moiety was perhaps being preferentially *hydrogenated* through a metal methoxide intermediate, rather than being *hydroformylated* through a hydroxymethyl complex (Equation 20). (Although the

$$
\begin{array}{c}
\overset{\overset{\text{H}}{|}}{(CO)_4Ru-OCH_3} \xrightarrow{\text{CO}} HOCH_3 \\
\nearrow \\
\overset{\overset{\text{H}\ \ \text{H}}{\diagdown\ /}}{\underset{\ }{(CO)_3Ru}}\overset{\text{CO}}{-}\overset{\overset{\text{O}}{\|}}{\underset{\overset{\diagup\diagdown}{\text{H}\ \ \text{H}}}{C}} \qquad (20) \\
\searrow \\
\overset{\overset{\text{H}}{|}}{(CO)_4Ru-CH_2OH} \xrightarrow{\text{CO}} \overset{\overset{\text{H}\ \ \text{O}}{|\ \ \|}}{(CO)_4Ru-CCH_2OH}
\end{array}
$$

hydroxymethyl complex shown in Equation 20 would also be able to reductively eliminate methanol, it has the opportunity to grow to a two-carbon product, which the methoxy complex cannot do.)

Reactions of formaldehyde with the ruthenium catalyst were therefore investigated under catalytic conditions (ca. 340 atm, 1:1 H_2/CO, 230°C). Paraformaldehyde depolymerizes at temperatures above 100°C,[30] so it was used as a conveniently handled precursor. It was found that in inert solvents such as sulfolane the formaldehyde was converted essentially quantitatively to methanol; i.e., it was simply hydrogenated. However, reactions in acetic solvent converted the formaldehyde to both C_1 and C_2 products, as shown in TABLE 1.

TABLE 1

EFFECT OF ADDING PARAFORMALDEHYDE TO A RUTHENIUM-CATALYZED
CO HYDROGENATION REACTION*

$(CH_2O)_x$ Added	C_1 Formed (mmol)	C_2 Formed (mmol)	C_1/C_2
—	52.2	1.37	38
167 mmol	162	7.53	21

* Conditions: 50 mL acetic acid solvent, 2.35 mmol Ru, 340 atm, $H_2:CO = 1$, 230°C, two hours.

These results are very similar to those obtained with $HMn(CO)_5$, as described above, suggesting that $H_2Ru(CO)_4$ reacts analogously with CH_2O (Equation 21).

$$(CO)_4RuH_2 + CH_2O \begin{array}{c} \nearrow (CO)_4HRu-OCH_3 \longrightarrow HOCH_3 \\ \\ \searrow_{\substack{HOAc \\ -H_2O}} (CO)_4HRu-CH_2OAc \xrightarrow[HOAc]{H_2} \overset{(AcO)}{} HOCH_2CH_2OAc \end{array} \quad (21)$$

The route by which formaldehyde would be converted to an acyloxymethyl complex is not certain. The formaldehyde may first be coordinated to the metal, and then be acylated by acetic acid (Equation 22). Approximate second-order dependence on acetic acid concentration is observed for the C_2-producing pathway

$$(CO)_3\overset{\displaystyle H \;\; H}{\underset{\displaystyle C}{Ru}} - \overset{O}{\underset{/\backslash}{\|}} \xrightarrow[-H_2O]{HOAc, CO} (CO)_4\overset{\displaystyle H}{\underset{}{Ru}}-CH_2OAc \quad (22)$$

(from H_2/CO),[22] suggesting that an acetic acid dimer or protonated acetic acid molecule could be the acylating agent. A formaldehyde ligand coordinated to vanadium has recently been shown to be acylated, in a very similar manner, by benzoyl chloride (Equation 23).[31]

$$(C_5H_5)_2V-\overset{O}{\underset{/\backslash}{\underset{C}{\|}}} + C_6H_5CCl \longrightarrow (C_5H_5)_2V\begin{array}{c}\nearrow CH_2O\overset{O}{\overset{\|}{C}}C_6H_5 \\ \searrow Cl\end{array} \quad (23)$$

It is also possible that the carboxylic acid reacts with the formaldehyde in an uncoordinated form, particularly in the reaction involving added paraformaldehyde. Methylene diacetate (see Reference 30, pages 345 and 350), for example, would be expected as an equilibrium component of such a system in acetic acid (Equation 24). This reactive ester could add to a ruthenium hydride, and could

$$2 AcOH + CH_2O \rightleftharpoons AcOCH_2OAc + H_2O \quad (24)$$

generate the acyloxymethyl intermediate by a second route (Equation 25). In fact,

$$(CO)_4RuH_2 + AcOCH_2OAc \longrightarrow (CO)_4HRuCH_2OAc + HOAc \quad (25)$$

TABLE 2
MOLE RATIO OF ESTERS (BY CHAIN LENGTH) PRODUCED IN CATALYSIS*

C_1/C_2	C_2/C_3
30–45	200–300

* Conditions: 50 mL acetic acid solvent, 2.35 mmol Ru, 340 atm, $H_2:CO = 1$.

it is possible that both pathways to this intermediate are involved. The existence of two routes to the acyloxymethyl complex could explain the observation that a different C_1/C_2 ratio is observed in the CO hydrogenation reaction when paraformaldehyde is added (TABLE 1).

An acyloxymethyl intermediate such as that of Equation 25 could reductively eliminate methyl ester product, but also has the opportunity to undergo chain extension by alkyl migration to coordinated CO (Equation 26). The initial product

$$(CO)_4HRu-CH_2OAc \underset{CO}{\overset{CH_3OAc}{\diagup\diagdown}} \qquad (26)$$
$$(CO)_4HRu-\overset{O}{\overset{\|}{C}}CH_2OAc \longrightarrow H\overset{O}{\overset{\|}{C}}CH_2OAc$$

expected after CO insertion and reduction elimination might be the glycolaldehyde ester shown, and this product was observed in the manganese model system described above. Pathways similar to those involved in the hydrogenation/hydroformylation of formaldehyde would also probably be operative in the further reaction of glycolaldehyde esters. Thus, the aldehyde could be hydrogenated to alcohol, or could be hydroformylated to the longer-chain aldehyde (Equation 27).

$$H\overset{O}{\overset{\|}{C}}CH_2OAc + (CO)_4RuH_2 \diagup\diagdown$$

$$(CO)_4HRu-OCH_2CH_2OAc \longrightarrow HOCH_2CH_2OAc$$

$$\overset{HOAc}{\underset{-H_2O}{}}$$

$$(CO)_4HRu-\overset{OAc}{\overset{|}{C}}HCH_2OAc \overset{CO}{\longrightarrow} H\overset{OOAc}{\overset{\|\,|}{C}}CHCH_2OAc \qquad (27)$$

Since this higher aldehyde could also be hydrogenated or hydroformylated, the process becomes a chain-growth sequence which proceeds through aldehyde intermediates. In fact, small amounts of the three-carbon product glycerol triacetate are observed to be formed by this system (TABLE 2). Since the three-carbon product is formed in such a small proportion, it is probable that the selectivity of branching to the metal-carbon-bonded intermediate in Equation 27 is considerably lower than at the formaldehyde stage, as in Equation 21. This may be rationalized on grounds of both the increased steric bulk of a higher aldehyde and the electronic differences in the aldehydes.

Since it appeared that the key to controlling selectivity to the two-carbon product lay in converting a formaldehyde-containing intermediate to the metal-carbon-bonded species, other means to induce this transformation were sought. One possibility appeared to be increasing the nucleophilicity of the metal, so that

TABLE 3
IONIC PROMOTERS IN RUTHENIUM-CATALYZED CO REDUCTION*

Salt	CH_3OH (mmol/hour)	$(CH_2OH)_2$ (mmol/hour)
None	13.9	undetected
KO_2CCH_3	55.8	1.5
K_3PO_4	60.0	1.9
KF	44.7	1
KCl	39.7	6.9
KBr	81.3	17.2
KI	202	43.5

* Conditions: 75 mL sulfolane solvent, 6 mmol Ru, 18 mmol salt, 850 atm, H_2:CO = 1, 230°C. (Adapted from Reference 32 with permission. Copyright 1982 American Chemical Society.)

it would add more preferentially to the (electrophilic) aldehyde carbon atom. Anionic promoters that could possibly coordinate to the metal were therefore investigated, and inert solvents were chosen since ester products are less desirable than free alcohols. Results of these experiments were indeed encouraging, although perhaps not for the reasons that led to their conception.

REDUCTION OF CO BY LEWIS BASE–PROMOTED RUTHENIUM CATALYSTS

Reactions of $Ru_3(CO)_{12}$ with ionic promoters, and especially halide salts, did give improved yields of the glycol product, and also provided substantially improved catalytic activities (TABLE 3). Best results were obtained with iodide salts.[32] and it was found that the cation normally had little effect on the process. Major products are methanol, ethylene glycol, and ethanol, along with smaller amounts of glycerol, acetaldehyde, methane, and other minor products.

It was soon concluded that this catalytic system was very different from that observed in the absence of ionic promoters, either in carboxylic acid or inert solvents. Studies of catalytic solutions, both during and after reaction, showed the presence of mainly two ruthenium complexes, $HRu_3(CO)_{11}^-$ and $Ru(CO)_3I_3^-$. Iodide stoichiometry studies also support the presence of mainly these complexes during catalysis.[32] The existence of the $HRu_3(CO)_{11}^-$ cluster under high pressures of H_2/CO was initially quite surprising, since many metal carbonyl clusters [including $Ru_3(CO)_{12}$][25] are easily fragmented by CO under relatively mild conditions.

It was found that these two complexes could be formed from $Ru_3(CO)_{12}$ and ionic iodides under H_2 at mild conditions (Equation 28). Neither of these two complexes alone was found to possess the catalytic activity provided by their

$$\tfrac{7}{3}Ru_3(CO)_{12} + 3\,I^- + H_2 \longrightarrow 2\,HRu_3(CO)_{11}^- + Ru(CO)_3I_3^- + 3\,CO \quad (28)$$

mixtures. No interaction between these complexes could be observed spectroscopically under mild conditions.

One possible way in which these complexes could interact under catalytic conditions that initially appeared plausible is by hydride transfer from the cluster to a carbonyl ligand on $Ru(CO)_3I_3^-$. Shore and co-workers have shown that under the right conditions, the cluster hydride behaves as a hydride donor.[33] Since the negative charge on the $Ru(CO)_3I_3^-$ complex would not favor such a reaction, prior conversion to $Ru(CO)_4I_2$ (Equation 29) might be required (although the

$$Ru(CO)_3I_3^- + CO \rightleftharpoons Ru(CO)_4I_2 + I^- \qquad (29)$$

equilibrium apparently lies toward the ionic complex). A reaction does occur between $HRu_3(CO)_{11}^-$ and $Ru(CO)_4I_2$ under mild conditions (25°C, 1 atm N_2), but no organic products or metal formyl intermediates are observable. It appears that this reaction is a redox process and not a conversion directly related to catalysis. Furthermore, kinetic observations made on the catalytic system indicate that the simple interaction of these two complexes in the glycol-forming process is unlikely.

A relatively high pressure dependence of the catalytic reaction (third to fourth order for the glycol-forming reaction) initially suggested that cluster fragmentation might be involved in the process. Declusterification of $HRu_3(CO)_{11}^-$ by CO would lead to $HRu(CO)_4^-$ (Equation 30), a mononuclear hydride that has recently been

$$HRu_3(CO)_{11}^- + 3\ CO \rightleftharpoons HRu(CO)_4^- + 2\ Ru(CO)_5 \qquad (30)$$

isolated and studied by Ford and Walker.[34] Analogous declusterification of $HFe_3(CO)_{11}^-$ has been previously described,[35] and we have found that this iron cluster is totally fragmented under 500 psi of CO at 60°C. If such a process does occur with $HRu_3(CO)_{11}^-$ under our catalytic conditions, the cluster appears to be favored by the equilibrium since it is the major species observed by high-pressure IR during catalysis.

Reaction of $HRu(CO)_4^-$ with $Ru(CO)_4I_2$ has not yet been observed to produce a metal formyl product, even when followed by ^1H (or ^2H) NMR at low temperatures. For example, combination of the two complexes at $-60°C$ in acetone leads to gradual disappearance of the $HRu(CO)_4^-$ complex and the production of several metal hydride resonances. No signal could be observed in the region of δ 12–17, where metal formyl resonances normally are seen.[26] A reaction between $Ru(CO)_4I_2$ and $LiDB(CH_2CH_3)_3$ in THF at $-60°C$ (Equation 31) did immediately generate a signal at δ 13.65, which disappeared within five minutes at this temperature.

$$Ru(CO)_4I_2 + LiDB(CH_2CH_3)_3 \xrightarrow{-60°C} Ru(CO)_3I_2(CDO)^-\ (?) \qquad (31)$$

If this signal is indeed due to the formyl depicted in Equation 31, the stability of this complex must be very low under these conditions. The fact that such a product could not be observed in the slower reaction between $Ru(CO)_4I_2$ and $HRu(CO)_4^-$ and $Ru(CO)_4I_2$ at $-60°C$ is therefore not surprising.

Because the ruthenium formyl could not be observed in this reaction, we chose to investigate the reaction between $HRu(CO)_4^-$ and a model for $Ru(CO)_4I_2$ that would produce a more stable formyl product. The rhenium complex $[(C_5H_5)Re(CO)_2(NO)]PF_6$ was chosen since it has been reported to give a relatively stable formyl upon borohydride reduction.[36–38] The reaction of Equation 32 does indeed generate the known rhenium formyl product, as well as $(C_5H_5)Re(CO)(NO)H$.

$$3\ HRu(CO)_4^- + 2\ (C_5H_5)Re(CO)_2(NO)^+$$
$$\xrightarrow{25°C} HRu_3(CO)_{11}^- + 2\ (C_5H_5)Re(CO)(NO)(CHO) + CO \qquad (32)$$

The yield of the formyl can be as high as 30%. The rhenium hydride appears

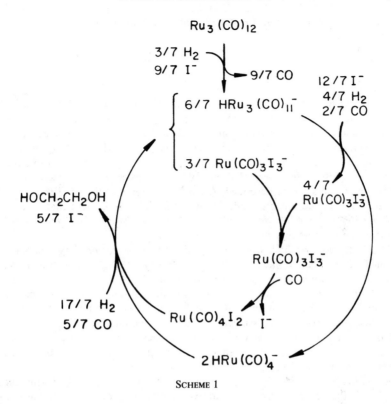

SCHEME 1

to be formed by a separate pathway not involving the formyl complex. This pathway could include electron-transfer processes (for related examples, see Reference 13), and such a pathway is probably involved in the reaction of $HRu_3(CO)_{11}^-$ with $(C_5H_5)Re(CO)_2(NO)^+$. This reaction does not produce a detectable amount of formyl product, but produces only the rhenium hydride and $Ru_3(CO)_{12}$ (Equation 33). This reaction is slower than that of $HRu(CO)_4^-$, and

$$HRu_3(CO)_{11}^- + (C_5H_5)Re(CO)_2(NO)^+$$
$$\xrightarrow{\text{25°C}} Ru_3(CO)_{12} + (C_5H_5)Re(CO)(NO)H \quad (33)$$

is retarded by a CO atmosphere. It therefore appears that the mononuclear hydride is considerably more reactive than the cluster, and is a substantially better hydride donor.

A catalytic cycle based on the reaction of $Ru(CO)_4I_2$ with $HRu(CO)_4^-$ is shown in SCHEME 1. Such a scheme appears to be consistent with kinetic and other experimental observations on the system, assuming that the rate-determining step occurs soon after the reaction of two $HRu(CO)_4^-$ ions with $Ru(CO)_4I_2$. Although the precise nature of this interaction is not certain, the generation of a ruthenium formyl by intermolecular hydride transfer from $HRu(CO)_4^-$ appears to be a plausible first step in the CO reduction process. The possibility exists that this pathway to a formyl intermediate, rather than an intramolecular hydride migration, could be responsible for the high activity of this system for CO hydrogenation.

CONCLUSIONS

It is evident that metal hydrides are involved in many aspects of CO hydrogenation. Of course, in order for a system to be catalytic, the metal hydrides involved must be regenerable from molecular hydrogen. Once on the metal, the hydrogen atom must be transferred to a carbonyl ligand to initiate CO reduction. Pathways from metal hydrides to formyls involving intramolecular hydride migration and intermolecular hydride transfer are implicated in two different catalytic systems described. Although the reasons that one route would be preferred over the other are not entirely clear, it seems evident that a reaction involving ionic complexes, specifically anionic metal hydrides, will have a better opportunity to proceed by the hydride transfer process.

Hydrogenation or hydroformylation of a possible formaldehyde-containing intermediate can determine the product selectivity of a CO reduction process. At this stage, it would appear that the more acidic hydrides will be more inclined to lead to hydroformylation, and the hydridic hydrides will more probably hydrogenate an aldehyde intermediate. Finally, an organic product must be released from the metal complex. Bimolecular reductive elimination by reaction of a transition metal hydride with a metal alkyl or acyl derivative may be involved in some CO reduction processes, and is a probable step in the stoichiometric model system involving manganese alkyls described above.

Metal hydrides, it appears, must serve in many roles in a catalytic reaction as complex as the conversion of CO and H_2 to organic products. It is the search for new systems containing metal hydrides that will accomplish all of these steps, and the study of why known systems are successful that makes this area of research such an interesting and challenging pursuit.

EXPERIMENTAL METHODS

Manganese acyloxymethyl complexes were prepared as described previously.[10] The hydride $HMn(CO)_5$ was obtained by the procedure of King.[39] Kinetics of hydrogenolysis were followed on reactions under pressure in Fischer-Porter bottles; samples were periodically withdrawn through a sample tube and analyzed by VPC and IR. The complexes $PPN[HRu_3(CO)_{11}]$,[40] $PPN[HRu(CO)_4]$,[34] $PPN[Ru(CO)_3I_3]$ [PPN = bis(triphenylphosphine)iminium],[41] and $[(C_5H_5)Re-(CO)_2(NO)]PF_6$[42] were prepared by modifications of literature procedures. Commercial paraformaldehyde was dried over P_4O_{10} in a vacuum desiccator for at least seven days before use.

High-pressure infrared spectra were recorded under reaction conditions by use of the apparatus described elsewhere.[43]

Reactions in carboxylic acids and related catalytic studies were performed in 500 mL stainless steel bombs containing glass liners of approximately 125 mL capacity. Agitation was accomplished by rocking during reaction. Reactions with Lewis base–promoted Ru catalysts were carried out in stainless steel autoclaves of 125 mL nominal volume (Autoclave Engineers) stirred by a magnetically driven turbine.

Analyses for organic products were carried out by VPC on Chromosorb 101 or Tenax columns. Identity of products was confirmed by IR, NMR, and mass spectrosocopy. Samples were also hydrolyzed (when necessary) and converted to derivatives such as glycol dibenzoate. Several acetic acid solutions were examined for glycerol triacetate content. The mixtures were filtered, diluted with H_2O, neutralized with $NaHCO_3$, and extracted into ether. Concentration of this solution gave a mixture rich in glycol diacetate, which also contained traces of the glycerol ester. Analyses were performed by VPC.

A reaction with 1.0 g $Mn_2(CO)_{10}$, 5 g paraformaldehyde, and 10 mL of acetic acid in 30 mL of sulfolane under 310 atm of 1:1 H_2/CO at 210°C (three hours) gave 4.69 g of methyl acetate and 0.10 g of ethylene glycol diacetate. An identical reaction using 10 mL of acetic anhydride instead of acetic acid yielded 2.70 g of methyl acetate and 0.57 g of glycol diacetate. Similar experiments in the absence of acetic acid or anhydride produced no ethylene glycol

$(CO)_3(dpe)MnOCH_2CH_3$ [dpe = 1,2-bis(diphenylphosphino)ethane]. This complex was prepared by a three-step procedure from $Mn_2(CO)_{10}$ through $(CO)_3(dpe)MnH$ and $(CO)_3(dpe)MnOSO_2CF_3$. Reaction of $Mn_2(CO)_{10}$ (4.2 g, 10.8 mmol) with dpe (8.0 g, 20.1 mmol) in 100 mL of refluxing n-propanol (16 hours) gave a yellow-orange solid after cooling. Extraction into hot acetone and evaporation of this solution gave 6.50 g (60%) of quite pure $(CO)_3(dpe)Mn(CO)_3H$: IR (CH_2Cl_2) 2000 m, 1918vs cm^{-1}. *Analysis.* Calculated: C, 64.56; H, 4.64%. Observed: C, 64.61; H, 4.91%.

This complex (4.0 g, 7.4 mmol) was dissolved in CH_2Cl_2 (60 mL) and anhydrous ethanol (30 mL). To this solution was added dropwise a solution of CF_3SO_3H (1.10 g, 7.4 mmol) in 4 mL of dichloromethane. After stirring for 15 minutes the solution was concentrated under vacuum, which produced 4.29 g (87%) of bright yellow crystalline $(CO)_3(dpe)MnOSO_2CF_3$: IR (CH_2Cl_2) 2020s, 1979vs, 1936vs cm^{-1}. *Analysis.* Calculated: C, 52.40; H, 3.49. Observed: C, 52.79; H, 3.64.

A suspension of $(CO)_3(dpe)MnOSO_2CF_3$ (5 g, 7.1 mmol) in 450 mL of hexane was stirred with $NaOCH_2CH_3$ (8.5 mmol) in 20 mL ethanol for five hours at 25°C. This solution was filtered through diatomaceous earth and evaporated to dryness to yield 2.86 g (69%) of $(CO)_3(dpe)MnOCH_2CH_3$: IR (hexane) 2016vs, 1948s, 1908s cm^{-1}. NMR (acetone-d_6)δ 7.68, 7.40 (10) C_6H_5; 2.9 m (4)—CH_2CH_2—; 2.87 q (2)—CH_2CH_3; 0.57 t (3)—CH_2CH_3; (J = 6.8 Hz). *Analysis.* Calculated: C, 63.92; H, 4.98. Observed: C, 63.63; H, 5.08.

ACKNOWLEDGMENTS

I am thankful to all of the individuals who were involved in this research, especially C. P. Wolfe, R. B. James, T. D. Myers, and N. Poppelsdorf.

REFERENCES

1. FEDER, H. M. & J. W. RATHKE. 1980. Ann. N.Y. Acad. Sci. **333**: 45.
2. PRUETT, R. L. 1977. Ann. N.Y. Acad. Sci. **295**: 239.

3. MUETTERTIES, E. L. & J. STEIN. 1979. Chem. Rev. **79**: 479.
4. FAHEY, D. R. 1981. J. Am. Chem. Soc. **103**: 136.
5. MASTERS, C. 1979. Adv. Organomet. Chem. **17**: 61.
6. GLADYSZ, J. A., J. C. SELOVER & C. E. STROUSE. 1978. J. Am. Chem. Soc. **100**: 6766.
7. CASEY, C. P., M. A. ANDREWS, D. R. MCALISTER, W. D. JONES & S. G. HARSY. 1981. J. Mol. Catal. **13**: 43.
8. CALDERAZZO, F. 1977. Angew. Chem. Int. Ed. Engl. **16**: 299.
9. KING, R. B., A. D. KING, M. Z. IQBAL & C. C. FRAZIER. 1978. J. Am. Chem. Soc. **100**: 1687.
10. DOMBEK, B. D. 1979. J. Am. Chem. Soc. **101**: 6466.
11. BYERS, B. H. & T. L. BROWN. 1977. J. Organomet. Chem. **127**: 181.
12. BYERS, B. H. & T. L. BROWN. 1977. J. Am. Chem. Soc. **99**: 2527.
13. JONES, W. D., J. M. HUGGINS & R. G. BERGMAN. 1981. J. Am. Chem. Soc. **103**: 4415.
14. JONES, W. D. & R. G. BERGMAN. 1979. J. Am. Chem. Soc. **101**: 5447.
15. NAPPA, M. J., R. SANTI, S. P. DIEFENBACH & J. HALPERN. 1982. J. Am. Chem. Soc. **104**: 619.
16. ROTH, J. A. & M. ORCHIN. 1979. J. Organomet. Chem. **172**: C27.
17. STERNBERG, H. W., I. WENDER, R. A. FRIEDEL & M. ORCHIN. 1953. J. Am. Chem. Soc. **75**: 2717.
18. SCHUNN, R. 1971. *In* Transition Metal Hydrides. E. L. Muetterties, Ed.: 239. Marcel Dekker. New York, N.Y.
19. BRADLEY, J. S. 1979. J. Am. Chem. Soc. **101**: 7419.
20. KING, R. B., A. D. KING & K. TANAKA. 1981. J. Mol. Catal. **10**: 75.
21. DOMBEK, B. D. 1979. European Patent Application 13008, to Union Carbide Corp.
22. DOMBEK, B. D. 1981. J. Am. Chem. Soc. **102**: 6855.
23. DOMBEK, B. D. 1981. Am. Chem. Soc. Symp. Ser. **152**: 213.
24. KNIFTON, J. F. 1982. J. Catal. **76**: 101.
25. WHYMAN, R. 1973. J. Organomet. Chem. **56**: 339.
26. GLADYSZ, J. A. 1982. Adv. Organomet. Chem. **20**: 1.
27. FAGAN, P. J., K. G. MOLOY & T. J. MARKS. 1981. J. Am. Chem. Soc. **103**: 6959.
28. WAYLAND, B. B., B. A. WOODS & R. PIERCE. 1982. J. Am. Chem. Soc. **104**: 32.
29. NORTON, J. R. 1979. Acc. Chem. Res. **12**: 139.
30. WALKER, J. F. 1964. Formaldehyde: 44. Reinhold Publishing Corp. New York, N.Y.
31. GAMBAROTTA, S., C. FLORIANI, A. CHIESI-VILLA & C. GUASTINI. 1982. J. Am. Chem. Soc. **104**: 2019.
32. DOMBEK, B. D. 1981. J. Am. Chem. Soc. **103**: 6508.
33. BRICKER, J. C., C. C. NAGEL & S. G. SHORE. 1982. J. Am. Chem. Soc. **104**: 1444.
34. WALKER, H. W. & P. C. FORD. 1981. J. Organomet. Chem. **214**: C43.
35. WADA, F. & T. MATSUDA. 1973. J. Organomet. Chem. **61**: 365.
36. CASEY, C. P., M. A. ANDREWS, D. R. MCALISTER & J. E. RINZ. 1980. J. Am. Chem. Soc. **102**: 1927.
37. TAM, W., W. WONG & J. A. GLADYSZ. 1979. J. Am. Chem. Soc. **101**: 1589.
38. SWEET, J. R. & W. A. G. GRAHAM. 1982. J. Am. Chem. Soc. **104**: 2811.
39. KING, R. B. 1965. Organometallic Synthesis. **1**: 158. Academic Press, Inc. New York, N.Y.
40. JOHNSON, B. F. G., J. LEWIS, P. R. RAITHBY & G. SUSS. 1979. J. Chem. Soc. Dalton Trans.: 1356.
41. CLEARE, M. J. & W. P. GRIFFITH. 1969. J. Chem. Soc. A: 372.
42. FISCHER, E. O. & H. STRAHMETZ. 1968. Z. Naturforsch. B **23**: 278.
43. WALKER, W. E., L. A. COSBY & S. T. MARTIN. 1975. U.S. Patent 3886364, to Union Carbide Corp.

NICKEL HYDRIDES: CATALYSIS IN OLIGOMERIZATION AND POLYMERIZATION REACTIONS OF OLEFINS

W. Keim

Institute for Technical and Petroleum Chemistry
Rhine-Westphalia Technical University
D-5100 Aachen, Federal Republic of Germany

The oligomerization of olefins is of considerable academic and industrial interest for the synthesis of linear and branched higher monoolefins. The linear olefins are key intermediates for detergents, plasticizers, and a variety of fine chemicals. The branched ones, which can also be used for plasticizers, represent potential candidates for improving octane numbers. Various mechanisms for the oligomerization are discussed:[1] a mechanism based on cyclic intermediates (Equation 1); a mechanism based on bi- or multimetallic intermediates (Equation 2); a mechanism based on 1,2-hydrogen shift equilibria (Equation 3); and a mechanism based on β-hydrogen elimination (FIGURE 1).

$$L_nM + 2CH_2{=}CHR \longrightarrow L_nM \underset{}{\Big\langle} \!\!\!\!{-}R_2 \longrightarrow \text{oligomers} \qquad (1)$$

$$L_nM{-}ML_n + nCH_2{=}CHR \longrightarrow L_nM\,(-CH_2{-}\overset{\overset{\textstyle R}{|}}{C}H)_n{-}ML_n$$
$$\longrightarrow \text{oligomers} \qquad (2)$$

$$\overset{\overset{\textstyle R_2CH}{|}}{M}CH_2{=}CHR \longrightarrow \overset{\overset{\textstyle CR_2\ \ H}{\diagdown\!\!\diagup}}{M}CH_2{=}CHR \longrightarrow M{-}CH_2{-}CHR{-}CR_2 \quad (3)$$
$$\downarrow$$
$$\text{oligomers}$$

The β-hydrogen elimination mechanism shown in FIGURE 1 is the most generally

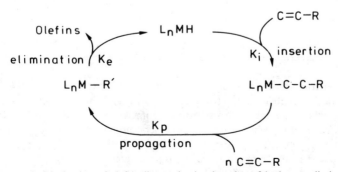

FIGURE 1. Mechanism of olefin oligomerization based on β-hydrogen elimination.

191

accepted one. The first step comprises the insertion of an olefin into a metal-hydrogen bond. Further insertion of olefins into the metal-alkyl bond leads to growth products (propagation step). The elimination of the thus formed oligomers occurs by a β-hydrogen elimination returning to L_nM-H, which reenters the catalytic cycle. Owing to the ease of β-hydrogen elimination, complexes of Type 1 could not be isolated, but in complexes of norbornene, in which a β-hydrogen elimination is more difficult, 2 could be isolated. An analogous complex 3, in which $\eta^2-\eta^1$ bonds are present, has been published.

The distribution of olefins in FIGURE 1 is determined by Equation 4:

$$x_p = \frac{\beta}{(1 + \beta)^p} \tag{4}$$

where x_p = share of olefins of polymerization amount p, and β = quotient of reaction rates of growth and displacement.

Based on Equation 4, TABLE 1 gives C-number distributions for the ethylene oligomerization at various β-values.

TABLE 1
C-NUMBER DISTRIBUTION FOR ETHYLENE OLIGOMERIZATION

	$\beta = 2.0$ Weight (%)	$\beta = 0.25$ Weight (%)
C_4	52.33	6.67
C_6	26.56	8.0
C_8	11.84	8.53
C_{10}	4.94	8.53
C_{12}	1.98	8.19
C_{14}	0.77	7.65
C_{16}	0.29	6.99
C_{18}	0.11	6.29
C_{20}	—	5.59
C_{20}^+	—	33.56

In general, three types of distribution can be considered: $k_p \gg k_e$, when a considerable number of insertion steps occur and higher olefins are formed; $k_p \approx k_e$, when oligomers with a broad geometric molecular weight distribution result; and $k_p \ll k_e$, when dimers and trimers dominate. Which of these reactions prevails is dependent on the nature of the metal and its oxidation state, on the nature of the ligand, and on the reaction parameters.

The simplified catalytic cycle illustrated in FIGURE 1 does not distinguish between

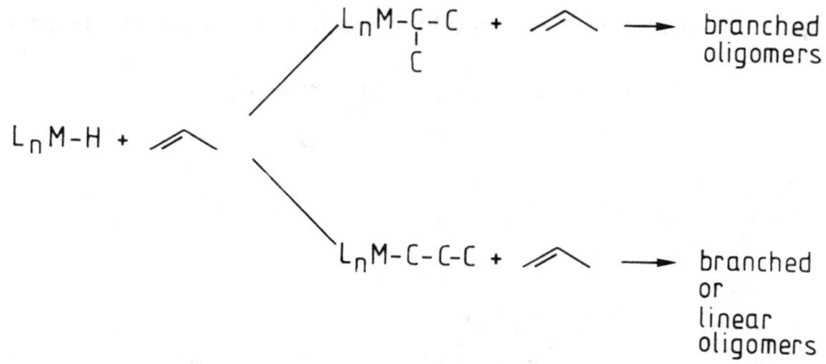

$$L_nM\text{-}\underset{\underset{C}{|}}{C}\text{-}C \ + \ \diagup\!\!\!\diagdown \ \longrightarrow \ \text{branched oligomers}$$

$$L_nM\text{-}H \ + \ \diagup\!\!\!\diagdown$$

$$L_nM\text{-}C\text{-}C\text{-}C \ + \ \diagup\!\!\!\diagdown \ \longrightarrow \ \text{branched or linear oligomers}$$

FIGURE 2. Markownikov addition of metal hydrides to propylene.

the two possible modes of addition of a metal-carbon or a metal-hydrogen bond to an unsymmetrically substituted olefin, namely, the Markownikov and anti-Markownikov addition (Equation 5). The direction of this addition is very important for obtaining linear or branched products. This is exemplified for propene in FIGURE 2.

$$L_nM-H + C{=}C-R \overset{\text{anti-Mark.}}{\underset{\text{Mark.}}{\Bigg[}} \quad \begin{array}{l} \longrightarrow L_nM-C-C-R \\[4pt] \longrightarrow L_nM-\underset{\underset{C}{|}}{C}-R \end{array} \tag{5}$$

The course of the metal hydride addition can be significantly influenced by the nature of the metal, the ligands applied, and the reaction conditions. Catalysts that yield exclusively linear products with α-olefins are unknown.

Some years ago, we initiated a program to investigate oligomerizations of α-olefins by nickel complexes. For our catalyst selection the model shown in FIGURE 3 was

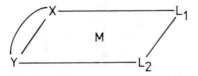

FIGURE 3. Model for α-olefin oligomerization. Square planar. X = Y; X = donor; Y = acceptor; M = Ni. L_1 and L_2 can dissociate easily.

chosen. Square planar nickel complexes provide appropriate orbitals to interact with incoming olefins. Chelate ligands favor square planar structures. In addition, the chelate should minimize the chance of coordinating olefins formed during the catalytic cycle. X and Y were chosen among soft and hard ligands. The ligands L_1 and L_2 should easily dissociate, thus providing empty coordination sites for the olefins to be oligomerized. The effect of the change on the metal and the attainment of an 18-valence shell of electrons are also two strong forces in determining preferred coordination number. Low coordination numbers are common for the latter transition elements in their low oxidation states.

Complex 4 (Equation 6) possesses all the requirements discussed.[2] In analogy

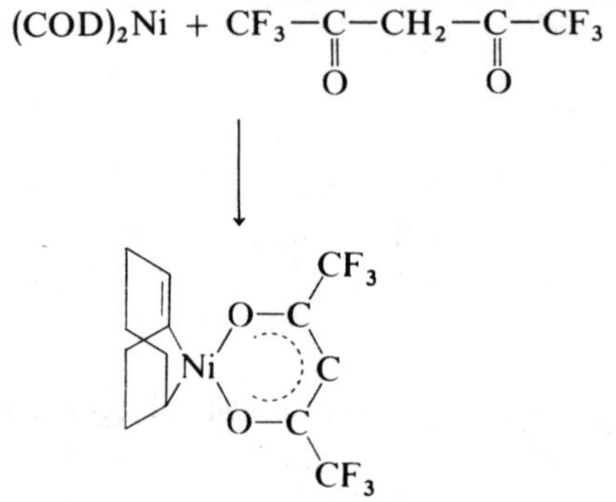

$$(COD)_2Ni + CF_3-\underset{\underset{O}{\|}}{C}-CH_2-\underset{\underset{O}{\|}}{C}-CF_3$$

(6)

4

to the methyl derivative of **4**, Complex 4 should also be square planar.[3]

Assuming a cyclooctadiene ligand dissociation, a nickel hydride complex should be formed. Indeed, reaction of **4** with $P(C_6H_{11})_3$ gave Complex 5, namely, $[(PC_6H_{11})_3]_2(CF_3COCH-COCF_3)NiH$,[4, 5] which upon reaction with $NaBH_4$ could be converted into the known complex $[(C_6H_5)_3P]_2NiH_3BH_2$. Reaction of **4** with ethylene produced oligomers as shown in TABLE 2.[2] The olefins are highly

TABLE 2
OLIGOMERIZATION OF ETHYLENE WITH NICKEL COMPLEX 4

C-Number	Weight (%)	Linearity (%)
4	21	100
6	38	82
8	20	77
10	10	75
12	4.8	75
14	2.6	77
16	1.5	78
18	1.2	80
20	0.7	—
22	0.3	—
24	0.1	—

isomerized, supporting the general observation that acidic nickel hydrides isomerize double bonds.

Complex 4 also reacts with α-olefins such as propene, 1-butene, 1-pentene, and 1-octene. The well-known order of reactivity ethylene $\gg$ propylene > 1-butene $\sim$ 1-pentene > 1-octene was confirmed. TABLE 3 summarizes data obtained with 1-butene.[5] At high conversions the linearity goes down, indicative of cooligomerizations with 2-butene formed via isomerization.

A mechanism similar to FIGURE 1 is supported by the ^1H NMR of **4** with ethylene, which shows a signal at -16.3 ppm characteristic of an Ni—H bond. In addition, the ^1H NMR shows a doublet at 0.6 ppm, indicative of an Ni-alkyl

TABLE 3
1-BUTENE OLIGOMERIZATION WITH COMPLEX 4

Conversion (%)	Selectivity (%)		Linearity (%)	
	C_8	C_{12}	C_8	C_{12}
20	95	5	82	65
40	93	7	73	63
50	90	10	70	60
70	87	12	67	50
80	85	14	65	40
95	82	17	60	35

group.

In a small pilot plant, the potential technical feasibility of the 1-butene oligomerization to 60% linear dimers is supportable.[4]

In order to obtain further insight into the function of the β-diketonate and the 4-enyl ligand of 4, ligand variations were carried out. A great variety of α-diketonate and β-diketonate ligands were tested.[6, 7] A direct relationship to the acidity of the diketonate applied could be demonstrated, as shown in FIGURE 4.

Interestingly, the 4-enyl part in 4 is also of importance in obtaining active catalysts. Complexes 6, 7, and 8 are practically inactive in the olefin oligomerization. All attempts to isolate a nickel hydride complex similar to 5 by adding $P(C_6H_{11})_3$ failed (FIGURE 5).

These results support the idea that catalytically active systems are found among complexes that easily form an Ni—H bond. Catalytic activity for oligomerizations

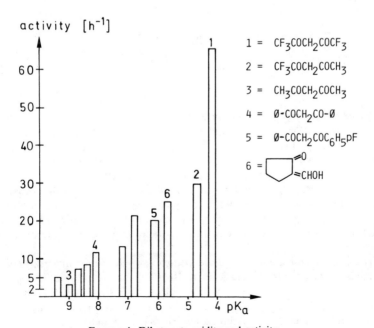

FIGURE 4. Diketonate acidity and activity.

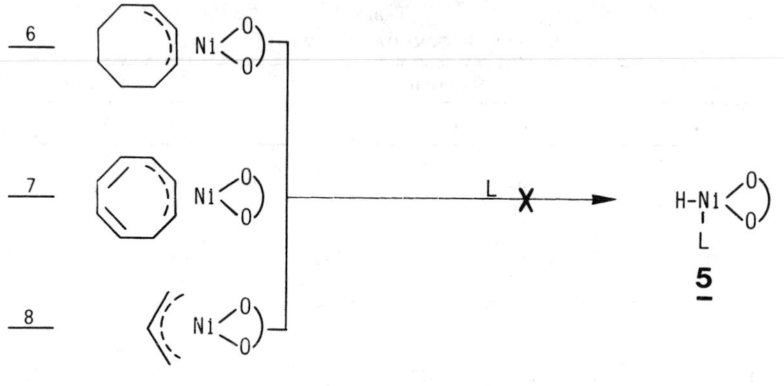

L = alkyl phosphine

0 0 = hexafluoroacetyl acetonate

FIGURE 5. Reaction of η^3-allyl complexes with $P(C_6H_{11})_3$.

may be related to the ease of metal hydride formation. Thus **8** is inactive because the elimination of allene to form a nickel hydride is unfavorable.

So far, only chelates in which X = Y = 0 have been discussed. Complexes **9**,[8] **10**,[9,10] and **11**[11] contain chelates of a P O combination (Ø = Ph). An x-ray analysis

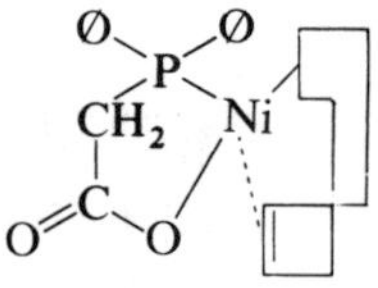

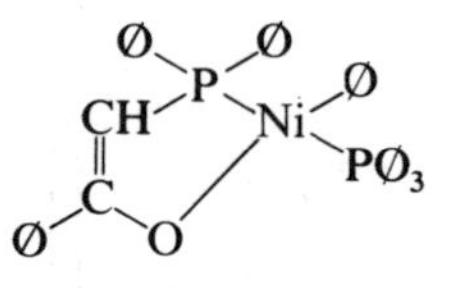

confirmed the square planar structure of **10**.

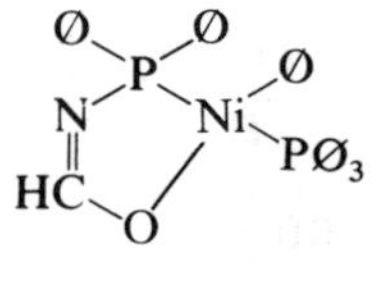

Toluene solutions of **9**, **10**, and **11** react at 90°C and 20 bar pressure with ethylene, yielding oligomers of 99% linearity and >98% α-olefin content. All three complexes are active in the Shell higher olefin plant (SHOP) process, which is used industrially in various plants to manufacture α-olefins.[12-14]

It is noteworthy that complexes of Types 9 and 10 practically do not isomerize double bonds. This could be confirmed by reacting 1-octene with **10** at 70°C for 24 hours. Only 10% of the 1-octene was isomerized to 2-octene. A self-explanatory nickel hydride mechanism elucidated in FIGURE 6 is proposed to account for the α-olefins obtained.

The isolation of a nickel hydride complex would be of interest in view of mechanistic considerations, and work to isolate nickel hydrides is in preparation. However, support for this mechanism is gained from the following observations: upon reacting **9** with ethylene, a ^{1}H NMR signal at -15.2 ppm characteristic of a nickel hydride is found; a set of three high-field ^{1}H NMR signals (0.7 ppm, 0.15 ppm, and -0.5 ppm) can be attributed to a nickel-alkyl group.

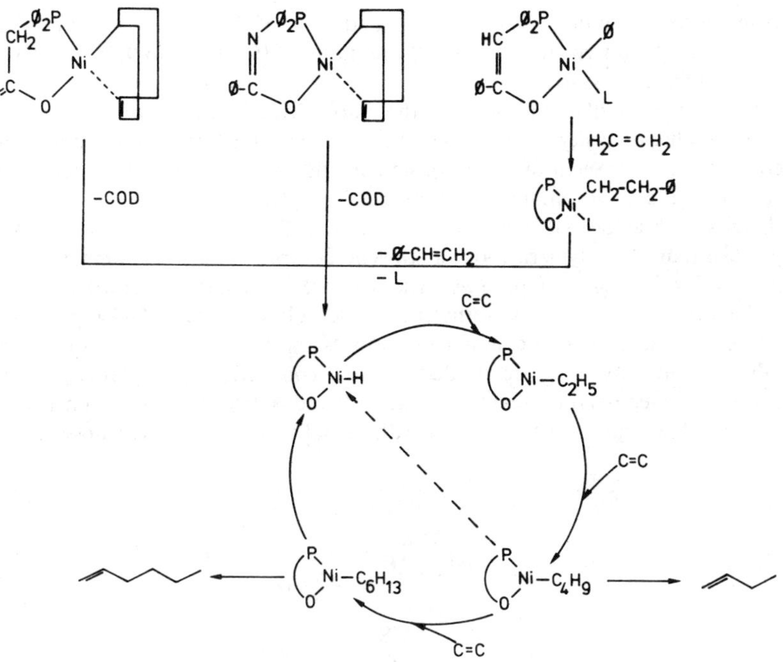

FIGURE 6. Nickel hydride mechanism for ethylene oligomerization.

A key step in the industrial utilization of ethylene oligomerization is the carbon-number distribution. In general the geometric carbon-number distribution can be altered by lowering the temperature and by increasing the pressure. A further handle to tailor the carbon-number distribution is given by adding various amounts of Ph_3P, as is evident from TABLE 4. By increasing the amount of Ph_3P added, the β-values increase indicating a shift of the carbon-number distribution toward $<C_{10}$ α-olefins.

There is some discussion in the literature dealing with β-elimination. Generally, vacant coordination sites ease β-elimination, and the dissociation of a ligand can

TABLE 4
β-VALUES AS A FUNCTION OF Ph_3P*

PPh₃/Complex 10 (molar)	β-Value
Pure complex 10	0.3
0.2	0.8
1	2.0
2	4.0
4	7.4
6	8
10	9

* Reaction conditions: 50°C, 11 hours, 50 bar, 140 mg complex.

become rate determining. This implies that addition of strong donor ligands should suppress the oligomerization. Indeed, the addition of $(C_2H_5)_3P$ to **9**, **10**, or **11** yields completely inactive systems.

Interestingly, molar additions of the bulky ligand (α-naphthyl)$_3$P yield predominantly high-density linear polyethylene of a molecular weight up to one million. Here steric reasons seem to be responsible for the rate of propagation being much greater than the rate of elimination. Linear polyethylene is also obtained by working in liquids such as hexane, which do not dissolve Complexes 9, 10, and 11. It can be argued that the polymerization of ethylene occurs on the solid complexes where β-elimination is largely suppressed giving rise predominantly to growth.

The formation of linear polyethylene via Complexes 9, 10, and 11 is remarkable because they represent one-component catalysts and no cocatalysts as, for instance, in Ziegler-Natta systems are needed. In this connection, Complex 12 must be mentioned, which polymerizes ethylene at 70°C and 50 bar to a short-chain branched polymer.[15] Spectral comparison of **12** with its palladium analogue, whose structure

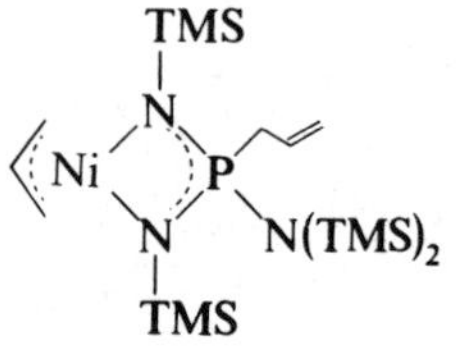

12

has been established by x-ray analysis, confirms a square planar configuration.

Activities of $> 1,000$ mol ethylene per mol complex are achieved. The physical properties of the polymer lie between those of high-pressure polyethylene and ethylene-propylene diene monomer (EPDM). Characteristic features of the polymer are: long linear chains, isolated methyl branching, and isolated larger chain branching.[16,17] The ratio of vinyl to internal double bond varies from 0.2 to 0.08.

The structure of the polyethylene can be understood by considering a nickel hydride mechanism, as shown below:

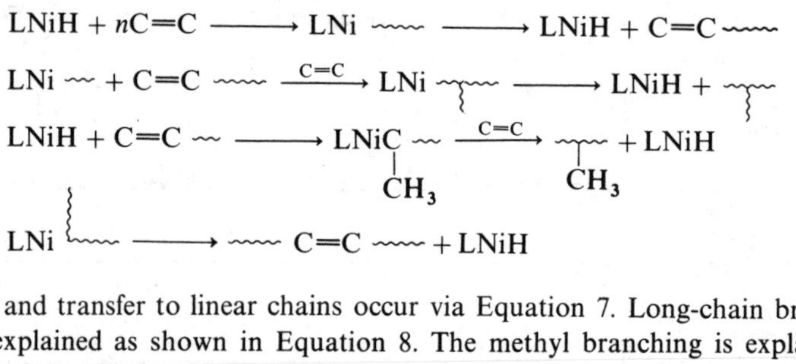

$$\text{LNiH} + n\text{C}{=}\text{C} \longrightarrow \text{LNi} \sim\sim\sim \longrightarrow \text{LNiH} + \text{C}{=}\text{C}\sim\sim\sim \qquad (7)$$

$$\text{LNi} \sim + \text{C}{=}\text{C} \sim\sim \xrightarrow{\text{C}=\text{C}} \text{LNi} \sim\gamma\sim \longrightarrow \text{LNiH} + \sim\gamma\sim \qquad (8)$$

$$\text{LNiH} + \text{C}{=}\text{C} \sim \longrightarrow \underset{\overset{|}{\text{CH}_3}}{\text{LNiC}} \sim \xrightarrow{\text{C}=\text{C}} \underset{\overset{|}{\text{CH}_3}}{\sim\gamma\sim} + \text{LNiH} \qquad (9)$$

$$\text{LNi} \overbrace{}^{} \longrightarrow \sim\sim\text{C}{=}\text{C}\sim\sim + \text{LNiH} \qquad (10)$$

Growth and transfer to linear chains occur via Equation 7. Long-chain branching can be explained as shown in Equation 8. The methyl branching is explained in Equation 9. In all cases, β-elimination accounts for the vinyl groups observed. The internal double bonds can be explained from β-elimination, as shown in Equation 10. Of course, double bond isomerization of the vinyl group could also give rise to internal double bonds. But this route is unlikely because 1-octene is practically not isomerized by **12**. The mechanism discussed invokes the idea that a cooligomerization of ethylene with the α-olefins formed during the reaction occurs (Equation 9) and that the catalyst dimerizes (oligomerizes) the intermediate α-olefins (Equation 8). This could be confirmed by reacting **12** with propylene. In a very slow reaction, predominantly 4-methyl-1-pentene is formed, a product that is expected from a reaction scheme similar to Equations 7–10. Catalysts of Type 12 may be of interest because of their potential for the production of high-pressure-type polyethylenes under low-pressure working conditions.

SUMMARY

Evidence is provided that chelate ligands offer unusual ligand properties. The majority of papers dealing with homogeneous transition metal catalysts relate to monodentate ligands. The effect of chelates has found only scattered interest. This is surprising when considering the impact chelating ligands play in nature. Chelates with an O O combination dimerize α-olefins to predominantly linear products. Chelates with P O combinations are selective for ethylene, thus providing only highly linear olefins. Their lack of ability to isomerize double bonds gives rise to the formation of only α-olefins. Chelates with P N combinations combine features of the O O and P O ligands. The linear oligomerization of ethylene prevails, but from time to time cooligomerization or dimerization (oligomerization) with the intermediate α-olefins occurs.

REFERENCES

1. KEIM, W., A. BEHR & M. RÖPER. 1982. Alkene and alkyne oligomerization, cooligomerization and telomerization reactions. *In* Comprehensive Organometallic Chemistry. **5**: 371. Pergamon Press. New York, N.Y.
2. KEIM, W., B. HOFFMANN, R. LODEWICK, M. PEUCKERT & G. SCHMITT. 1979. J. Mol. Catal. **6**: 796.
3. MILLS, O. S. & E. PAULUS. 1966. J. Chem. Soc. Chem. Commun. **20**: 738.
4. DESPEYROUX, B. 1981. Doctoral Dissertation. Rhine-Westphalia Technical University. Aachen, Federal Republic of Germany.
5. DESPEYROUX, B., R. LODEWICK & W. KEIM. 1980/81. Erdöl Kohle Erdgas Petrochem. Compendium DGMK: 123.
6. KRAUS, G. 1982. Doctoral Dissertation. Rhine-Westphalia Technical University. Aachen, Federal Republic of Germany.
7. KEIM, W., A. BEHR & G. KRAUS. 1982. Poster, Third International Symposium on Homogeneous Catalysts, Milan, Italy, Aug. 30–Sept. 3.
8. PEUCKERT, M. 1980. Doctoral Dissertation. Rhine-Westphalia Technical University. Aachen, Federal Republic of Germany.
9. KEIM, W., F. H. KOWALDT, R. GODDARD & C. KRÜGER. 1978. Angew. Chem. Int. Ed. Engl. **17**: 466.
10. KOWALDT, F. H. & W. KEIM. 1978/79. Erdöl Kohle Erdgas Petrochem. Compendium DGMK: 453.
11. GRUBER, B. 1983. Doctoral Dissertation. Rhine-Westphalia Technical University. Aachen, Federal Republic of Germany.
12. KEIM, W., *et al.* 1972. Shell Development Co. U.S. Patents No. 3,635,937, 3,644,563, 3,647,914, 3,647,915, 3,686,159.
13. KEIM, W. & F. H. KOWALDT. 1978/79. Erdöl Kohle Erdgas Petrochem. Compendium DGMK: 453–462.
14. FREITAS, E. R. & C. R. GUM. 1979. Chem. Eng. Prog. (January): 73.

15. KEIM, W., R. APPEL, A. STORECK, C. KRÜGER & R. GODDARD. 1981. Angew. Chem. Int. Ed. Engl. **20**: 116.
16. FINK, G. & W. KEIM. Unpublished results.
17. MÖHRING, V. Doctoral Dissertation. Rhine-Westphalia Technical University. Aachen, Federal Republic of Germany. (In preparation.)

HOMOGENEOUS NICKEL-CATALYZED OLEFIN HYDROCYANATION

W. C. Seidel and C. A. Tolman

E. I. du Pont de Nemours & Company
Experimental Station
Wilmington, Delaware 19898

Adiponitrile has *traditionally* been manufactured by the reaction of chlorine with butadiene (FIGURE 1). Dichlorobutene is reacted with two equivalents of sodium cyanide in a substitution reaction to make dicyanobutene. Dicyanobutene is hydrogenated to form adiponitrile.

A much more efficient synthesis would be the direct addition of hydrogen cyanide to butadiene (FIGURE 2). In this technology, hydrogen cyanide is added to butadiene in two steps. In the first step, hydrogen cyanide reacts with butadiene to form 3-pentenenitrile and the branched 2-methyl-3-butenenitrile. These mononitriles are separated from the catalyst by distillation. The branched 2-methyl-3-butenenitrile is then isomerized in the "isom" step to the linear 3-pentenenitrile. Unsaturated mononitriles can be isomerized from branched to linear, because of the presence of the double bond, via a π-allyl isomerization mechanism. In Step II, the 3-pentenenitrile

$$2NaCl + 2H_2O \xrightarrow{\text{electrolysis}} Cl_2 + 2NaOH + H_2^{\uparrow}$$

$$2NaOH + 2HCN \longrightarrow 2NaCN + 2H_2O$$

$$C_4H_6 + Cl_2 \longrightarrow C_4H_6Cl_2$$

$$C_4H_6Cl_2 + 2NaCN \longrightarrow C_4H_6(CN)_2 + 2NaCl$$

$$C_4H_6(CN)_2 + H_2 \longrightarrow NCC_4H_8CN$$

$$\text{Net:} \quad C_4H_6 + 2HCN \longrightarrow NCC_4H_8CN$$

Byproducts: chloroprene, HCl

FIGURE 1. Adiponitrile—old chlorination technology.

201

$$C_4H_6 + 2HCN \xrightarrow[\text{NiL}_4 \text{ cat}]{} NCC_4H_8CN$$

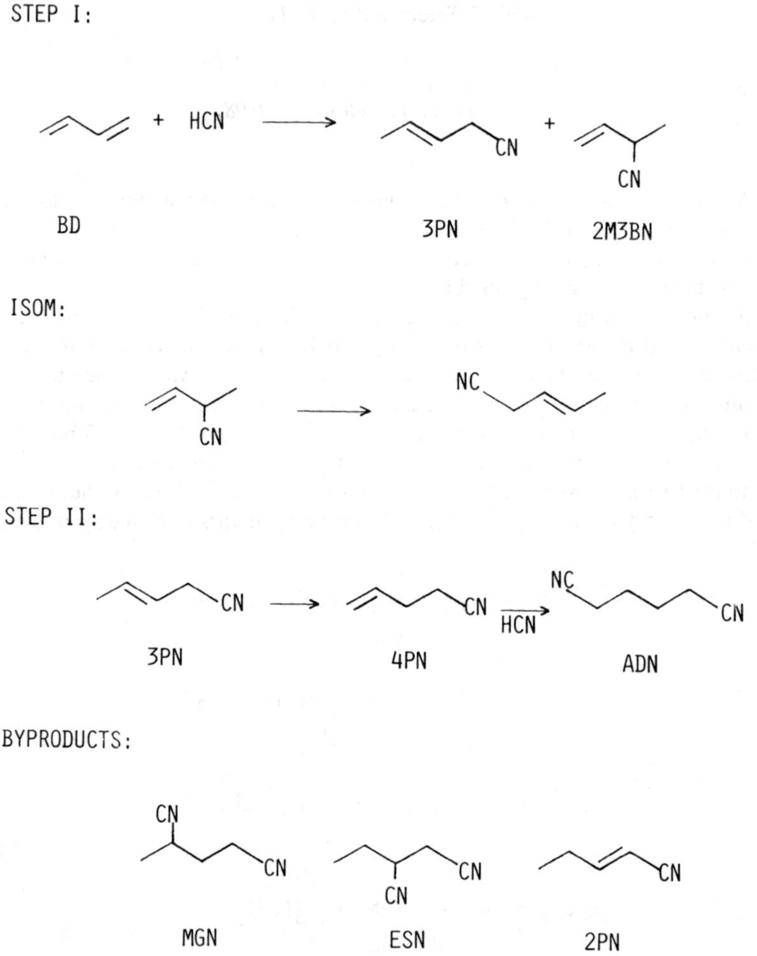

STEP I:

BD   3PN   2M3BN

ISOM:

STEP II:

3PN   4PN   ADN

BYPRODUCTS:

MGN   ESN   2PN

FIGURE 2. New hydrocyanation technology.

is reacted with hydrogen cyanide to form adiponitrile. The 3-pentenenitrile must first be isomerized to 4-pentenenitrile before the addition of the second equivalent of hydrogen cyanide. In Step II, the by-products are methylglutaronitrile and ethylsuccinonitrile. They cannot be isomerized to the linear dinitrile because there is no double bond. 2-Pentenenitrile, the conjugated mononitrile, is also formed in Step II. It cannot be isomerized to the linear 3-pentenenitrile because the 2-pentenenitrile is favored by conjugation.

Olefin hydrocyanation follows the 16- and 18-electron rule (FIGURE 3). Organometallic reactions proceeding by elementary steps involving only complexes with

$$NiL_4 \; \rightleftharpoons \; NiL_3 + L$$
$$18 16$$

$$NiL_3 + HCN \; \rightleftharpoons \; HNiL_3CN$$
$$16 18$$

$$NiL_4 + H^+ \; \rightleftharpoons \; HNiL_4^+$$
$$18 18$$

$$HNiL_3(01)^+ \; \rightleftharpoons \; RNiL_3^+$$
$$18 16$$

FIGURE 3. The 16- and 18-electron rule. Organometallic reactions proceed by elementary steps involving only 16 or 18 metal valence electrons. For the most part, successive steps involve a simple alternation between 16 and 18.

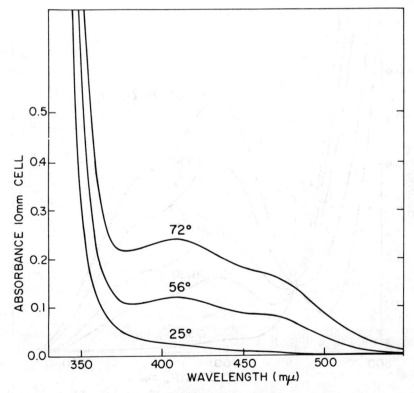

FIGURE 4. Optical spectrum of 0.019 M Ni[P(Op-tolyl)$_3$]$_4$ at various temperatures in benzene.

$$NiL_4 \underset{\longleftarrow}{\overset{K_d}{\longrightarrow}} NiL_3 + L \text{ in benzene } (25°)$$

L	K_d (M)	θ (deg)
$P(O\underline{o}tolyl)_3$	4.0×10^{-2}	141°
$P(O\underline{i}Pr)_3$	2.7×10^{-5}	~130°
$P(O\underline{p}tolyl)_3$	6×10^{-10}	128°
$P(OEt)_3$	$<10^{-10}$ at 70°	109°

FIGURE 5. Ligand dissociation from NiL_4 complexes.

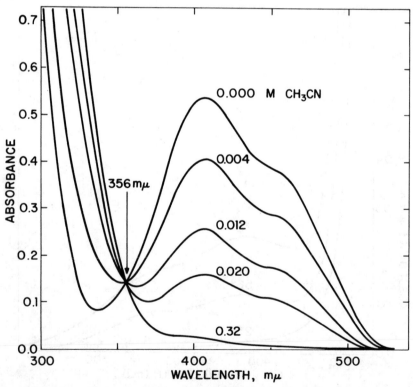

FIGURE 6. Optical spectrum of 0.11 M $Ni[P(Op\text{-tolyl})_3]_3$ in benzene at 25°C, 0.1 mm cell with CH_3CN added.

$$NiL_4 \xrightleftharpoons{K_d} NiL_3 + L$$

$$Ol + NiL_3 \xrightleftharpoons{K_1} (Ol)NiL_2 + L$$

$$Ol + (Ol)NiL_2 \xrightleftharpoons{K_2} (Ol)_2NiL_2$$

$$RCN + NiL_3 \xrightleftharpoons{K_N} (RCN)NiL_3$$

Isolation of $(C_2H_4)Ni[P(O\underline{o}tolyl)_3]_2$ $\left.\begin{array}{c}\\\\\end{array}\right\}$ x-ray

$(ACN)Ni[P(O\underline{o}tolyl)_3]_2$

$(MA)Ni[P(O\underline{p}tolyl)_3]_2$

For $Ni[P(O\underline{p}tolyl)_3]_4$ at 25° $k_d = 1.8 \times 10^{-4}$ sec^{-1} ($t_{1/2} = 1$ hr)

FIGURE 7. Olefin and nitrile complexes.

16 and 18 electrons are very common in the chemical literature. For example, the NiL_4 complex, containing 18 electrons, is dissociated to form an NiL_3, which is now a 16-electron complex. The addition of hydrogen cyanide to the 16-electron NiL_3 complex forms the 18-electron hydridocyanonickel complex. The protonation of an NiL_4 complex to form the $HNiL_4^+$ is an example of a reaction without a change in the number of valence electrons.

We will start the discussion of the mechanism of olefin hydrocyanation by first describing some of the NiL_4 complexes we have examined. The optical spectrum of $Ni[P(Op\text{-tolyl})_3]_4$ in benzene is temperature dependent (FIGURE 4). As the temperature is increased from 25°C to 72°C, the solution takes on a distinct red-orange color. This is indication of the formation of the NiL_3 complex by ligand dissociation.

The dissociation constant for ligand dissociation from NiL_4 complexes varies as a function of the ligand (FIGURE 5). As the ligand cone angle is increased from 109° to 141°, the dissociation constant increases from 10^{-10} for triethyl phosphite to only 10^{-2} for o-tolyl phosphite. Thus the bulkier the ligand the more ligand dissociation is favored. This is an important factor in explaining some hydrocyanation rate data.

Olefins and nitriles are present in hydrocyanation reaction mixtures, and they

Olefin/Nitrile	Abbrev.	K_1	$K_N (M^{-1})$
$CH_2=CHCN$	ACN	4.0×10^4	$\neq$
$CH_2=CH_2$	C_2H_4	2.5×10^2	
cis-$CH_3CH_2CH=CHCN$	C2PN	17	~500
trans-$CH_3CH_2CH=CHCN$	T2PN	17	$\neq$
$CH_2=CHCH_2CN$	3BN	10	$\neq$
$CH_2=CH\overset{\displaystyle CH_3}{\underset{\displaystyle |}{C}}HCN$	2M3BN	6.0	$\neq$
$CH_2=CHCH_2CH_2CN$	4PN	3.6	$\neq$
$CH_2=CH(CH_2)_3CN$	5HN	2.7	$\neq$
$CH_2=CH(CH_2)_3CH_3$	1-hexene	0.5	
trans-$CH_3CH=CHCH_2CN$	T3PN	1.7×10^{-2}	~200
CH_3CN			200

$\neq$ Not Determined

FIGURE 8. Equilibrium constants for olefin and nitrile reactions with Ni[P(Oo-tolyl)$_3$]$_3$ in benzene at 25°C.

both interact with nickel(0). The optical spectrum of Ni[P(Oo-tolyl)$_3$]$_3$ varies dramatically as small amounts of acetonitrile are added to the solution (FIGURE 6). In the absence of acetonitrile, there is a strong adsorbance at 400 mμ. Addition of acetonitrile bleaches the strong red color, evidence for the formation of an NiL$_3$ nitrile complex.

Of course olefins also form stable complexes with nickel(0), as evidenced by the x-ray structure determination of the ethylene and acrylonitrile complexes (FIGURE 7).

When a molecule contains both an olefin and a nitrile, for example 3-pentene-nitrile, there is a competition for the open coordination site on nickel. Equilibrium constants for olefin and nitrile reactions with Ni[P(Oo-tolyl)$_3$]$_3$ were measured in benzene at 25°C (FIGURE 8). Many of the molecules had both an olefin and a nitrile function. With acrylonitrile, only olefin coordination was observed. Most of the entries in this figure can bind either as olefins or as nitriles. For instance, cis-2-pentenenitrile has a nitrile formation constant of 500 and an olefin formation

$$NiL_4 + HCN \xrightleftharpoons{K} HNiL_3CN + L \text{ in } CH_2Cl_2 \text{ at } 25°$$

L	$K=K_d K_{HCN}$	$\nu_{CN} (cm^{-1})$
$P\phi(OEt)_2$	0.03	2113
$P(OEt)_3$	0.005	2120
$P(OCH_2CH_2Cl)_3$	0.0001	2130
$P(O\underline{o}tolyl)_3$	$>10^4$	2126
$P(O\underline{p}tolyl)_3$	4×10^{-4} (benzene)	

FIGURE 9. Equilibrium constants for $HNiL_3CN$ formation.

constant of 17. The numbers 17 and 500 are not directly comparable because of a difference in units.

Among the olefins that are present in the hydrocyanation, 2-pentenenitrile is the strongest for olefin binding to nickel, but 4-pentenenitrile does bind more tightly than 3-pentenenitrile. This is important because when an equivalent of hydrogen cyanide is added to pentenenitriles, only 4-pentenenitrile can give the linear adiponitrile. If 3-pentenenitrile is hydrocyanated, only branched dinitriles can be formed. Notice that 3-pentenenitrile binds as a nitrile with a formation constant of 200. We think that all of the nitriles bind to nickel with similar formation constants, but the amount of olefin binding varies dramatically.

We formed hydridocyanonickel complexes from different nickel(0) compounds and observed their CN stretching frequency in the infrared. We also measured the hydridocyanonickel formation constant (FIGURE 9). Very bulky $Ni[P(Oo\text{-tolyl})_3]_3$ has a high formation constant for the formation of the hydridocyanonickel complex. However, the p-tolyl phosphite complex has a formation constant of only 10^{-4}. This is a factor of 10^8 difference in formation constants that depends only on the steric size of the ligand. Cyanohydride formation is important in determining the rate of hydrocyanation. We observe faster hydrocyanation rates with $Ni[P(Oo\text{-tolyl})_3]_3$ than we do with the $Ni[P(Op\text{-tolyl})_3]_4$. This is partially explained by the higher concentration of hydridocyanonickel present in solution when orthotolyl phosphite nickel is used as a hydrocyanation catalyst.

We have already discussed hydridocyanonickel formation and olefin complex formation. We believe that ethylene hydrocyanation follows the 16- and 18-electron

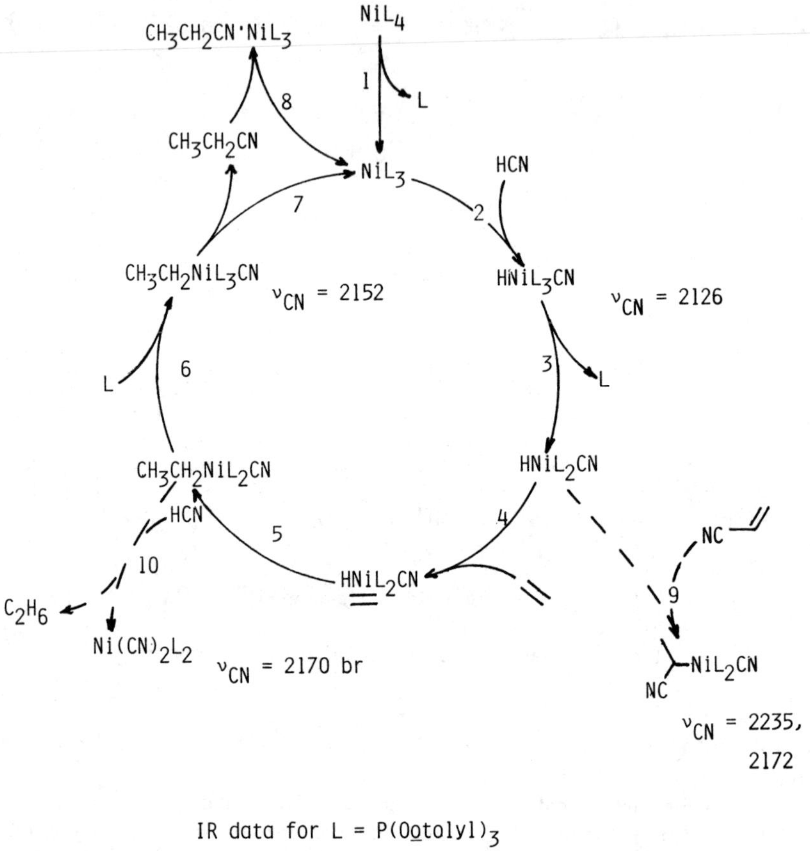

IR data for L = P(Oₒtolyl)₃

FIGURE 10 Mechanism of ethylene hydrocyanation.

rule. The first step is the dissociation of ligand from the NiL_4 complex to form NiL_3 (FIGURE 10). This is followed by oxidative addition of HCN to form $HNiL_3CN$ accompanied by its infrared stretch at 2126 wave numbers. Dissociation of ligand forms $HNiL_2CN$ followed by addition of olefin and rapid insertion to produce the cyanoethylnickel complex, which has been observed in the proton NMR. Now, addition of another ligand followed by collapse of the cyanoethylnickel complex forms propionitrile and regenerates NiL_3, which complexes with the nitrile to form the NiL_3 nitrile complex, and the cycle is repeated.

The combination of nickel triethyl phosphite and sulfuric acid in methanol is an extremely efficient butene isomerization catalyst (FIGURE 11). The use of a strong acid without the presence of cyanide allowed us to study olefin isomerization without hydrocyanation. When isomerization is carried out using deuterated sulfuric acid we found that 99% of the 2-butene formed contained no deuterium incorporation. Thus the mechanism appears to be formation of nickel deuteride $DNiL_4^+$, which then reacts with 1-butene to form 2-butene without regenerating an NiL_4 complex. We

- NiL_4, $HNiL_4^+$, and Ni(II) only Ni species observed

- Very fast: $t_{1/2}$ ~20 sec at 25°

    $$k_1 = 1500 \; M^{-1} \; sec^{-1} \quad K_1 = 50 \; M^{-1}$$

    $$k_2 = 10^{-3} \; sec^{-1} \quad\quad K_2 \; < 10^{-4} \; M$$

- D_2SO_4 in CH_3OD gives ~99.4% d_0-2-butene and 0.5% d_1-1-butene in 15 sec exposure (40% isomerized)

- Statistical scrambling on long exposure

- 3000 cycles at 0° and 300 cycles at 25°

FIGURE 11. Butene isomerization by $Ni[P(OEt)_3]_4$ and H_2SO_4 in CH_3OH.

then have $HNiL_4^+$ present, which continues to isomerize 1-butene to 2-butene. We observed 3,000 catalyst cycles at 0°C.

The 16- and 18-electron rule also describes butene isomerization (FIGURE 12). In this case instead of dissociation of NiL_4 to NiL_3 we believe the first step is protonation to form $HNiL_4^+$. This complex can dissociate ligand to form $HNiL_3^+$, which reacts with 1-butene to form the 18-electron $HNiL_3$olefin followed by insertion of the olefin to form the alkyl nickel. Alkyl collapse to form 2-butene, *cis* or *trans*, is followed by dissociation of the 2-butene and pickup of another 1-butene. 1-Butene has a much higher olefin formation constant than 2-butene; thus the isomerized 2-butene is readily replaced by another equivalent of 1-butene to repeat the cycle. When butadiene is added, a π-allyl is formed and the reaction is effectively stopped.

A competitive hydrocyanation between ethylene and propylene with $Ni[P(Oo\text{-tolyl})_3]_3$ occurs at 0°C (FIGURE 13). We observed that ethylene rapidly hydrocyanated to form propylnitrile in just 10 minutes followed by a much slower hydrocyanation of propylene to form the mixture of linear and branched butylnitriles.

We have a similar experiment at 35°C, the hydrocyanation of butadiene, where we follow the reaction in the infrared (FIGURE 14). We observe the formation of a nickel π-allyl intermediate at 2110 wave numbers followed by a slow decrease in the concentration of the intermediate and the formation of 3-pentenenitrile and 2-methyl-3-butenenitrile.

The mechanism of butadiene hydrocyanation is more complicated than that of ethylene hydrocyanation. We believe the dissociation of NiL_4 to form the NiL_3

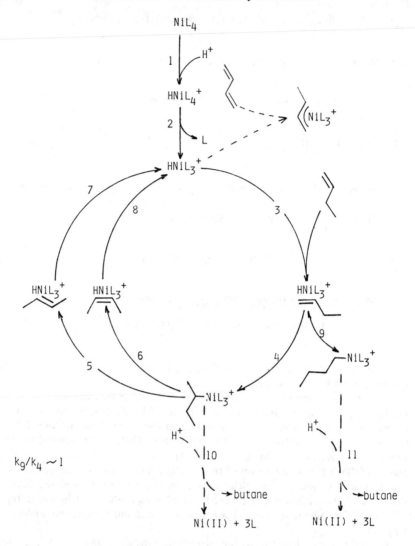

FIGURE 12. Mechanism of butene isomerization.

complex is the first step (FIGURE 15), followed by HCN oxidative addition to form the hydridocyanonickel complex. Now butadiene attacks the cyanohydride to form the π-allylcyanonickel complex. This π-allyl decomposes in two ways to form either the linear 3-pentenenitrile or the branched 2-methyl-3-butenenitrile. We find that 3-pentenenitrile is formed about two and one-half times faster than 2-methyl-3-butenenitrile.

Because an oxidative addition to form a π-allyl nickel complex can be reversed, 2-methyl-3-butenenitrile can be charged to a reactor containing NiL_4 and isomerized to 3-pentenenitrile. Heating at 110°C forms the thermodynamic mixture, 85%

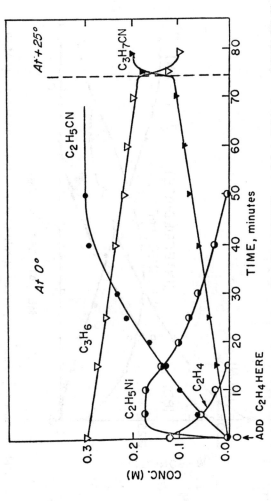

FIGURE 13. Competitive hydrocyanation of ethylene and propylene with 0.175 M Ni[P(Oo-tolyl)$_3$]$_3$ in 75% toluene-d$_8$/25% CH$_2$Cl$_2$ by ^{1}H NMR.

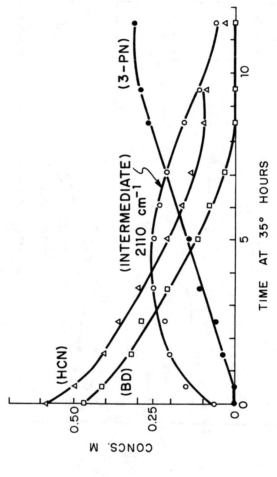

FIGURE 14. Hydrocyanation of butadiene (BD) with 0.5 M Ni[P(OEt)$_3$]$_4$ in CH$_2$Cl$_2$, 0.5 M BD + 0.6 M HCN.

STEP I AND ISOM

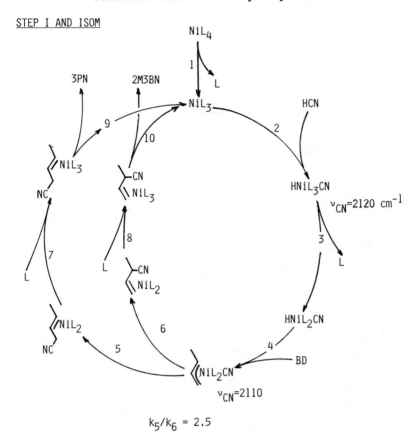

$k_5/k_6 = 2.5$

IR Data for L = P(OEt)$_3$ in CH$_2$Cl$_2$

FIGURE 15. Mechanism of butadiene hydrocyanation.

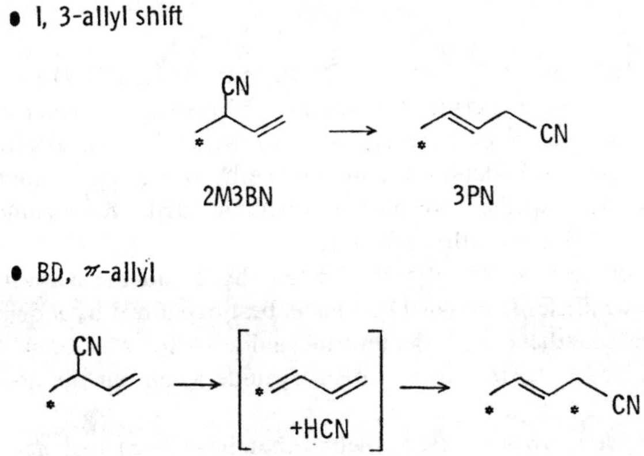

● 1, 3-allyl shift

2M3BN   3PN

● BD, π-allyl

FIGURE 16. Mechanisms of 2M3BN ⟶ 3PN isomerization. NiL$_4$, 110°C.

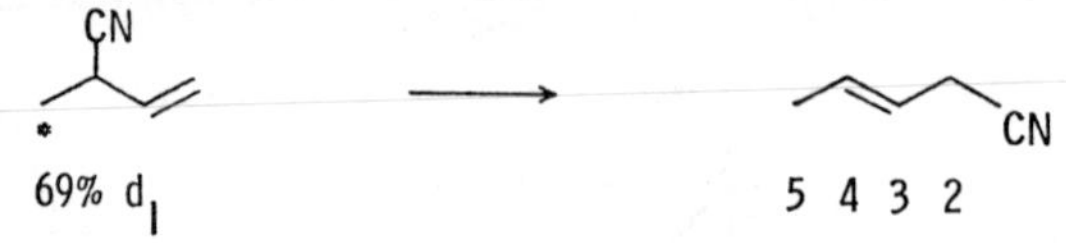

Isom. Int.	Relative Proton Integrals			
	5	3+4	2	5/2
I, 3 allyl shift	2.31	2.0	2.0	1.15
BD, π-allyl	2.54	2.0	1.77	1.43
Observed	2.23	2.0	1.57	1.42

Conclusions:

π-allyl intermediates involved

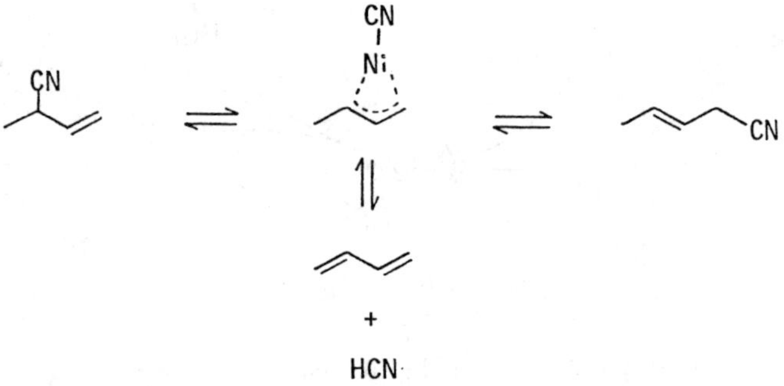

FIGURE 17. Two mechanisms for 2M3BN isomerization.

3-pentenenitrile (3PN) and 15% 2-methyl-3-butenenitrile (2M3BN).

A deuterium labeling experiment (FIGURE 16) provided further mechanistic information on isomerization of 2-methyl-3-butenenitrile. We started with 2-methyl-3-butenenitrile labeled with deuterium in the methyl position (containing about 69% deuterium). We expected to find a 3-pentenenitrile containing the same amount of deuterium in the methyl position.

We observed deuterium scrambling between the 2 and the 5 positions in the recovered 3-pentenenitriles (FIGURE 17). This is best explained by a dehydrocyanation to form free butadiene and deuteriumcyanide. After the dehydrocyanation step, we can now add butadiene to hydrogen cyanide again but the deuterium will be scrambled to the other position.

In FIGURE 18 we show a series of olefins that have been hydrocyanated with NiL_3 at 25°C in the absence of Lewis acids. When Lewis acids are added to the

Olefin

Products

C_2H_4

CH_3CH_2CN

$CH_2CH=CH_2$

$CH_3CH_2CH_2CN$

CH_3CHCH_3
$\quad\quad|$
$\quad\quad CN$

2   :   1

$CH_3(CH_2)_3CH=CH_2$

$CH_3(CH_2)_4CH_2CN$

$CH_3(CH_2)_3CHCH_3$
$\quad\quad\quad\quad|$
$\quad\quad\quad\quad CN$

$CH_3(CH_2)_2CHCH_2CH_3$
$\quad\quad\quad\quad\quad|$
$\quad\quad\quad\quad\quad CN$

3   :   1   :   0.5

$CH_3C=CH_2$
$\quad|$
$\quad CH_3$

CH_3CHCH_2CN
$\quad\quad|$
$\quad\quad CH_3$

$\quad\quad CH_3$
$\quad\quad|$
$CH_3SiCH=CH_2$
$\quad\quad|$
$\quad\quad CH_3$

$\quad\quad CH_3$
$\quad\quad|$
$CH_3SiCH_2CH_2CN$
$\quad\quad|$
$\quad\quad CH_3$

$PhCH=CH_2$

$PhCHCH_3$
$\quad|$
$\quad CN$

$PhCH_2CH_2CN$

10   :   1

$cis\text{-}CH_3CH_2CH=CHCN$

$trans\text{-}CH_3CH_2CH=CHCN$

$CH_2=CHF$

CH_3CHF
$\quad\quad|$
$\quad\quad NiL_nCN\ (2143)$

$CH_2=CHCO_2CH_3$

$CH_3CHCO_2CH_3$
$\quad\quad|$
$\quad\quad NiL_nCN\ (2152)$

$CH_2=CHCN$

CH_3CHCN
$\quad\quad|$
$\quad\quad NiL_nCN\ (2164)$

FIGURE 18. Steric and electronic effects of olefins in hydrocyanations by $Ni[P(Oo\text{-tolyl})_3]_3$ at 25°C.

$$HCN + NiL_4 \rightleftharpoons HNiL_3CN + L \quad K \sim 4 \times 10^{-4}$$

$$HCN + NiL_4 + B\phi_3 \rightleftharpoons HNiL_3CN \cdot B\phi_3 + L \quad K = 20\ M^{-1}$$

$$t_{1/2} = 1\ hr\ at\ 25°$$

$$For\ HNiL_3CN + B\phi_3 \rightleftharpoons HNiL_3CN \cdot B\phi_3$$

$$K_{calc} = 5 \times 10^4\ M^{-1}$$

FIGURE 19. HCN and $B\phi_3$ with $Ni[P(Op\text{-tolyl})_3]_4$.

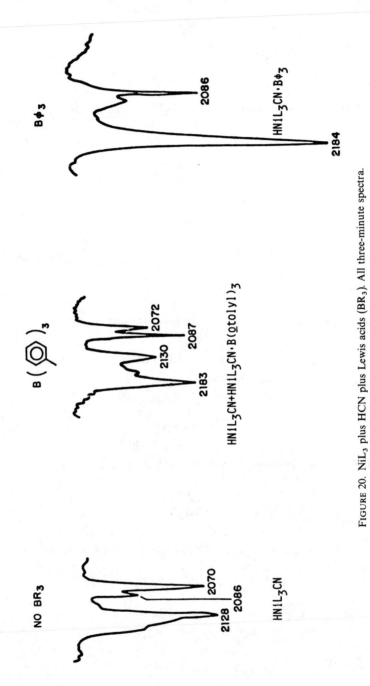

FIGURE 20. NiL₃ plus HCN plus Lewis acids (BR₃). All three-minute spectra.

Lewis Acid	% Linear Nitrile Product	
	Styrene	4PN
Bφ₃	33	98
B(otolyl)₃	9	74
BCy₃	11	72
None	9	77
B(oφ)₃	1	70

FIGURE 21. Lewis acid effects on product distribution with P(Oo-tolyl)₃ at 25°C.

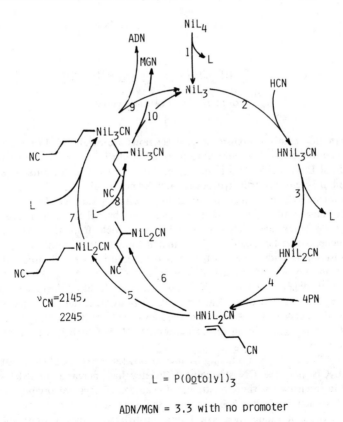

L = P(Ootolyl)₃

ADN/MGN = 3.3 with no promoter

= 50 with Bφ₃

FIGURE 22. Mechanism of 4PN hydrocyanation.

- INCREASES THE EQUILIBRIUM CONSTANT FOR HYDRIDE

FORMATION FROM NiL_4

$$NiL_4 + HCN \rightleftharpoons HNiL_3CN + L$$

$$NiL_4 + HCN + B \rightleftharpoons HNiL_3CN \cdot B + L$$

- DECREASES THE EQUILIBRIUM CONSTANT FOR OLEFIN

COORDINATION/INSERTION

$$HNiL_3CN + S \rightleftharpoons RNiL_3CN$$

$$HNiL_3CN \cdot B + S \rightleftharpoons RNiL_3CN \cdot B$$

FIGURE 23. $B\phi_3$ Lewis acid effects.

hydrocyanation catalyst mixture, several reactions are catalyzed. In particular, the hydrocyanation of 3-pentenenitrile to form adiponitrile is catalyzed relative to its rate without Lewis acids. This is very important for a successful butadiene hydrocyanation. If 3-pentenenitrile hydrocyanation did not require a Lewis acid cocatalyst, then butadiene hydrocyanation would go to the dinitrile directly. This would result in the formation of large amounts of undesired branched methylglutaronitrile. 2-Methyl-3-butenenitrile, which is formed in 30% yield from the hydrocyanation of butadiene, would be directly hydrocyanated to methylglutaronitrile.

Because Lewis acids are important in hydrocyanation, we will discuss the interaction of one Lewis acid, triphenylboron, with HCN and NiL_4 (FIGURE 19). We observe, with $Ni[P(Op\text{-tolyl})_3]_4$, that the cyanohydride has a formation constant of only 10^{-4}. Now remember that with o-tolyl phosphite, the formation constant was 10^4, a difference of 10^8. Addition of the Lewis acid, triphenylboron, to a mixture of HCN and NiL_4 results in the formation of $HNiL_3CN \cdot B\phi_3$ with a formation constant of 20.

$HNiL_3CN$ has a CN stretching frequency of 2128 wave numbers (FIGURE 20). In $HNiL_3CN \cdot B\phi_3$ the CN stretching frequency has increased to 2184 with an increase in intensity for the CN stretch of cyanide. o-Tolyltriphenylboron has the same CN stretch, 2183, but does not have the increase in intensity. We think this is because of steric hindrance. The bulky o-tolyltriphenylboron does not have a high formation constant for forming the Lewis acid adduct.

In addition to catalyzing the hydrocyanation of 3-pentenenitrile and other

- CHANGES RATE OF HYDROCYANATION

  [INCREASE FOR P(O$\underline{p}$tolyl)$_3$, DECREASE FOR P(O$\underline{o}$tolyl)$_3$]

- INCREASES PERCENTAGE OF LINEAR PRODUCT

- INCREASES NUMBER OF CATALYTIC CYCLES

FIGURE 24. Bϕ_3 Lewis acid effects.

unactivated olefins, the presence of Lewis acids also affects the amount of linear products formed (FIGURE 21). When 3-pentenenitrile was hydrocyanated at 25°C in the absence of Lewis acid, 77% of the linear adiponitrile was formed. However, when the hydrocyanation was carried out in the presence of triphenylboron, the amount of desired linear product increased to 98%. Other boranes, like tricyclohexyl borane, do not have this ability to increase the selectivity to linear products. Thus Lewis acids catalyze the formation of cyanohydrides, increase the rate of hydrocyanation of unactivated olefins like 3-pentenenitrile, and increase the selectivity to desired linear product.

We will discuss the mechanism of 4-pentenenitrile hydrocyanation in the absence of Lewis acid first (FIGURE 22). Once again we have dissociation of NiL$_4$ to form NiL$_3$ followed by oxidative addition of hydrogen cyanide to form a cyanohydride and formation of a 4-pentenenitrile complex. The 4-pentenenitrile was formed from 3-pentenenitrile by a butene isomerization type mechanism. Now the cyanohydride can hydrocyanate the olefin to form either a branched or a linear alkyl. If the linear alkyl decomposes we form adiponitrile, but the branched alkyl decomposes to give methylglutaronitrile. Triphenylboron dramatically catalyzes the formation of the linear dinitrile.

Triphenylboron has two important effects on hydridocyanonickel (FIGURE 23). It increases the stability of HNiL$_3$CN · Bϕ_3 and decreases the equilibrium constant for olefin coordination and insertion. This is probably because HNiL$_3$CN · Bϕ_3 is so sterically hindered that it prefers to be in the form of the free nickel hydride instead of the cyanoalkylnickel complex.

Triphenylboron has three Lewis acid effects in the hydrocyanation (FIGURE 24). First it changes the rate of hydrocyanation for unactivated olefins; second, it increases the percentage of linear product; and third, it gives a dramatic increase in the number of catalytic cycles. In effect it protects the nickel from oxidative addition with another equivalent of HCN to form a nickel(II)cyanide.

The complete mechanism for the borane-promoted hydrocyanation including both the non–Lewis acid catalyzed loop, previously discussed, and the triphenylborane-catalyzed loop is very complex (FIGURE 25). Once again, dissociation of ligand from NiL$_4$ to NiL$_3$ starts the process, followed by oxidative addition of hydrogen cyanide. At this point, Bϕ_3 can attach itself to the cyanide to form HNiL$_3$CN · Bϕ_3. Now olefin insertion into the nickel hydride bond forms an alkyl,

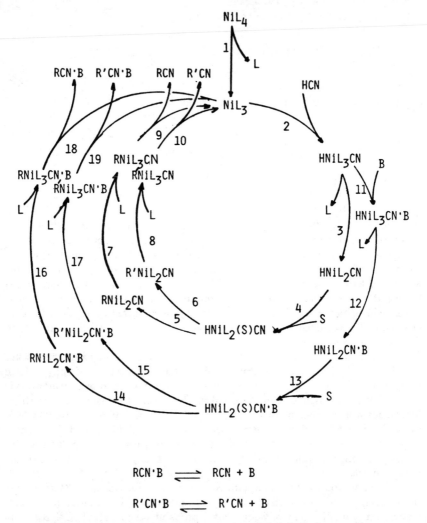

FIGURE 25. Mechanism of a borane-promoted olefin hydrocyanation.

but in the presence of triphenylboron only the linear alkyl is formed, which decomposes to form adiponitrile. The inner loop to form branched dinitrile is slow. The inner loop addition of $HNiL_3CN$ without Lewis acid should form the linear and branched cyanoalkyls. We know that the ratio of linear to branched product on the inner loop is 3:1. Thus we conclude, in the presence of triphenylboron, all of the chemistry follows the outer loop because of the high ratio of linear to branched products observed.

In conclusion, olefin hydrocyanation follows the 16- and 18-electron rule. Activated olefins like ethylene and butadiene add hydrogen cyanide without Lewis acids at high rates to give good yields of the nitriles. However unactivated olefins, in particular 4-pentenenitrile, require both Lewis acids and nickel(0) complexes for efficient linear hydrocyanation. The Lewis acid affect is important for the success of a process for the addition of hydrogen cyanide to butadiene.

ACKNOWLEDGMENTS

We would like to thank our colleagues William C. Drinkard, J. Douglas Druliner, Larry W. Gossner, and Lloyd S. Guggenberger for their participation in this work.

CATALYTIC TRANSITION METAL HYDRIDES IN THE ABSENCE OF EXTRANEOUS HYDROGEN: STUDIES OF HYDROGEN TRANSFERS IN THE REACTIONS OF DIPHENYLACETYLENE WITH (A) METHYL- OR BENZYLMAGNESIUM CHLORIDE AND TRIS(TRIPHENYLPHOSPHINE)COBALT CHLORIDE $(Ph_3P)_3CoCl$ AND (B) TRIS(TRIPHENYLPHOSPHINE)METHYL COBALT $(Ph_3P)_3CoCH_3$*

M. Michman, B. Steinberger, and A. Scherz

Institute of Chemistry
The Hebrew University of Jerusalem
Jerusalem, Israel

H. Schwarz and G. Höhne

Institute for Organic Chemistry
Technical University of Berlin
D-1000 Berlin 12, Federal Republic of Germany

Free radical mechanisms in hydrogenation chemistry have recently been comprehensively reviewed.[1] Several distinctive reaction pathways have been outlined, while it was stressed that in spite of their general nature they have received considerably less attention as compared with nonradical reactions. This seems to be especially true in the case of organometallic hydrogenation reactions. The significance of such processes has recently been demonstrated by a very detailed study.[2]

In the present study we wish to describe the stepwise mechanism of what we believe to be a free radical hydrogenation reaction of an alkyne as inferred from electron spin resonance (ESR) and mass spectrometry (MS) studies. Although ESR spectra are difficult to interpret, we think that some of our results reveal defined steps involving complexation of the substrate to the metal catalyst as well as the initial process of alkyne hydrogenation. These investigations are complemented by isotopic labeling as analyzed by MS, which clearly indicate that radicals are formed and terminated during the hydrogenation reaction.

The widespread capacity of organometallic reagents to bring about hydrogen abstraction and transfer in organic media is by now well known, and a considerable number of hydrogenation catalysts have been developed on the basis of this reactivity.[3-5] Although with homogeneous catalysts it is the activation of molecular hydrogen that is of greatest importance, it is quite reasonable that initiation of hydrogenation reactions should be attributed to the formation of transient metallohydride species from reactions with solvent or ligand molecules, or a

* Financial support of the Fonds der Chemischen Industrie, Frankfurt, and the Technical University (exchange program TU Berlin/HU Jerusalem) is kindly acknowledged.

cocatalyst.† Indeed, more attention has recently been given to hydrogenation reactions in which hydrogen is transferred from a donor molecule only to be added to the acceptor via a metallic hydride and without using molecular hydrogen at all.[6-9] We have found that in many cases the chemistry of transition metal alkyl compounds is quite often closely connected with hydrogen-transfer reactions, due to their tendency to form active hydrides by α- or β-elimination. Such compounds often act as hydrogenation catalysts as well and as a consequence yield alkylation and hydrogenation products at the same time. In particular with regard to the activation of alkylmagnesium halides by transition metal salts, the various products obtained suggest at least a formal similarity between hydrogenation and alkylation reactions.[10] In the specific case in which we used $(Ph_3P)_3CoCl$ together with a Grignard reagent or $(Ph_3P)_3CoCH_3$ alone to alkylate and hydrogenate diarylalkynes, e.g., $PhC\equiv CPh$[11,12] (SCHEME 1), an unusual effect was observed in the presence of oxygen, in that hydrogenation was completely inhibited without having any measurable effect on the alkylation reaction.[11,12] Inasmuch as the latter is concerned, we could not find explanatory results; however, it seems from the following that the hydrogenation is a radical process that is very sensitive to the presence of oxygen.

We have studied by means of ESR and MS the reaction of $PhC\equiv CPh$ with both CH_3MgCl and $C_6H_5CH_2MgCl$ in the presence of $(Ph_3P)_3CoCl$ (SCHEME 1) as well as the interaction of $PhC\equiv CPh$ with $(Ph_3P)_3CoCH_3$ alone. In all three reactions, no hydrogenation of the alkyne takes place when argon is replaced by oxygen.

Slurries of $(Ph_3P)_3CoCl$ in tetrahydrofuran (THF) were charged with CH_3MgCl or $C_6H_5CH_2MgCl$ in THF, at $-70°C$, in sealed vessels equipped with tubes to fit the ESR cavity. All this was done under Ar. These mixtures show a complex pattern of ESR signals to be designated as background signals. Although they are better defined with the benzyl reagent than with the methyl reagent and some of the signals become more intense on warming to $0°C$, an interpretation of these ESR spectra has not been attempted. It should be mentioned that the reactions of other Grignard reagents with phosphinecobalt complexes have been reported to yield ESR signals (doublets), which were attributed to cobalt hydrides;[13] the presence of a doublet in our case may have the same or a similar origin. With both methyl and benzyl reagent, the addition of $PhC\equiv CPh$ causes immediate buildup of a new signal (FIGURES 1 and 2, respectively). The signals are temperature dependent, an additional absorption appears as a shoulder at 277 K (FIGURE 2), and the signals are identical.

Due to high sensitivity it is sometimes possible that ESR signals may have little to do with the actual reaction at hand, but in the present case the spectrum is relevant to the hydrogen-transfer reaction for the following reasons: (a) the signal appears instantaneously upon addition of $PhC\equiv CPh$; (b) an identical signal forms in both systems and these are similar as far as hydrogenation is concerned; (c) the signal is instantaneously extinguished at the addition of oxygen, in close coincidence with the hydrogenation reactions and in contrast to the background signal, which is retained.

It is difficult to interpret the ESR results unequivocally, yet a rationalization will be attempted. The g value of 2.0027 points to the formation of an organic

---

† These often become catalytic hydrogenation systems in the presence of molecular hydrogen. See however References 3–5.

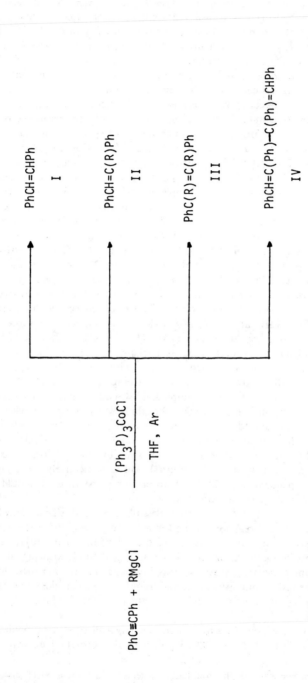

SCHEME 1. Products of reaction of **RMgCl** (R = CH$_3$, C$_6$H$_5$CH$_2$), PhC≡CPh, and (Ph$_3$P)$_3$CoCl under Ar. The presence of O$_2$ inhibits selectively the formation of **I** and **II**. Although we have separated stereochemical isomers of **I** and **II** by GCMS analysis, the detailed discussion of this is of no relevance for the present study.

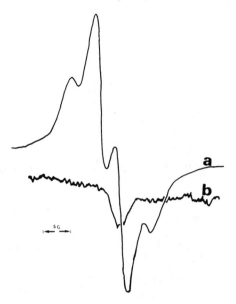

FIGURE 1. ESR spectrum of CH_3MgCl and $(Ph_3P)_3CoCl$ with $PhC{\equiv}CPh$. (**a**) Signal appears immediately after the addition of excess $PhC{\equiv}CPh$ to the mixture of $(Ph_3P)_3CoCl$ and CH_3MgCl, replacing a weak broad background signal. $g = 2.0027$. (**b**) Signal in **a** disappears instantaneously upon addition of O_2. The new absorption is probably due to a cobalt oxygen adduct. $T = 255$ K, $g = 1.9994$.

radical. The form of the multiplet suggests a set of two doublets, one with $a = 16$ G (the two external lines in FIGURES 1 and 2), the other with $a = 6$ G. The ratio of intensity of the two doublets is temperature dependent, indicating an equilibrium between two species. This is substantiated by obtaining a linear plot of ln (intensity) vs. $1/T$ (FIGURE 3). The given description is supported by a computer simulation (FIGURE 4) with a, g, and ΔH (line width) as variables (TABLE 1). Best fit at four different temperatures was obtained when treating the external doublet as a Lorenzian and the internal doublet as a Gaussian shaped signal. This could mean that the two species differ in their relaxation mechanisms.

Such signals could fit an organic radical or radical anion with a vinylic structure,

e.g.,
$$\begin{array}{c} H \\ \diagdown \\ Ph \diagup \end{array} C{=}\dot{C}{-}Ph,$$
which in some way may still be complexed with cobalt; the notation of the species is not meant to be a free radical, a free vinyl radical is presumably too short-lived to give rise to a persistent ESR signal. Incidentally, radical anions of different stereochemistry have been described for hydrogenations in liquid ammonia.[14] Such species should also have E and Z configurations leading eventually to *cis* and *trans* olefins.

The question of coordination to the cobalt complex is indirectly reflected in the results of the reactions between $PhC{\equiv}CPh$ and $(Ph_3P)_3CoCH_3$. Whereas the latter is diamagnetic,[15] addition of the alkyne immediately generates a paramagnetic

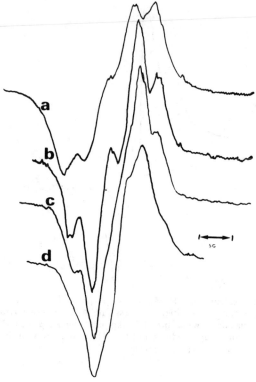

FIGURE 2. ESR spectrum of the mixture of $C_6H_5CH_2MgCl$, $PhC{\equiv}CPh$, and $(Ph_3P)_3CoCl$ under Ar, at (a) 193 K; (b) 239 K; (c) 259 K; and (d) 277 K. $g = 2.0027$.

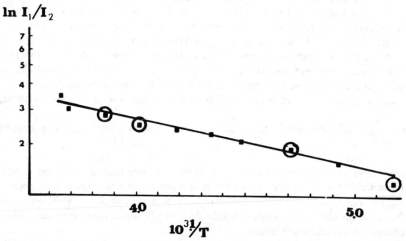

FIGURE 3. Plot of the ln values of the ratios of ESR signal intensities of the two doubtlets formed by the reaction of $PhCH_2MgCl$, $(Ph_3P)_3CoCl$, and $PhC{\equiv}CPh$ versus $1/T$. Circles indicate relations to FIGURE 2 and FIGURE 4.

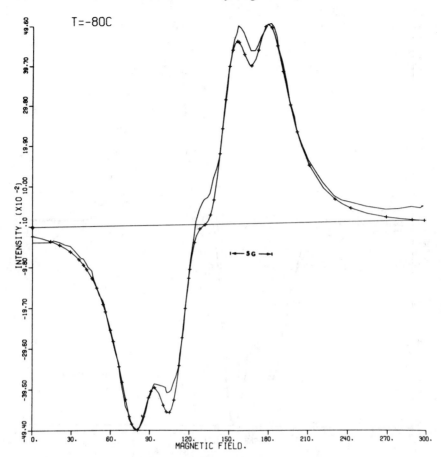

FIGURE 4. Simulation of ESR absorption of the mixture PhCH$_2$MgCl, (Ph$_3$P)$_3$CoCl, and PhC≡CPh at 259°K (FIGURE 4a, above); 249°K (FIGURE 4b); 212°K (FIGURE 4c); and 193°K (FIGURE 4d). Solid line, experimental; hatched line, simulated.

TABLE 1

VALUES FOR ESR BANDS OBTAINED BY COMPUTER SIMULATION
(FIGURE 4) AT FOUR TEMPERATURES

	193 K		212 K		239 K		259 K	
	G	L	G	L	G	L	G	L
Band form*	G	L	G	L	G	L	G	L
Band width (Gauss)	3.61	4.00	3.33	3.33	3.42	2.83	3.67	2.67
Band amplitude†	1.36	1	1.89	1	2.94	1	4.17	1
Band splitting factor‡ (Gauss)	4.83	14.33	4.83	14.33	4.83	14.50	4.83	15.00

* Inner band Gaussian (G). External band Lorenzian (L).
† Arbitrary numbers.
‡ Direct graphical analysis yielded values of ∼6 and ∼16 G.

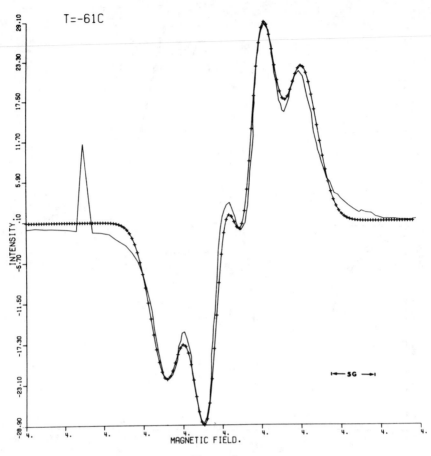

FIGURE 4b

species, which is shown by the octet in the ESR spectrum (FIGURE 5). The splitting is caused by $^{59}Co^1$ (I = 7/2). A g value of 2.0010 and a splitting factor of a_{Co} = 4.5 G were determined, somewhat less than half the value reported for a_{Co} = 11.46 G.[16] and this signal is superimposed by a broad doublet, which becomes more prominent as the temperature is raised (FIGURE 6). $(Ph_3P)_3CoCH_3$ is shown by ^{31}P NMR[17] to dissociate in solution, and the alkyne may easily be accommodated as a ligand. Whether as an alkyne complex the cobalt compound reverts to the high-spin form, which is the stable form of $(Ph_3P)_3CoCl$ for example, or whether it adds to the alkyne with formation of a sigma compound followed by oxidation is difficult to judge. The spectrum suggests that the odd electron resides away from the cobalt atom as an organic radical.

Investigation of the reaction products by combined gas chromatography/mass spectrometry yields conclusive evidence for the existence and termination of a radical entity of the type suggested above. The reactions were run in the presence of small quantities of $C_6D_5CD_3$. Even with a ratio of THF to perdeuteriotoluene of

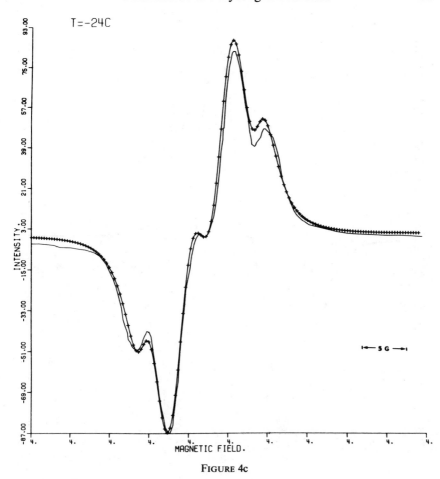

FIGURE 4c

20:1, we obtained considerable yields of monodeuterated stilbene (TABLE 2) and, significantly, no dideuterated product. If hydridocobalt derivatives alone were the cause of hydrogenation, one should expect the formation of d_0, d_1, and d_2 stilbene. This situation is indeed observed in other cases.[18] The implication of hydrogen abstraction from the solvent by a radical entity is ultimately borne out by the detection of considerable amounts of $C_6D_5CD_2CD_2C_6D_5$.

Our results imply a stepwise radical hydrogenation with addition of the first hydrogen atom (perhaps likewise an alkyl group), possibly from a Co—H unit with the formation of the intermediary radical adduct. This is eventually terminated by reaction with solvent, releasing the weakly coordinated olefin. Termination is also possible by radical-radical dimerization, and the corresponding product is observed, namely, PhCH=C(Ph)—C(Ph)=CHPh. All this holds for both Grignard reagents, which indeed differ only with regard to their alkylation products. Of these we detected PhC(R)=CHPh and PhC(R)=CDPh (R = CH_3, $C_6H_5CH_2$). Bibenzyl is present in the reaction mixture as a result of the Grignard preparation. There is no

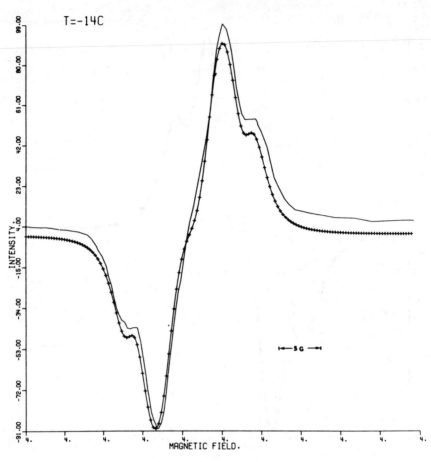

FIGURE 4d

TABLE 2

DEUTERIUM CONTENT OF OLEFIN PRODUCTS FROM THE REACTION OF PhC≡CPh AND
PhCH$_2$MgCl/(Ph$_3$P)$_3$CoCl IN THE PRESENCE OF ·C$_6$D$_5$CD$_3$*

Product	Percent d$_0$	Percent d$_1$	Percent d$_2$
cis Stilbene	79	19	2
trans Stilbene	55	44	1
PhCH=C(CH$_2$Ph)Ph†	87	13	—

* Bibenzyls containing both d$_0$ and d$_{14}$ were also present, but no other isotopomers of this compound were found.

† No separation of the stereoisomers was attempted.

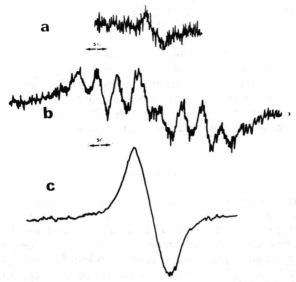

FIGURE 5. (a) ESR spectrum of $(Ph_3P)_3CoCH_3$ in toluene. (b) As in a, but following addition of $PhC\equiv CPh$. (c) Spectrum b vanishes and new absorption appears upon addition of O_2; probably $Co-O_2$ adduct. $T = 213$ K; $g = $ (a) 1.9996; (b) 2.0010; and (c) 1.9993.

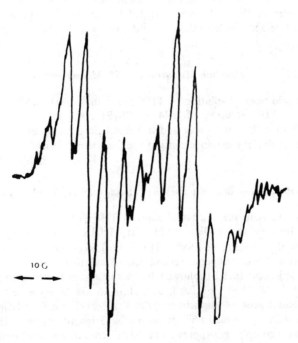

FIGURE 6. ESR spectrum of $(Ph_3P)_3CoCH_3$ and $PhC\equiv CPh$ at 250 K. (Sample in FIGURE 4b warmed up.)

formation of it during the reaction, neither is there any cross product with deuteriotoluene. This proves the absence of free alkyl radicals derived from Grignard compounds.

Thus, although much remains to be clarified as to the nature of the active intermediate and clearly the system is a very complicated one, our results provide some information regarding the reaction pathway. As has been stated in earlier work,[19] these pathways lose their importance once the hydrogenation is run under hydrogen pressure; however, they may still be significant in the initiation of hydrogen transfer and in activating a catalytic system.

EXPERIMENTAL

ESR measurements were carried out on a Varian X Band E-104 spectrometer. Temperatures were varied by a standard procedure using a V-4540 varian temperature controller. Solvents were dried in accordance with previously described procedures.[11] All preparations of solutions, reactions, and ESR tests were carried out under argon, unless stated otherwise. $PhC\equiv CPh$ of commercial grade (Fluka) was used without any further purification. $(Ph_3P)_3CoCH_3$[15] and $(Ph_3P)_3CoCl$[20] were prepared by standard methods.

Mass spectra were recorded in combination with on-line capillary gas chromatography on a Varian MAT 44 mass spectrometer using chemical ionization with NO^+ as reagent gas; data were taken on-line to a SS 200 computer. The numbers given in TABLE 2 were corrected for natural ^{13}C isotope contribution; they are the average of at least three separate runs, and the relative errors are $\pm 5\%$.

Reaction Mixtures for ESR Measurements

A Grignard reagent solution in THF (10^{-2} molar) was added in $\times 20$ molar excess to a THF solution of 0.44 g $(Ph_3P)_3CoCl$ (5×10^{-4} mol); 0.089 g $PhC\equiv CPh$ (5×10^{-4} mol) was added either before measurements started or at later stages during the measurements at specified temperatures (as in FIGURE 3).

Reaction of Benzylmagnesium chloride with Diphenylacetylene

Diphenylacetylene 0.178 g (10^{-3} mol) in THF (10 ml) was kept under Ar at $-62°C$. To this, a 0.6 molar solution of $PhCH_2MgCl$ was added (33 ml, 2×10^{-2} mol) followed by the addition of 0.880 g (10^{-3} mol) of $(Ph_3P)_3CoCl$. The slurry was warmed up to room temperature and stirred overnight. Hydrolysis with 40 ml aqueous acetic acid (pH 3) followed by extraction with diethylether yielded the organic products. These consisted, according to GC analysis, of *trans* stilbene, bibenzyl, and traces of diphenylacetylene. A check on the Grignard reagent showed that bibenzyl formed while preparing the Grignard reagent. All hydrogenation reactions were completely suppressed in the presence of oxygen. Products were also identified and deuterium incorporation positions were verified with 100 MHz H NMR spectroscopy. Reactions with CH_3MgCl were likewise carried out.

Reaction of Benzylmagnesium chloride with Diphenylacetylene,
$(Ph_3P)_3CoCl$ in the Presence of $C_6D_5CD_3$

A mixture consisting of PhC≡CPh (0.089 g, 0.5 mmol), $(Ph_3P)_3CoCl$ (0.440 g, 0.5 mmol), and $C_6D_5CD_3$ (1 ml, 9.4 mmol) was kept under Ar at −60°C. Freshly prepared $PhCH_2MgCl$ (0.61 molar in THF) was added (16 ml, 9.8 mmol). The mixture was stirred and brought to room temperature within 20 minutes. After continuous stirring for 24 hours reaction was terminated by hydrolysis; organic products were extracted with diethylether. Reaction with CH_3MgCl was carried out likewise.

Determination of Concentration of Grignard Reagents

Concentration of Grignard reagents was determined by acid-base back titration of freshly prepared solutions.

SUMMARY

The reactions of RMgCl (R = CH_3; $C_6H_5CH_2$) with PhC≡CPh yield products of hydrogenation, e.g., stilbene, and alkylation, e.g., alkylated stilbenes, when cobalt compounds such as $(Ph_3P)_3CoCl$ and $(Ph_3P)_3CoCH_3$ are used as catalysts. The role of these reagents as hydrides and hydrogen-transfer agents has been studied by ESR, isotopic labeling, and MS to reveal the operation of a radical stepwise hydrogenation. The system described here is basically different from another hydrogenating reagent, i.e., $NaBH_4 + (Ph_3P)_3CoCl$, the properties of which have recently been discussed.[18]

ACKNOWLEDGMENTS

The authors wish to thank Professor R. West from the University of Wisconsin for his encouragement, Professor C. Levanon from the Hebrew University of Jerusalem for discussion of the ESR results, and M. Uzi Eliav, Jerusalem, for running his computer program and the simulation tests.

REFERENCES

1. HALPERN, J. 1979. Pure Appl. Chem. **51**: 2171.
2. KINGLER, R. J., K. MOCHIDA & J. K. KOCHI. 1979. J. Am. Chem. Soc. **101**: 6626.
3. JAMES, B. R. 1973. Homogeneous Hydrogenation. John Wiley. New York, N.Y.
4. MUETTERTIES, E. L. 1971. Transition Metal Hydrides. Marcel Dekker. New York, N.Y.
5. JAMES, B. R. 1979. Adv. Organomet. Chem. **17**: 319.
6. BLUM, J., S. SHTELZER, P. ALBIN & Y. SASSON. 1982. J. Mol. Catal. **16**: 67.
7. BAR, R., Y. SASSON & J. BLUM. 1982. J. Mol. Catal. **16**: 175.
8. CRABTREE, R. H. 1982. Chem. Tech.: 507.
9. PARSHALL, G. W., D. L. THORN & T. H. TULIP. 1982. Chem. Tech.: 571.
10. MICHMAN, M. & H. H. ZEISS. 1970. J. Organomet. Chem. **25**: 161.
11. MICHMAN, M., B. STEINBERGER & S. GERSHONI. 1976. J. Organomet. Chem. **113**: 293.

12. MICHMAN, M. & L. MARCUS. 1976. J. Organomet. Chem. **122**: 77.
13. HENRICI-OLIVE, G. & S. OLIVE. 1969. J. Chem. Soc. Chem. Commun.: 1482.
14. HOUSE, H. O. 1965. Modern Synthetic Reactions: 71. Benjamin Inc. New York, N.Y.
15. YAMAMOTO, A., S. KITAZUME, L. S. PU & S. IKEDA. 1971. J. Am. Chem. Soc. **93**: 371.
16. GIANNOTTI, C., G. MERLE & J. R. BOLTON. 1975. J. Organomet. Chem. **99**: 145.
17. MICHMAN, M., V. R. KAUFMAN & S. NUSSBAUM. 1979. J. Organomet. Chem. **182**: 547.
18. MICHMAN, M., B. STEINBERGER, H. SCHWARZ & G. HÖHNE. 1983. J. Organomet. Chem. **244**: 283.
19. JAMES, B. R., L. D. MARKHAM & D. K. W. WANG. 1974. J. Chem. Soc. Chem. Commun.: 439.
20. ARESTA, M., M. ROSSI & A. SACCO. 1969. Inorg. Chim. Acta **3**: 227.

THE HYDROGENATION OF CARBONYL COMPOUNDS CATALYZED BY ANIONIC RUTHENIUM HYDRIDE COMPLEXES AND ALKALI-DOPED SUPPORTED GROUP VIII METALS

Roger A. Grey

Technology and Development
ARCO Chemical Company
Newtown Square, Pennsylvania 19073

Guido P. Pez

Corporate Science Center
Air Products and Chemicals, Inc.
Allentown, Pennsylvania 18105

Andrea Wallo and Jeff Corsi

Corporate Research and Development
Allied Corporation
Morristown, New Jersey 07960

INTRODUCTION

Main group metallic hydrides, such as $LiAlH_4$, are well known as excellent reagents for the reduction of a wide variety of unsaturated polar organic substrates such as aldehydes, ketones, and carboxylic acid esters.[1] One disadvantage of these reagents for large-scale processes is that they must be used in stoichiometric quantities to obtain high yields of desired products. We felt that this disadvantage might be overcome by preparing certain anionic transition metal hydride complexes of formula $A^+[L_nM-H_x]^-$, formally analogous to $LiAlH_4$, wherein A^+ is an alkali-metal cation, M is a group VIII metal, and L a tertiary phosphine ligand. In the presence of hydrogen gas, therefore, the reduction could be catalytic with respect to the metal.

A hypothetical, possible mechanism for the catalytic reduction of acetone using an anionic hydride catalyst is illustrated in SCHEME 1. The key step here (step 2) is the transfer of the hydride ion from the ruthenium to the coordinated ketone. We expected that any factor that favored this transfer would make a more efficient catalyst (i.e., cationic assistance, electron-withdrawing groups on the ketone, nucleophilicity of the hydride).

The anionic formulation was chosen in an attempt to impart the maximum hydridic and nucleophilic character to the hydride atom on the metal. Ruthenium was chosen as the central metal atom because of its inherent capacity to react reversibly with alkoxy ligands. Tertiary phosphines were chosen as stabilizing ligands for ruthenium to minimize the dispersal of the negative charge of the complex.

Hydridometalates containing strongly π-acid ligands which tend to stabilize the negative charge such as $HFe(CO)_4^-$ and $HCo(CN)_5^{3-}$ are well known (see Reference 2 and references therein). Hydridometalates containing only weakly

235

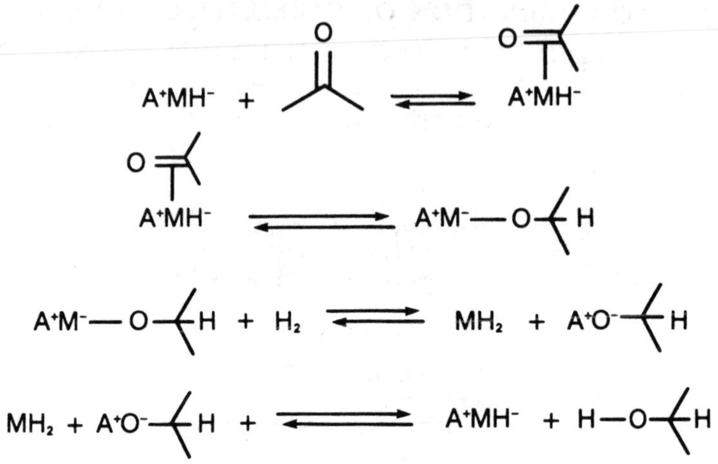

A^+MH^- = cation; M = transition metal with associated ligands.

SCHEME 1. A possible mechanism for the catalytic hydrogenation of acetone with an anionic hydride catalyst. (Adapted from Reference 7 with permission. Copyright 1981 by the American Chemical Society.)

π-acceptor ligands are rarer. However, Chatt and Davidson have previously reported a phosphine hydridoruthenate complex prepared by the sodium naphthalene reduction (2 equiv) of trans-[Ru(PP)$_2$HBr] [PP = 1,2-bis(dimethylphosphino)ethane].[3] Although this complex was characterized only by its reaction with D_2O to give cis-[Ru(PP)$_2$HD], this report suggested that an anionic ruthenium hydride complex stabilized by only phosphine ligands could be prepared. Since we felt that the catalytic activity of hydridometalates would be maximized with less than four phosphorus ligand atoms per ruthenium, we pursued the synthesis of tris(triphenylphosphine)- and bis(triphenylphosphine)hydridoruthenates by the reduction of (Ph$_3$P)$_3$RuHCl and [(Ph$_3$P)$_2$RuHCl]$_2$, respectively.

DISCUSSION

Preparation and Catalytic Reactions of
$K^+[(Ph_3P)_2Ph_2PC_6H_4RuH_2]^- \cdot C_{10}H_8 \cdot (C_2H_5)_2O$ (Complex 1)

The reaction of (Ph$_3$P)$_3$RuHCl·C$_6$H$_5$CH$_3$[4] with potassium naphthalene (2 equiv) at $-78°C$ in tetrahydrofuran gave Complex 1, which can be isolated as a yellow crystalline solid from ether solutions in the presence of excess naphthalene.[5] Although the precursor to 1 possesses only one hydride ligand, spectroscopic evidence indicates that 1 possesses two hydride ligands. The IR spectrum of 1 exhibits two distinct bands at 1,825 and 1,735 cm^{-1} attributed to the two hydrides. As would be expected for an anionic complex, these bands are of distinctly lower frequency than the original neutral complex [ν_{MH}(Ph$_3$P)$_3$RuHCl = 2,010 cm^{-1}]. The ^{1}H NMR of 1 exhibited two multiplets centered at -7 and -11 ppm (δ_{TMS}). The presence of two

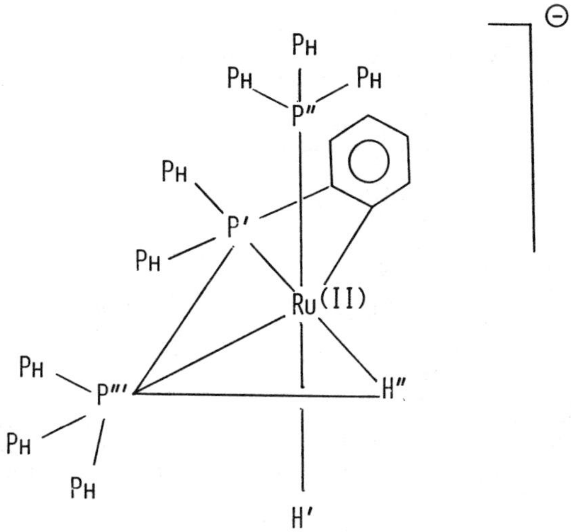

FIGURE 1. Solution structure for the $[(Ph_3P)_2Ph_2PC_6H_4RuH_2]^-$ anion.

hydrides is also supported by the reaction of **1** with excess methyl iodide and HCl to form two equivalents of methane and hydrogen, respectively, per ruthenium atom. It was reasoned that the additional hydride ligand arises from the *ortho*-metalation of a phenyl ring of a triphenyl phosphine ligand. From the ^{1}H and ^{31}P NMR data, the structure shown in FIGURE 1 was deduced. The structure was confirmed in the solid state by an x-ray crystal structure.*

With a well-characterized anionic ruthenium hydride in hand, we proceeded to test our hypothesis that it would function as a catalyst for the hydrogenation of organic substrates containing polar unsaturated functional groups.[6] We found that **1** readily catalyzes the hydrogenation of ketones, and these results are summarized in TABLE 1. Acetone, for example, is readily hydrogenated to isopropanol (95% selectivity) at 85°C and 620 kPa gauge of hydrogen. Minor aldol products of acetone believed to be caused by the anionic nature of the catalyst were observed as the only by-products. The parent neutral complex, $(Ph_3P)_3RuHCl$, under the same reaction conditions gives less than 4% reduction of the acetone to isopropanol. We suggest that the enhanced activity of **1** for this reduction is due to its anionic nature. The presence of 18-crown-6 in the reaction mixture causes a much faster reduction of the acetone (see TABLE 1). Since 18-crown-6 is a known complexing agent of alkali cations this result indicates that the degree of association of the potassium ion with the anionic complex or the carbonyl group of the ketone has an effect on the catalytic activity of **1**. Hexafluoroacetone is hydrogenated at a considerably faster initial rate than acetone although the hydrogenation does not go to completion, possibly due to the relatively acidic hexafluoroisopropanol product. The initial rapid hydrogenation rate could be due to the electron-with-drawing effect of the two trifluoromethyl groups, thus activating the carbonyl group for hydride transfer. The relative rates of hydrogenation of the cyclic ketones cyclo-pentanone, cyclohexanone, and cycloheptanone are consistent with a mechanism

---

* The hydride ligands were not observed in the x-ray crystal structure.

TABLE 1
CATALYTIC HYDROGENATION OF KETONES*

Substrate	Percent Conversion	Time (hours)	Remarks
Acetone	98	16	
Acetone	100	6	18-crown-6
Acetone	98	16	THF solvent
Acetone	78	16	1 equiv Ph_3P
2-Butanone	91	16	
Benzophenone	85	16	2.7 mmol
Hexafluoroacetone	60	2	
Cyclopentanone	93	16	
Cyclohexanone	100	1	
Cycloheptanone	55	16	

* Conditions: toluene (3 mL); ketone (13.9 mmol); catalyst (0.035 mmol); hydrogen (620 kPa gauge); 85°C. (Table adapted from Reference 7 with permission. Copyright 1981 by the American Chemical Society.)

involving hydride addition to the carbonyl group. Of these ketones, cyclohexanone has the most favorable energy difference of the carbonyl groups in going from an "sp^2 hybridized" ground state to an "sp^3" transition state[7,8] and is also hydrogenated the fastest with 1 as the catalyst. H. C. Brown has reported a similar relative reactivity of cyclopentanone and cyclohexanone using $NaBH_4$ as the reductant.[9]

Complex 1 also catalyzed the hydrogenation of aldehydes. For example, propanal was converted to n-propanol (94% conversion, 95% selectivity) and aldol condensation products at 100°C and 620 kPa of H_2. The α,β-unsaturated aldehyde acrolein was also investigated. At low acrolein conversions (~35%), allyl alcohol and propanol were obtained as the two major products. Tetrahydrofuran (THF) solutions of 1 gave allyl alcohol in 10% selectivity, while in the presence of 18-crown-6 no allyl alcohol was observed. Allyl alcohol was formed in selectivities up to 20% if the lithium analogue of 1 is used [prepared by the reacton of $Li^+C_{10}H_8^-$ and $(Ph_3P)_3RuHCl$]. In the hydrogenation of acrolein, therefore, the nature of the cation has an effect on the selectivity of 1.

After having demonstrated 1 as an effective catalyst for the hydrogenation of aldehydes and ketones, we investigated 1 as a catalyst for the hydrogenation of carboxylic acid esters. The first ester investigated, ethyl formate, however, instead of being hydrogenated was catalytically decarbonylated to ethanol and carbon monoxide (~1,280 turnovers).

$$HCO_2Et \xrightarrow{1, 90°C} CO + EtOH \qquad (1)$$

The ability of 1 to catalytically decarbonylate formate esters may be related to its basicity since it is known that alkyl formates can be catalytically decomposed by alkoxides.[10] In order to avoid the decarbonylation complication, methyl acetate was investigated as a hydrogenation substrate. However, methyl acetate could not be hydrogenated using 1 as the catalyst.

Since we had found earlier that trifluoromethyl groups apparently activated a ketone carbonyl for hydrogenation, we next investigated esters containing electron-withdrawing groups adjacent to the carboxyl group. Methyl trifluoroacetate in toluene with **1** as catalyst reacted under 620 kPa of hydrogen at 90°C to give a 10% conversion (16 catalyst turnovers) to trifluoroethanol and methanol (Equation 2). The presence of 18-crown-6 in the reaction mixture, in contrast to the acetone

$$CF_3CO_2CH_3 \xrightarrow[\Delta]{H_2, \text{ cat.}} CF_3CH_2OH + CH_3OH \qquad (2)$$

reduction, slows the hydrogenation of this fluorinated ester.

Trifluoroethyl trifluoroacetate was even more readily hydrogenated than methyl trifluoroacetate. This ester was completely converted to trifluoroethanol (162 catalyst turnovers) after 12 hours at 90°C with 620 kPa of H$_2$.

$$CF_3CO_2CH_2CF_3 \xrightarrow[H_2]{1, 90°C} 2\ CF_3CH_2OH \qquad (3)$$

Dimethyl oxalate, which contains two adjacent electron-withdrawing carboxyl groups, was hydrogenated in THF solution at 90°C and 620 kPa of H$_2$ in the presence of **1** to methanol and methyl glycolate (10% conversion, 16 catalyst turnovers). The product, methyl glycolate (an unactivated ester), was not reduced further to ethylene glycol.

The Preparation and Catalytic Reactions of $[(Ph_3P)_3(Ph_2P)Ru_2H_4]^{2-}K_2^+ \cdot 2C_6H_{14}O_3$ (Complex 2)

In order to prepare an anionic ruthenium hydride catalyst more active than **1** that could be used to hydrogenate even unactivated esters such as methyl acetate, we attempted to synthesize a more unsaturated analogue of **1** that would contain only two triphenyl phosphine ligands. The analogous potassium naphthalene reduction of $[(Ph_3P)_2RuHCl]_2 \cdot 2$ toluene,[11] however, gave two major potassium phosphine ruthenate products. The material that gave the most interesting catalytic properties was crystallized as the diglyme adduct **2**. This material, characterized by chemical and spectroscopic means, does not have a structure completely analogous to **1**. The ^{1}H NMR and ^{31}P NMR of **2** show four nonequivalent hydrides at -13.0, -13.7, -15.6, and -16.7 ppm (δ_{TMS}) and four nonequivalent phosphorus resonances at 110.6, 76.4, 69.8, and 69.1 ppm (δ 85% H$_3$PO$_4$), respectively. The IR spectrum of **2** shows complex medium intensity bands between 1,700 and 1,900 cm^{-1} assigned to the metal hydrides and medium to weak absorptions at 1,403 and 1,550 cm^{-1} indicative of *ortho*-metalation.[12] The reaction of THF solutions of **2** with HCl gave two valuable facts concerning this complex. First, four moles of H$_2$/mole of **2** were evolved, indicating two hydrides per ruthenium atom. Secondly, the ruthenium complex formed after protonation was characterized as having the formula $(Ph_3P)_3(Ph_2PH)Ru_2Cl_4$. The diphenyl phosphine ligand is believed to have arisen by the protonation of a diphenyl phosphido ligand, which is unexpectedly present in **2**. The diphenyl phosphido ligand in **2**, which is believed to bridge two ruthenium atoms, is probably formed by a reductive cleavage of triphenyl phosphine during the potassium naphthalene reduction. The ^{31}P NMR resonance at 110.6 ppm, which is greatly shifted from the other three phosphorus resonances, is assigned to the bridging Ph$_2$P ligand.[13] Based on these data, **2** is empirically formulated as $K_2^+[(Ph_3P)_3(Ph_2P)Ru_2H_4]^{2-} \cdot 2C_6H_{14}O_3$.

TABLE 2

ACTIVITY OF $[(Ph_3P)_2Ph_2PC_6H_4RuH_2]^-$ (COMPLEX 1)
VERSUS $[(Ph_3P)_3(Ph_2P)Ru_2H_4]^-$ (COMPLEX 2)*

Substrate	Catalyst	Percent Conversion	Time (hours)
$HCO_2CH_2CH_3$	1	92	4
$HCO_2CH_2CH_3$	2	5	20
$(CH_3)_2CO$	1	98	16
$(CH_3)_2CO$	2	100	4†
$CF_3CO_2CH_3$	1	10	20
$CF_3CO_2CH_3$	2	88	20
$CH_3O_2CCO_2CH_3$	1	10	20
$CH_3O_2CCO_2CH_3$	2	70	20
$CF_3CO_2CH_2CF_3$	1	100	12
$CF_3CO_2CH_2CF_3$	2	100	4
$CH_3CO_2CH_3$	1	0	20
$CH_3CO_2CH_3$	2	30	20

* Conditions: toluene (3 mL); ketone (13.9 mmol); ester (5.7 mmol); 1 (0.035 mmol); 2 (0.017 mmol); H_2 (620 kPa); 90°C.
† Two and a half hours with added 18-crown-6.

As we had anticipated, 2 was indeed a more active catalyst than 1 for the hydrogenation of carbonyl containing substrates (see TABLE 2). Acetone, methyl trifluoroacetate, dimethyl oxalate, and trifluoroethyl trifluoroacetate were hydrogenated faster and with higher conversions using 2 as the catalyst. The presence of 18-crown-6 in the reaction mixture also had an effect on the catalytic activity of 2. For the hydrogenation of acetone the rate was accelerated (2.5 vs. 4 hours for complete conversion), whereas for esters (i.e., dimethyl oxalate) the rate was slowed (52 vs. 70% conversion).

With 2 as the catalyst, the unactivated ester methyl acetate was successfully hydrogenated. Toluene solutions of methyl acetate were hydrogenated to a mixture of ethanol, methanol, and ethyl acetate at 90°C and 620 kPa of H_2 (35 catalyst turnovers). The ethyl acetate comes from the *trans*-esterification of the starting ester with the product ethanol. The cationic assistance effect of the potassium is critical for this hydrogenation since no hydrogenation of the methyl acetate was observed in the presence of 18-crown-6.

Anionic Hydrides—Solid-State Analogues

In an attempt to extend the concept of anionic hydride catalysts to heterogeneous systems, we investigated various alkali-doped, supported metal catalysts. We envisioned that a highly dispersed group VIII transition metal on an appropriate support such as carbon in the presence of an alkali metal could react with hydrogen to form an anionic hydride *in situ* at the catalyst surface.

M = group VIII transition metal
A = alkali metal

The anionic hydride thus produced could then react with the carboxyl carbon of an ester that is susceptible to nucleophilic attack.

Group VIII metals supported on carbon doped with alkali metals are known and have been demonstrated by Ozaki to be effective catalysts for the hydrogenation of molecular nitrogen to ammonia at 300–350°C.[14] To our knowledge we are the first to demonstrate the use of this type of catalyst for the hydrogenation of esters.[15]

We have found that active alkali-doped catalysts can be prepared in several ways:

1. Deposition of alkali metal vapor. In the same manner as described by Ozaki, 5% ruthenium on carbon was doped with potassium metal by heating a mixture of the supported transition metal and potassium metal at 400°C under dynamic vacuum. Potassium-doped ruthenium catalysts (K/Ru/C) were prepared containing 4 to 12% potassium by weight. Since the vapor pressures of other alkali metals (i.e., Li, Na) are not suitable for this method of deposition, we investigated other catalyst preparations that would have general utility for all of the alkali metals.

2. Alkali metals dissolved in liquid ammonia. We found that active catalysts could be prepared by treating the transition metal on carbon with liquid ammonia solutions containing the desired quantity of lithium, sodium, or potassium at −65°C. After the ammonia solutions become colorless, the ammonia is removed by heating the catalyst in vacuum. A catalyst prepared in this manner consisting of rhodium and potassium is designated as K(NH$_3$)/Rh/C. The ammonia in parenthesis is used to indicate the manner of preparation of the catalyst and does not designate ammonia as part of the composition.

3. Alkali metal radical anion solutions. Treatment of group VIII transition metals supported on carbon with radical anion solutions was found to be another useful technique for preparing alkali metal doped catalysts. Tetrahydrofuran solutions of the radical anion potassium naphthalene (KNp) were prepared by adding THF to equal molar amounts of potassium metal and naphthalene and allowing them to react at room temperature. The potassium naphthalene solution is then added to 5% rhodium on carbon and allowed to react at room temperature for one hour to give potassium naphthalene on rhodium on carbon (KNp/Rh/C). Enough potassium naphthalene is added to give a catalyst with a potassium content between 4 and 16% by weight. These catalysts can be used *in situ* or isolated by filtering and drying the solids in vacuum. In a similar manner, rhodium on carbon was doped with lithium naphthalene, sodium naphthalene, and cesium naphthalene. Potassium naphthalene–doped catalysts were also prepared from 5% ruthenium, palladium, and platinum on carbon.

4. Alkoxide solutions. Active catalysts can also be prepared by treating 5% rhodium on carbon or 5% ruthenium on carbon with THF solutions containing alkali alkoxides of one of the alcohols to be produced during hydrogenation. For example, if a ruthenium catalyst is to be used for the hydrogenation of methyl acetate, 5% Ru/C is mixed with potassium methoxide in THF to give a catalyst designated as K$^+$ $^-$OCH$_3$/Ru/C.

We have found that the above alkali-doped catalysts in THF solutions are active for the hydrogenation of several carboxylic acid esters to their corresponding alcohols under mild conditions of pressure and temperature (25°C, 620 kPa of H$_2$) (see Table 3). These conditions are in contrast to most heterogeneous catalysts for

TABLE 3
THE HYDROGENATION OF ESTERS WITH ALKALI-DOPED CATALYSTS*

Catalyst	Substrate	Reaction Time (hours)	Conversion (%)
K/Ru/C	MeAc	20	20
KNp/Ru/C	MeAc	20	49
KNp/Rh/C	MeAc	20	55
KOCH$_3$/Rh/C	MeAc	20	46
KNp/Rh/C	CF$_3$CO$_2$CH$_3$	12	100
KNp/Rh/C	CF$_3$CO$_2$CH$_2$CF$_3$	3	100
KNp/Rh/C	Me benzoate	20	10

* Conditions: substrates to metal = 80:1 in THF, 25°C, 620 kPa H$_2$.

ester hydrogenation, which operate at conditions as high as 3,000 psi of H$_2$ and 250–300°C (see Reference 16 and references therein). Esters that were hydrogenated include methyl acetate, methyl trifluoroacetate, trifluoroethyl trifluoroacetate, and methyl benzoate. The ease of hydrogenation followed the same sequence as the homogeneous anionic hydride catalysts: trifluoroethyl trifluoroacetate > methyl trifluoroacetate > methyl acetate > methyl benzoate. Interestingly, the hydrogenation of methyl benzoate with KNp/Rh/C is selective to benzyl alcohol and methanol. This result is in contrast to undoped Rh/C, which under the same conditions yields the ring hydrogenated product carbomethoxycyclohexane.

The activities of the alkali-doped catalysts were found to be dependent upon which alkali metal was chosen as the dopant. Based on methyl acetate conversions, the order of catalyst activity follows the sequence KNp/Rh/C > NaNp/Rh/C > CsNp/Rh/C > LiNp/Rh/C (see TABLE 4). The order of activity of different group VIII metal catalysts with the same alkali dopant for the hydrogenation of methyl acetate is KNp/Rh/C > KNp/Ru/C > KNp/Pt/C > KNp/Pd/C.

SUMMARY AND CONCLUSIONS

We have prepared two new anionic phosphine ruthenium hydride complexes, **1** and **2**, and have demonstrated that they are effective catalysts for the hydro-

TABLE 4
HYDROGENATION OF METHYL ACETATE WITH ALKALI-DOPED CATALYSTS*

Catalyst	Percent Conversion
LiNp/Rh/C	11
NaNp/Rh/C	32
CsNp/Rh/C	24
KNp/Rh/C	55
KNp/Ru/C	49
KNp/Pt/C	20
KNp/Pd/C	11

* Conditions: substrates to metal = 80:1 in THF, 25°C, 620 kPa H$_2$.

genation of aldehydes, ketones, and carboxylic acid esters. We suggest that the ability to hydrogenate these substrates is related to the anionic nature of the complexes. The observation of cationic assistance effects on the rates and selectivity of the hydrogenation supports this hypothesis.

Group VIII metals supported on carbon doped with various forms of alkali (i.e., K^+Np^-, $K°$, etc.) are catalysts for liquid-phase ester hydrogenation under mild conditions (25°C, 620 kPa H$_2$). The corresponding undoped analogues are not active catalysts under these conditions.

REFERENCES

1. AUGUSTINE, R. L., Ed. 1968. Reduction. Marcel Dekker. New York, N.Y.
2. SCHUNN, R. A. 1971. *In* Transition Metal Hydrides. E. L. Muetterties, Ed.: 203–269. Marcel Dekker. New York, N.Y.
3. CHATT, J. & J. M. DAVIDSON. 1965. J. Chem. Soc.: 843–855.
4. HALLMAN, P. S., B. R. McGARVEY & G. WILKINSON. 1968. J. Chem. Soc. A: 3143–3150.
5. PEZ, G. P., R. A. GREY & J. CORSI. 1981. J. Am. Chem. Soc. 103: 7528–7535.
6. GREY, R. A., G. P. PEZ & A. WALLO. 1981. J. Am. Chem. Soc. 103: 7536–7542.
7. BROWN, H. C., R. S. FLETCHER & R. B. JOHANNESEN. 1951. J. Am. Chem. Soc. 73: 212–221.
8. BROWN, H. C., J. H. BREWSTER & H. SHECHTER. 1954. J. Am. Chem. Soc. 76: 467–474.
9. BROWN, H. C. 1957. J. Org. Chem. 22: 439–441.
10. ADICKES, F. & G. SCHAFER. 1932. Ber. Dtsch. Chem. Ges. B. 65: 950–955.
11. JAMES, B. R., A. D. RATTRAY & D. K. W. WANG. 1976. J. Chem. Soc. Chem. Commun: 792–793.
12. COLE-HAMILTON, D. J. & G. WILKINSON. 1977. J. Chem. Soc. Dalton Trans.: 797–804.
13. MOTT, G. N. & S. J. CARTY. 1979. Inorg. Chem. 18: 2926–2928.
14. OZAKI, A., K. AIKA, A. FURUTA & A. OKAGAMI. 1973. U.S. Patent 3,770,658.
15. GREY, R. A. & G. P. PEZ. 1982. U.S. Patent 4,346,240.
16. ADKINS, H. 1954. Org. React. 8: 1–27.

TRANSITION METAL HYDRIDES IN HOMOGENEOUS CATALYTIC HYDROGENATION*

Jack Halpern

Department of Chemistry
The University of Chicago
Chicago, Illinois 60637

INTRODUCTION

Transition metal hydrides (e.g., CuH^+) were postulated as intermediates in homogeneous catalytic hydrogenation reactions more than 25 years ago, before the existence of stable transition metal hydride complexes was widely recognized (see Reference 1 and references therein). Subsequently, many such hydride complexes have been prepared and characterized[2] and their roles in homogeneous catalytic hydrogenation reactions have been extensively studied.

Interest in the development of new transition metal hydride catalysts continues to be high, notably in respect of catalysts exhibiting distinctive selectivities and/or higher activities as reflected in higher rates or in the ability to catalyze hydrogenation of less reactive substrates such as ketones, nitriles, and arenes. Recent attention in this context has focused particularly on cationic[3-11] and anionic[12-16] transition metal hydrides, which have been found to exhibit distinctive coordination chemistry and catalytic properties. This paper describes some recent studies on such systems notably involving anionic ruthenium hydrides and cationic rhodium hydrides.

ANIONIC RUTHENIUM HYDRIDES

The starting point of our studies on anionic ruthenium hydride complexes was the reported synthesis by Pez, Grey, *et al.* of the orthometallated complex $[RuH_2(PPh_3)_2(PPh_2C_6H_4)]^-$ (1) and their finding that this complex was effective as a homogeneous catalyst or catalyst precursor for the selective hydrogenation of polynuclear aromatic compounds, for example of anthracene to 1,2,3,4-tetrahydroanthracene.[12-15] We were interested in the origin of this selectivity, particularly in the light of earlier reports of different selectivities for other homogeneous hydrogenation catalysts, for example $HCo(CO)_4$, which catalyzes the hydrogenation of anthracene exclusively to 9,10-dihydroanthracene.[17,18] This prompted us to undertake an investigation of the mechanism of the $[RuH_2(PPh_3)_2(PPh_2C_6H_4)]^-$-catalyzed hydrogenation of anthracene. As the first stage of such an investigation it proved necessary to examine the background coordination chemistry of $[RuH_2(PPh_3)_2(PPh_2C_6H_4)]^-$ and related anionic ruthenium complexes and the stoichiometric reactions of such complexes with possible relevance to their catalytic chemistry.

*Supported by the National Science Foundation.

Reaction of $[\overline{RuH_2(PPh_3)_2(PPh_2C_6H_4)}]^-$ *with* H_2: *Formation of*
$fac\text{-}[RuH_3(PPh_3)_3]^-$ (2)

$K[\overline{RuH_2(PPh_3)_2(PPh_2C_6H_4)}]^{12-15}$ reacted with H_2 (1 atm) in tetrahydrofuran (THF) at 25°C according to Equation 1 to form, after 24 hours in ca. 85% yield, $fac\text{-}[RuH_3(PPh_3)_3]^-$ (2), which was isolated as the yellow K^+ salt. The ^1H NMR signal at δ −9.53 ppm, due to the three Ru-bonded protons, corresponded to a six-peak multiplet resembling that previously reported for $fac\text{-}[IrH_3(PPhEt_2)_3]$ and analyzed by computer simulation as an AA'A"XX'X" pattern,[19] with a corresponding ^{31}P multiplet at 65.9 ppm. The IR spectrum of $K[RuH_3(PPh_3)_3]$ exhibited bands, assignable to ν_{Ru-H}, at 1,857 cm^{-1} and 1,815 cm^{-1} (Nujol), 1,835 cm^{-1} (THF). The formation of **2** from **1** was essentially irreversible, i.e., **1** could not be regenerated by pumping off H_2.

$$[\overline{RuH_2(PPh_3)_2(PPh_2C_6H_4)}]^- + H_2 \longrightarrow fac\text{-}[RuH_3(PPh_3)_3]^- \qquad (1)$$
$$\mathbf{1} \qquad\qquad\qquad\qquad\qquad\qquad \mathbf{2}$$

Reaction of $fac\text{-}[RuH_3(PPh_3)_3]^-$ *with Anthracene: Formation of*
$[RuH(PPh_3)_2(anthracene)]^-$ (3)

$fac\text{-}[RuH_3(PPh_3)_3]^-$ reacted with an excess of anthracene in THF according to Equation 2 to form a new red complex, $[RuH(PPh_3)_2(anthracene)]^-$ (3), which was isolated as the $[Ph_3P{=}N{=}PPh_3]^+$ and K^+ salts. The reaction, which was monitored by ^1H and ^{31}P NMR, went to completion in ca. 24 hours at 65°C. The ^1H NMR spectrum of 3^{16} resembles that previously reported for $[Fe(CO)_3(anthracene)]^{20}$ and is interpreted in terms of an analogous structure on which the present assignments are based. A structurally related compound, $[IrH(PPr_3^i)_2(butadiene)]$,[21] has been characterized crystallographically.[22]

$$fac\text{-}[RuH_3(PPh_3)_3]^- + 1.5 \text{ anthracene} \longrightarrow [RuH(PPh_3)_2(anthracene)]^-$$
$$\mathbf{2} \qquad\qquad\qquad\qquad\qquad\qquad\qquad\qquad\qquad \mathbf{3}$$

$$+ 0.5 \text{ (1,2,3,4-tetrahydroanthracene)} + PPh_3 \qquad (2)$$

Reaction 2 exhibited the same rate law as the isotopic exchange of **2** with D_2 (Equation 3), i.e., $-d[2]/dt = k_4[2]$, where $k_4 = 7.6 \times 10^{-4}$ sec^{-1} at 65°C, independent of the H_2 (or anthracene) concentration. This implies that both reactions proceed through a common unimolecular rate-determining step, namely, the reductive elimination of H_2 to form the common intermediate $[RuH(PPh_3)_3]^-$ (4), an isomer of **1**, in accord with Equation 4. Reactions 2 and 3 are much faster than the phosphine exchange reactions of **2** (e.g., replacement of PPh_3 by PEt_3 or $P(OMe)_3$), ruling out PPh_3 dissociation as the rate-determining step.

$$fac\text{-}[RuH_3(PPh_3)_3]^- \xrightarrow{D_2} [RuHD_2(PPh_3)_3]^- (\xrightarrow{D_2} [RuD_3(PPh_3)_3]^-) \qquad (3)$$

fac-$[RuH_3(PPh_3)_3]^-$

$$\xrightarrow[-H_2]{k_4} [RuH(PPh_3)_3]^- \left\{ \begin{array}{l} \xrightarrow[-PPh_3]{\text{anthracene}} [RuH(PPh_3)_2(\text{anthracene})]^- \quad \textbf{(4a)} \\ \\ \xrightarrow{D_2} [RuHD_2(PPh_3)_3]^- \quad \textbf{(4b)} \end{array} \right.$$

4

Reaction of $[RuH(PPh_3)_2(anthracene)]^-$ with H_2: Formation of $[RuH_5(PPh_3)_2]^-$ (5)

$[RuH(PPh_3)_2(\text{anthracene})]^-$ reacted rapidly with H_2 in THF at 25°C in accord with Equation 5 to yield $[RuH_5(PPh_3)_2]^-$ (**5**), which was isolated as the white K^+ salt. The NMR and IR spectral data for **5**[16] are consistent with a pentagonal bipyramidal (or fluxional) structure analogous to that of the known compounds $[IrH_5(PR_3)_2]$ (R = Ph, Et, etc.).[23]

$$[RuH(PPh_3)_2(\text{anthracene})]^- + 4\,H_2 \longrightarrow \underset{\textbf{5}}{[RuH_5(PPh_3)_2]^-} + 1,2,3,4\text{-}H_4\text{anthracene} \quad \textbf{(5)}$$

Reactions of $[RuH_5(PPh_3)_2]^-$

$[RuH_5(PPh_3)_2]^-$ reacted with a stoichiometric amount (1:2) of anthracene in THF in accord with Equation 6 to form $[RuH(PPh_3)_2(\text{anthracene})]^-$ in quantitative yield (ca. 24 hours at 25°C; 0.5 hour at 65°C). The corresponding reaction with cyclohexadiene yielded the analogous diene adduct, $[RuH(PPh_3)_2(\text{cyclo-hexadiene})]^-$, together with cyclohexane and cyclohexene. Reaction of **5** with 1-hexene (Equation 7) resulted in partial dehydrogenation and formation of $[RuH_3(PPh_3)_2]^-$, whose NMR spectrum is consistent with Structure **6**. Reaction of **5** with ethylene yielded a new compound whose IR and NMR spectra contained no evidence for a hydride ligand but were consistent with the tentative formulation as the orthometallated complex $[\overline{Ru(PPh_3)(PPh_2C_6H_4)}(C_2H_4)_2]^-$, Complex **7** formed by Reaction 8, by analogy with the known and structurally characterized compound $[\overline{Ir(PPh_3)(PPh_2C_6H_4)}(C_2H_4)_2]$, formed by the analogous reaction of C_2H_4 with $[IrH_5(PPh_3)_2]$.[21,24]

$$[RuH_5(PPh_3)_2]^- + 2\,\text{anthracene} \longrightarrow [RuH(PPh_3)_2(\text{anthracene})]^- + 1,2,3,4\text{-}H_4\text{anthracene} \quad \textbf{(6)}$$

$$[RuH_5(PPh_3)_2]^- + 1\text{-hexene} \longrightarrow \underset{\textbf{6}}{[RuH_3(PPh_3)_2]^-} + \text{hexane} \quad \textbf{(7)}$$

$$[RuH_5(PPh_3)_2]^- + 5\,C_2H_5 \longrightarrow \underset{\textbf{7}}{[\overline{Ru(PPh_3)(PPh_2C_6H_4)}(C_2H_4)_2]^-} + 3\,C_2H_6 \quad \textbf{(8)}$$

The chemistry of hydridoruthenate complexes is summarized in SCHEME 1 and encompasses the synthesis and characterization of several new complexes. It is noteworthy that nearly every new anionic ruthenium complex that these

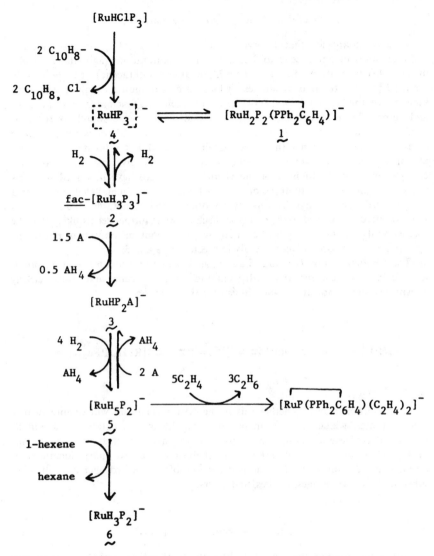

SCHEME 1. Summary of reaction chemistry. (P = triphenylphosphine, A = anthracene, AH_4 = 1,2,3,4-tetrahydroanthracene.)

studies have uncovered finds a parallel in a known (and, in most cases, structurally characterized) neutral iridium complex, $[IrH_2(PPh_3)_2(PPh_2C_6H_4)]$,[25] fac-$[IrH_3(PEt_2Ph)_3]$,[19] $[IrH(PPr^i_3)_2(C_4H_6)]$,[21,22] $[IrH_5(PEt_2Ph)_2]$,   and $[Ir(PPh_3)(PPh_2C_6H_4)(C_2H_4)_2]$[21,25] being direct analogues of 1, 2, 3, 5, and 7, respectively. We also have noted limited parallels between the chemistry of such corresponding species.

Catalytic Hydrogenation of Anthracene

We have confirmed that **1** serves as a catalyst or catalyst precursor for the hydrogenation of anthracene to 1,2,3,4-tetrahydroanthracene (and, more slowly, further hydrogenation to 1,2,3,4,5,6,7,8-octahydroanthracene) as previously reported.[12-15] Preliminary studies, which are continuing, suggest that the kinetics are approximately first order in an Ru complex, first order in anthracene, and zero order in H_2. Compounds 2, 3 and 5 also were found to serve as catalyst precursors for the hydrogenation of anthracene with rates that, in some cases, were initially higher than that obtained with **1** but ultimately leveled off at approximately the same rate, suggesting that they give rise to a common catalytic mechanism. In light of the chemistry that we have described it seems likely that, under the conditions of the reaction, the orthometallated precursor **1** is converted rapidly and irreversibly to other species (notably **3** and **5**) and so is not directly involved in the catalytic mechanism. It further seems likely that the species that are involved in the catalytic cycle contain only two phosphine ligands per Ru, as do the active catalytic precursors **3** and **5**.

The combination of Reactions 5 and 6, as depicted by Equation 9, corresponds to a catalytic cycle for the hydrogenation of anthracene and, thus, clearly constitutes one demonstrated mechanism for this reaction.

$$[RuH(PPh_3)_2(anthracene)]^- \rightleftharpoons [RuH_5(PPh_3)_2]^- \qquad (9)$$

4 H_2 H_4-anthracene

H_4-anthracene   2 anthracene

The determination of whether this is the only mechanism will require further kinetic studies, which are now in progress, on the overall catalytic reaction as well as on the several component steps that we have identified. It also remains to be established to what extent the chemistry that we have identified is relevant to the catalysis by **1**, or derivatives thereof, of the hydrogenation of other substrates such as ketones, nitriles, and esters.

CATIONIC RHODIUM COMPLEXES

Our conclusion that the activity and selectivity of anionic ruthenium complexes for the catalytic hydrogenation of arenes reflects the ability to bind arenes in the η^4 "diene" mode prompted us to look for related catalytic activity in the case of other hydrogenation catalysts with marked diene binding affinities. In line with this reasoning we have found that cationic rhodium phosphine complexes, notably $[Rh(DIPHOS)(MeOH)_2]^+$ [where DIPHOS = 1,2-bis (diphenyl-phosphino)ethane], which previously has been identified as a catalyst for the hydrogenation of alkenes, dienes, alkynes, and ketones[3-7] and which forms stable $[Rh(DIPHOS)(diene)]^+$ adducts,[8,9] do indeed form arene adducts and are effective homogeneous catalysts for the hydrogenation of anthracene to 1,2,3,4-tetrahydroanthracene.

Formation and Characterization of $[Rh(DIPHOS)(arene)]^+$ *Complexes*

The formation of $[Rh(DIPHOS)(benzene)]^+$ according to Reaction 10 and the characterization of this complex have previously been described.[8,9]

$$[Rh(DIPHOS)(MeOH)_2]^+ + benzene \underset{}{\overset{K_1}{\rightleftharpoons}} [Rh(DIPHOS)(benzene)]^+ \quad (10)$$

These studies were extended to the corresponding naphthalene and anthracene complexes including (a) determination of the ^1H and ^{31}P NMR spectra of the complexes and (b) determination of the equilibrium constants (K_1) for the formation of the complexes according to Equation 10.[11] Because of the low solubility of anthracene in methanol, the equilibrium constant K_1 was determined for the adduct of 9-methylanthracene.

The $[Rh(DIPHOS)(benzene)]^+$ adduct has previously been characterized as involving symmetrical η^6-coordination of the benzene ligand to Rh, in agreement with the x-ray structure of a closely related phenyl adduct.[8,9] Analogous coordination of the arene through one of the end rings in the corresponding $[Rh(DIPHOS)(naphthalene)]^+$ and $[Rh(DIPHOS)(anthracene)]^+$ adducts (Structures 8 and 9) can be inferred from the large upfield shifts of the 1,2,3,4-protons in each case.[11] The pattern of chemical shifts for $[Rh(DIPHOS)(anthracene)]^+$ resembles that for $[Cr(CO)_3(anthracene)]^+$ $(\eta^6)^{20}$ rather than for $[Fe(CO)_3-(anthracene)]$ $(\eta^4)^{20}$ or $[RuH(PPh_3)_2(anthracene)]^-$, (η^4),[16] suggesting that the mode of bonding of the arene is η^6 {as in $[Rh(DIPHOS)(benzene)]^+$} rather than η^4. This conclusion also is consistent with (a) the fact that Rh achieves an 18 valence electron configuration in these complexes through η^6 bonding of the arene and (b) the finding that the Rh—P coupling constants of $[Rh(DIPHOS)(naphthalene)]^+$ and $[Rh(DIPHOS)(anthracene)]^+$ are close to that of $[Rh(DIPHOS)(benzene)]^+$ rather than to those of the diene adducts $[Rh(DIPHOS)(1,3-cyclohexadiene)]^+$ and $[Rh(DIPHOS)(norbornadiene)]^{+}$.[11]

The equilibrium constants, K_1, for the formation of the $[Rh(DIPHOS)(arene)]^+$ complexes increase markedly along the sequence benzene (20 M^{-1}) < naphthalene $(2.6 \times 10^2 \ M^{-1})$ < 9-methylanthracene $(1.6 \times 10^3 \ M^{-1})$.[11] This reflects a trend of increasing binding with decreasing localization energy and, hence, is not unexpected.

Catalytic Hydrogenation

Under relatively mild conditions ($\gtrsim 60°C$, 1 atm H_2), the rate of hydrogenation of benzene in the presence of the corresponding $[Rh(DIPHOS)(arene)]^+$ complex was negligible. However, hydrogenation of anthracene, 9-methylanthracene, and 9-CF_3CO-anthracene was found to be catalyzed by the corresponding $[Rh(DIPHOS)(arene)]^+$ adducts at conveniently measurable rates. The product in each case was the 1,2,3,4-tetrahydroanthracene, together with a small amount ($\sim 1\%$) of the corresponding 1,2,3,4,5,6,7,8-octahydroanthracene. The catalyst system was stable, and no diminution of rate could be detected after ca. 20 turnovers. Under the same conditions, the hydrogenation of naphthalene (to 1,2,3,4-tetrahydronaphthalene) was considerably slower but still measurable. The hydrogenation of 9-CF$_3$CO-anthracene obeyed the second-order rate law corresponding to Equation 2 with k ($59.7°C$) = $(9.0 \pm 1.0) \times 10^{-2}$ M^{-1} sec^{-1}, $\Delta H^{\ddagger} = 16.6 \pm 2$ kcal/mol, and $\Delta S^{\ddagger} = -14 \pm 5$ cal/mol K (all for 9-CF$_3$CO-anthracene).

$$-d[\text{arene}]/dt = k\,[\text{Rh(DIPHOS)(arene)}^+][\text{H}_2] \qquad (11)$$

The rate law is similar to that found earlier for the [Rh(DIPHOS)(olefin)]$^+$-catalyzed hydrogenation of olefins[8,9,26] and may be interpreted in terms of a similar mechanism, depicted by Equations 12–14.

$$[\text{Rh(DIPHOS)(anthracene)}]^+ + \text{H}_2 \xrightarrow[\text{rate determining}]{k_2}$$

$$\text{1,2-dihydroanthracene} + [\text{Rh(DIPHOS)(MeOH)}_2]^+ \quad (12)$$

$$\text{1,2-dihydroanthracene} + \text{H}_2 \xrightarrow{[\text{Rh(diphos)(MeOH)}_2]^+} \text{1,2,3,4-tetrahydroanthracene} \quad (13)$$

$$[\text{Rh(DIPHOS)(MeOH)}_2]^+ + \text{anthracene} \xrightarrow{\text{fast}} [\text{Rh(DIPHOS)(anthracene)}]^+ \quad (14)$$

The following features of this interpretation warrant comment:

1. The stepwise hydrogenation, i.e., via 1,2-dihydroanthracene, is consistent with earlier demonstrations that other dienes (norbornadiene, 1,3-cyclohexadiene, etc.)[3-7] are hydrogenated in two stages via the intermediate monoenes. Since olefins are typically hydrogenated more rapidly than anthracene, the failure of the 1,2-dihydroanthracene to accumulate is expected. In line with this, the rate constant for the hydrogenation of 1,2-dihydronaphthalene to 1,2,3,4-tetrahydronaphthalene was found to be 2.2 M^{-1} sec^{-1} at 60°C (vs. 5×10^{-2} M^{-1} sec^{-1} for anthracene and 6×10^{-3} M^{-1} sec^{-1} for naphthalene).

2. In one set of comparisons, anthracene, 9-methylanthracene, and 9-CF$_3$CO-anthracene exhibited very similar rates of catalytic hydrogenation. From the relative rates of formation of 1,2,3,4-tetrahydroanthracene and 1,2,3,4,5,6,7,8-octahydroanthracene, the rate of hydrogenation of 1,2,3,4-tetrahydroanthracene can be estimated to be about 10^{-1}–10^{-2} that of anthracene. In accord with this the corresponding rate constant for hydrogenation of the closely related substrate naphthalene was found to be 6×10^{-3} M^{-1} sec^{-1} at 60°C (vs. 5×10^{-2} M^{-1} sec^{-1} for anthracene).

3. The hydrogenation of the [Rh(DIPHOS)(arene)]$^+$ adduct (Reaction 12) presumably proceeds through the following sequence, previously demonstrated[26] for olefinic substrates, in which step a is rate determining: (a) oxidative addition of H$_2$ to form [RhH$_2$(DIPHOS)(arene)]$^+$, (b) migratory insertion to form a hydridoalkyl complex [RhH($-$C$-$CH)(DIPHOS)S$_n$]$^+$, and (c) C$-$H bond-forming reductive elimination to yield the hydrogenated product and regenerate [Rh(DIPHOS)S$_2$]$^+$. For oxidative addition of H$_2$ to occur, the [Rh(DIPHOS)(arene)]$^+$ adduct

presumably must adopt the 16-electron η^4 configuration. Such a configuration is expected to be readily accessible for anthracene in view of the fairly low localization energy (1.16 β, Hückel approximation) involved in going from the "anthracene" to the "diene + naphthalene" configuration. The corresponding localization energies for naphthalene and, especially, for benzene are considerably higher (1.21 and 1.53 β, respectively), presumably contributing to the lower reactivities of these arenes.

The selectivity exhibited by this catalyst system for anthracene hydrogenation (reflected in the formation of 1,2,3,4-tetrahydroanthracene) differs from that previously found for the $HCo(CO)_4$-catalyzed hydrogenation, which yields 9,10-dihydroanthracene and which has been interpreted in terms of a free radical mechanism.[17,18] In this connection it is noteworthy that the same selectivity (i.e., to 1,2,3,4-tetrahydroanthracene) now has been observed for cationic, neutral $[Rh(\eta^5\text{-}C_5Me_5)Cl_2]_2$,[27] and anionic $[RuH_2(PPh_3)_2(PPh_2C_6H_4)]^-$ catalysts[8,9,18] and probably reflects the ability of the catalyst, in each case, to hydrogenate anthracene while coordinated in a "diene" (i.e., 1,2,3,4-η^4) mode.

Concluding Remarks

Notwithstanding their opposite charges, the anionic ruthenium and cationic rhodium complexes that we have examined exhibit marked similarities in their coordination chemistry and catalytic properties. Both classes of complex are unusually active hydrogenation catalysts as reflected in their abilities to hydrogenate rather unreactive substrates such as ketones and arenes (also esters and nitriles in the case of the anionic ruthenium catalysts)[12-15] and in the remarkably high rates of olefin hydrogenation with the cationic rhodium catalysts.[8,9,26] Both catalysts also exhibit binding affinities for arenes and similar selectivities for arene hydrogenation, e.g., anthracene to 1,2,3,4-tetrahydroanthracene. These similarities may be interpreted, in part, in terms of a common feature of the pathways through which the catalytic complexes are generated. Thus, in each case the catalytic species is generated by reductively induced irreversible elimination of a ligand (Cl^- and norbornadiene, respectively) from a stable catalyst precursor in a poorly coordinating solvent and in the absence of potentially bridging ligands or counterions. The resulting species in each case exhibit coordinative unsaturation and strong affinities for binding a variety of ligands including relatively weak ligands such as arenes. The marked activities of these complexes {i.e., $[RuH_2(PPh_3)_2(PPh_2C_6H_4)]^-$ and $[Rh(DIPHOS)(MeOH)_2]^+$} as catalysts or catalyst precursors may be interpreted, at least in part, in terms of these attributes. Other examples of this strategy for generating highly active catalytic transition metal hydrides undoubtedly will be found.

References

1. Halpern, J. 1980. J. Organomet. Chem. **200**: 133.
2. Kaesz, H. D. & R. G. Saillant. 1972. Chem. Rev. **72**: 231.
3. Schrock, R. R. & J. A. Osborn. 1970. Chem. Commun.: 567.
4. Schrock, R. R. & J. A. Osborn. 1971. J. Am. Chem. Soc. **93**: 2397.

5. SCHROCK, R. R. & J. A. OSBORN. 1976. J. Am. Chem. Soc. **98:** 2134.
6. SCHROCK, R. R. & J. A. OSBORN. 1976. J. Am. Chem. Soc. **98:** 2143.
7. SCHROCK, R. R. & J. A. OSBORN. 1976. J. Am. Chem. Soc. **98:** 4450.
8. HALPERN, J., R. P. RILEY, A. S. C. CHAN & J. J. PLUTH. 1977. J. Am. Chem. Soc. **99:** 8055.
9. HALPERN, J., A. S. C. CHAN, D. P. RILEY & J. J. PLUTH. 1979. Adv. Chem. Ser. **173:** 16.
10. CRABTREE, R. H., J. M. MIHELCIC & J. M. QUIRK. 1979. J. Am. Chem. Soc. **101:** 7738.
11. LANDIS, C. R. & J. HALPERN. 1983. Organometallics **2:** 840.
12. GREY, R. A., G. P. PEZ, A. WALLO & J. CORSI. 1980. J. Chem. Soc. Chem. Commun.: 783.
13. GREY, R. A., G. P. PEZ & A. WALLO. 1980. J. Am. Chem. Soc. **102:** 5948.
14. PEZ, G. P., R. A. GREY & J. CORSI. 1981. J. Am. Chem. Soc. **193:** 7528.
15. GREY, R. A., G. P. PEZ & A. WALLO. 1981. J. Am. Chem. Soc. **103:** 7536.
16. WILCZYNSKI, R., W. A. FORDYCE & J. HALPERN. 1983. J. Am. Chem. Soc. **105:** 2066.
17. FRIEDMAN, S., S. METLIN, A. SVEDI & I. WENDER. 1959. J. Org. Chem. **24:** 1287.
18. FEDER, H. M. & J. HALPERN. 1975. J. Am. Chem. Soc. **97:** 7186.
19. MANN, B. E., C. MASTERS & B. L. SHAW. 1971. J. Inorg. Nucl. Chem. **33:** 2195.
20. MANUEL, T. A. 1964. Inorg. Chem. **3:** 1794.
21. CLERICI, M. G., S. DI GIOACHINO, F. MASPERO, M. PERROTTI & A. ZANOBI. 1975. J. Organomet. Chem. **84:** 379.
22. DEL PIERO & M. CESARI. 1975. Gazz. Chim. Ital. **105:** 529.
23. MANN, B. E., C. MASTERS & M. CESARI. 1970. Chem. Commun.: 703.
24. PEREGO, G., G. DEL PIERO, M. G. CLERICI & E. PERROTTI. 1973. J. Organomet. Chem. **54:** C51.
25. MORANDINI, F., B. LONGATO & S. BRESADILA. 1977. J. Organomet. Chem. **132:** 291.
26. CHAN, A. S. C. & J. HALPERN. 1980. J. Am. Chem. Soc. **102:** 838.
27. RUSSELL, M. J., C. WHITE & P. M. MAITLIS. 1977. J. Chem. Soc. Chem. Commun.: 427.

HETEROGENIZED HOMOGENEOUS CATALYSTS*

Richard T. Smith, R. Kurt Ungar, Laura J. Sanderson,
and Michael C. Baird

Department of Chemistry
Queen's University
Kingston, Ontario, Canada, K7L 3N6

INTRODUCTION

Although the potential advantages of homogeneous over heterogeneous catalysis (selectivity, high activity) are well recognized, a major impediment to the industrial utilization of homogeneous systems is the general problem of separation of products and catalyst.[1] A common solution to this problem is to attach a normally soluble catalytic species to some inert, insoluble support, in an attempt to combine the virtues of both homogeneous and heterogeneous catalysts.[2-4] However, the anchoring of soluble metal catalyst precursors to, for instance, phosphine-functionalized polymeric supports has not in general led to useful catalysts; problems encountered include leaching of metal into solution, lowered activity or selectivity, and supervenient oxidation of the phosphorus atoms.[4-9] A quite different approach to the problem of separation of catalyst and product has involved the use of metal complexes of polar, water-soluble ligands such as sulfonated triaryl phosphines (see Reference 10 and references therein). Aqueous solutions of such species do catalyze reactions of water-immiscible substrates, possibly via some phase-transfer process, but little has been done in this area.

COMPLEXES OF THE WATER-SOLUBLE PHOSPHINE $Ph_2P(CH_2)_2NMe_3^+$ ($\equiv$ amphos)

We have earlier reported the metal carbonyl derivatives $[Fe(CO)_4amphos]^+$ and $[M(CO)_5amphos]^+$ (M = Mo, W).[11] Comparison of IR and NMR data of these compounds with data for similar compounds (TABLE 1) shows that amphos has net donor properties very similar to those of other tertiary phosphines, and that the effect of the positive charge must be almost completely attenuated along the aliphatic chain of the ligand. Thus it is to be expected that amphos will behave as a "normal" ligand so far as the metal in an amphos complex is concerned, although it is recognized that solvation effects on the chemistry of amphos complexes may well result in anomalous behavior.

The rhodium coordination chemistry was investigated initially via the reaction of $[norbornadieneRhCl]_2$ with amphos nitrate as it was found that similar reactions of $RhCl_3 \cdot 3H_2O$ and $[(C_2H_4)_2RhCl]_2$ gave rise to complex mixtures. Treatment of $[NBDRhCl]_2$ with two equivalents of amphos nitrate in methanol led to dissolution of the former within 15 minutes at room temperature. On

* We thank the Natural Sciences and Engineering Research Council of Canada for support of this research in the forms of an operating grant to M.C.B., a graduate scholarship to R.T.S., and a summer research award to L.J.S.

TABLE 1
IR AND NMR DATA FOR AMPHOS-METAL CARBONYL COMPLEXES

Complex Type	L	v_{CO}	^{31}P Coordination Shift (ppm)
Fe(CO)$_4$L	amphos	2054(m), 1983(w), 1945(vs), 1932(vs)	84.4
	PMePh$_2$	2048(m), 1974(w), 1934(vs)	84.2
	PPh$_3$	2049(m)m 1975(w), 1935(vs)	—
Mo(CO)$_5$L	amphos	2074(w), 1944(vs)	46.9
	PMePh$_2$	2069(w), 1940(vs)	43.0
	PPh$_3$	2070(w), 1940(vs)	43.5
W(CO)$_5$L	amphos	2073(w), 1938(vs)	27.2*
	PMePh$_2$	2069(w), 1930(vs)	24.2†
	PPh$_3$	2070(w), 1935(vs)	26.6‡

* $J_{WP} = 244$ Hz.
† $J_{WP} = 245$ Hz.
‡ $J_{WP} = 280$ Hz.

concentration of the solution and addition of ethyl ether, a 90% yield of [NBDRhCl(amphos)]NO$_3$ (I) was obtained. Complex I is a stable, yellow solid which can be recrystallized from acetonitrile-ethyl ether. It was characterized by elemental analyses, IR spectroscopy, and ^1H and ^{31}P NMR spectroscopy.

Some reactions of I were studied in methanol and water, as outlined in the following equations:

$$I + H_2 \longrightarrow Rh \text{ metal} + \text{norbornane} \qquad (1)$$

$$I + \text{amphos} \longrightarrow [NBDRh(\text{amphos})_2]^{3+} \qquad (2)$$
$$II$$

$$II + H_2 \longrightarrow [RhH_2(\text{amphos})_2(\text{solvent})_2]^{3+} \qquad (3)$$
$$III$$

$$II + H_2 + \text{amphos} \longrightarrow mer\text{-} [RhH_2(\text{amphos})_3(\text{solvent})]^{4+} \qquad (4)$$
$$IV$$

$$III \longrightarrow [Rh(\text{amphos})_2(\text{solvent})_2]^{3+} + H_2 \qquad (5)$$
$$V$$

Attempts to isolate Compounds II–V were generally unsuccessful. Thus a yellow material obtained by concentrating and cooling a solution of II turned gummy on warming to room temperature, while addition of tetrafluoroborate or hexafluorophosphate salts to reaction mixtures resulted only in precipitation of amphos fluoroborate or -phosphate. All compounds were therefore identified by comparisons of their ^1H and ^{31}P NMR spectral parameters, either at room temperature or (in methanol) at −70°C, with data for known compounds of other phosphines. As has been shown previously,[12] the coordination shift (Δ)† and

† The difference in ^{31}P chemical shifts between free and coordinated ligands.

TABLE 2
^{31}P NMR DATA (IN METHANOL)

Compound	$\delta_p^*(J_{RhP}, Hz)$	Δ
Amphos nitrate	-21.3	—
I NBDRhClPPh$_3$	25.3(176) 31.0(171)	46.6 37.0
II [NBDRh(PPh$_3$)$_2$]$^+$	15.4(156) 30.0(157)	36.7 36.0
III† [RhH$_2$(MeOH)$_2$(PPh$_3$)$_2$]$^+$	31.5(121) 42.0(121)	52.8 48.0
IV‡ mer-RhH$_2$Cl(PPh$_3$)$_3$	11.9(95), 31.2(117)§ 20.7(90), 40.3(114)	33.2, 52.5 26.7, 45.7
V [Rh(MeOH)$_2$(PPh$_3$)$_2$]$^+$	46.4(205) 51.2(207)	67.7 57.2

* In ppm relative to external H$_3$PO$_4$; downfield shifts positive.
† Hydride resonance at δ -20.1 (*m*).
‡ Hydride resonances at δ -10.4 (br d, J_{HP} = 168 Hz), δ -17.8 (br s).
§ J_{PP} = 20 Hz.

^{1}J(Rh-P) are usually very reliable criteria for the identification of solution species. Comparisons of data for **II–V** with the corresponding data for several PPh$_3$ compounds appear in TABLE 2, and there can be little doubt that the species in solution are as formulated.

While Reactions 1–5 generally proceeded much as expected on the basis of known chemistry of the triphenylphosphine system, subtle differences were observed. Thus treatment of **I** with hydrogen resulted in reduction of both the metal and the olefin, although NBDRhClPPh$_3$ is reported to be inert to hydrogen;[13] furthermore **I** did not react with a second equivalent of amphos, although NBDRhClPPh$_3$ reacts to give the five-coordinate NBDRhCl(PPh$_3$)$_2$.[14] Reductive elimination of hydrogen from **III** in methanol proceeded over one hour even under one atmosphere of hydrogen, in contrast to the PPh$_3$ analogue, which is stable with respect to loss of hydrogen even at 0.1 mm Hg pressure.[15] Interesting solvent effects were also observed for this reductive elimination reaction, as treatment of **II** with hydrogen in water gave [Rh(amphos)$_2$(H$_2$O)$_2$]$^{3+}$ quantitatively within two minutes, no aqua analogue of **III** being evident.

CATALYSIS STUDIES

Olefin Hydrogenation

It is clear from the previous section that the chemistry of the amphos-rhodium(I) compounds is very similar to that of the well-known, catalytically active PPh$_3$ rhodium(I) system. We were prompted, therefore, to investigate the catalytic activities of the new compounds.

Solutions of **II** were normally generated by treating a suspension of [NBDRhCl]$_2$ with four equivalents of amphos nitrate. Reactions were complete within 30 minutes, and the species in solution could be characterized by ^{31}P NMR spectroscopy. Preliminary experiments using the water-soluble olefins maleic and crotonic acids showed that **II** does indeed catalyze the hydrogenation of olefins in aqueous solution (catalyst concentration $\sim$0.01 M, olefin concentration $\sim$0.8 M). After one hour at 25°C and one atmosphere pressure of hydrogen, turnover numbers were 10–12 moles olefin per mole catalyst. Catalyst activities decreased on the addition of a third equivalent of amphos per rhodium, but increased dramatically (turnover number 100–125) in methanol, possibly because of the higher solubility of hydrogen in that solvent, but also perhaps because of the greater stability of the dihydride (see above). Interestingly, the amphos-rhodium catalyst system was slightly more active for the hydrogenation of maleic acid in methanol than was [NBDRh(PPh$_3$)$_2$]$^+$,[16] and about one-quarter as active for the hydrogenation of styrene in 1:1 ethanol-benzene as was RhCl(PPh$_3$)$_3$.[17] Rhodium complexes of the sulfonated triphenylphosphine (m-C$_6$H$_4$SO$_3$Na)PPh$_2$, on the other hand, are not catalytically active in aqueous solution.[18]

Having thus established the position of **II** in the hierarchy of homogeneous, rhodium olefin hydrogenation catalysts,[19] we turned to investigations of its utilization in two-phase systems. A very useful approach involved mixing an aqueous solution of **II** (20 ml solution, 2.5×10^{-3} M) with a water-immiscible olefin ($\sim$50 mmol styrene or 1-hexene), either neat or dissolved in an organic solvent (20 ml of n-pentane, methylene chloride, or ethyl ether). The mixtures were then shaken in a Parr hydrogenation apparatus for up to six hours under three atmospheres hydrogen.

Again good catalytic activities were observed, with turnover numbers in some runs comparable with those of the homogeneous catalyst system in methanol. Ether was found to be the best of the co-solvents tried, and 1-hexene was hydrogenated faster than styrene, although some isomerization to 2-hexene also occurred. Again addition of free amphos resulted in reduced activity, as did the use of water that had not been distilled. The aqueous phase containing the catalyst could be reused with little loss in activity, even after short periods of exposure to air. Indeed, hydrogenations could even be run with a 3:1 hydrogen-air mixture, although activity did decrease because of slow oxidation of amphos to amphos oxide. In contrast two-phase catalytic hydrogenation systems containing rhodium complexes of (m-C$_6$H$_4$SO$_3$Na)PPh$_2$ are said to be very air sensitive.[20]

Of even greater interest with the system discussed here is the observation that virtually no rhodium was leached from the aqueous phase. Analyses showed that rhodium concentrations in the organic phases were $<$0.25 ppm, corresponding to $<$0.1% of the catalyst used. Emulsification between liquid layers was rarely a problem, and thus separation of catalyst and product could be readily effected by simple decantation.

A second useful approach to the utilization of **II** as a catalyst in a two-phase system involved absorbing **II** from solution onto a cationic exchange resin. This approach has been taken elsewhere, using anionic complexes, but with mixed success.[21,22] Catalysts were prepared by mixing an aqueous solution of **II** (0.05–1.2 mmol, 50 ml) with 10 g of the sodium form of a macroreticular sulfonated polystyrene resin (200–400 mesh). Adsorption of **II** into the resin was generally complete within 45 minutes, as judged by the disappearance of the yellow color of the solution; analyses of the remaining solution showed that uptake of the rhodium was essentially quantitative. The water was then filtered off and the resin was washed well with methanol and ether.

The supported catalyst was transferred under nitrogen to a Parr Mini Reactor (300 ml) equipped with a glass liner. Solutions of olefin (styrene, 1-hexene, cyclohexene) in a solvent (ethyl ether, acetone), were added, and the reaction was sealed and pressured with hydrogen to 75–200 psi. The reactions were normally stirred overnight (17 hours), whereupon the reactor was opened and the liquid phases were analyzed by gas chromatography. While the reaction conditions were somewhat variable as various attempts were made to achieve optimization, yields of hydrogenated products of 85–100% could be achieved from 1-hexene; little isomerization to internal olefins was observed if the reactions were stopped short of completion. Acetone was much better as a solvent for the olefins than was ethyl ether, possibly because the latter did not solvate the highly polar surfaces of the resin beads. Used catalyst was found to exhibit little or no loss of activity when reused, but the organic phases were totally inactive. Analyses on the organic phases showed, in fact, that leaching of the rhodium was again minimal, although it can be eluted from the resin with 6 M perchloric acid. Recovery of precious metals from supported catalysts is rarely mentioned in the literature, presumably because it is rarely accomplished without destroying the supporting material. Most supported rhodium catalysts are formed by coordinating the metal to tertiary phosphine moieties incorporated onto the surface of the support,[2-4] and there is no well-developed methodology for the quantitative recovery, through substitution reactions, of rhodium from its complexes with tertiary phosphines.

Other Catalytic Reactions

The two two-phase catalyst systems already described are also effective for the hydroformylation of olefins. The best results were obtained using catalyst systems containing three amphos per rhodium, and 85–100% conversion of 1-hexene to heptanals could be effected by stirring an aqueous solution of the catalyst (0.1 mmol, buffered to pH 5.5–7) with a solution of 1-hexene (50 mmol in *n*-pentane, methylene chloride, or ethyl ether) for 24 hours at 90°C and a total pressure of 40 atm (1:1 CO-H$_2$). The *n*:iso product ratios varied between 1.7 and 4.6, and were not affected by added amphos. Somewhat better *n*:iso ratios were obtained (6.8) using the resin-supported system, but yields were erratic for reasons not yet clear. In both types of two-phase systems, however, metal leaching was again minimal, and thus these approaches to heterogenization show promise.

Finally, mention should be made of catalytic reactions that do not proceed. Although other rhodium compounds do catalyze the water-gas shift reaction,[23] the hydrogenation of ketones,[24] and the carbonylation of methanol,[25] little or no catalytic activity was observed with the amphos-rhodium system under seemingly appropriate conditions.

ACKNOWLEDGMENTS

This research was facilitated by a loan of RhCl$_3$·3H$_2$O from Johnson Matthey Ltd.

REFERENCES

1. PARSHALL, G. W. 1980. Homogeneous Catalysis. Wiley-Interscience. New York, N.Y.
2. HARTLEY, F. R. & P. N. VEZEY. 1977. Adv. Organomet. Chem. **15:** 189.
3. BAILEY, D. C. & S. H. LANGER. 1981. Chem. Rev. **81:** 109.
4. FRANCESCO, C., G. BRACA, C. CARLINI, G. SBRANA & G. VALENTINI. 1982. J. Mol. Catal. **14:** 1.
5. LANG, W. H., A. T. JUREWICZ, W. O. HAAG, D. D. WHITEHURST & L. D. ROLLMANN. 1977. J. Organomet. Chem. **134:** 85.
6. GRUBBS, R. H. & E. M. SWEET. 1977/78. J. Mol. Catal. **3:** 259.
7. TANG, S. C., T. E. PAXSON & L. KIM. 1980. J. Mol. Catal. **9:** 313.
8. DE CROON, M. H. J. M. & J. W. E. COENEN. 1981. J. Mol. Catal. **11:** 301.
9. BEMI, L., H. C. CLARK, J. A. DAVIES, C. A. FYFE & R. E. WASYLISHEN. 1982. J. Am. Chem. Soc. **104:** 438.
10. JOÓ, F. & Z. TÓTH. 1980. J. Mol. Catal. **8:** 369.
11. SMITH, R. T. & M. C. BAIRD. 1982. Inorg. Chim. Acta **62:** 135.
12. SLACK, D. A., I. GREVELING & M. C. BAIRD. 1979. Inorg. Chem. **18:** 3125.
13. SHAPLEY, J. R., R. R. SHROCK & J. A. OSBORN. 1969. J. Am. Chem. Soc. **91:** 2816.
14. VRIEZE, K., H. C. VOLGER & A. P. PRAAT. 1968. J. Organomet. Chem. **14:** 185.
15. BROWN, J. M., P. A. CHALMER & P. N. NICHOLSON. 1978. J. Chem. Soc. Chem. Commun.: 646.
16. SCHROCK, R. R. & J. A. OSBORN. 1976. J. Am. Chem. Soc. **98:** 2134.
17. OSBORN, J. A., F. H. JARDINE, J. F. YOUNG & G. WILKINSON. 1966. J. Chem. Soc. A: 1711.
18. DROR, Y. & J. MANASSEN. 1977. J. Mol. Catal. **2:** 219.
19. JAMES, B. R. 1973. Homogeneous Hydrogenation (Chapter 11). John Wiley & Sons. New York, N.Y.
20. BOROWSKI, A. F., D. J. COLE-HAMILTON & G. WILKINSON. 1978. Nouv. J. Chim. **2:** 137.
21. CHAUVIN, Y., D. COMMERENC & F. DAWANS. 1977. Prog. Polym. Sci. **5:** 95.
22. DRAGO, R. S., E. D. NYBERG, A. E. AMMA & A. ZOMBECK. 1981. Inorg. Chem. **20:** 641.
23. YOSHIDA, T., T. OKANO, Y. UEDA & S. OTSUKA. 1981. J. Am. Chem. Soc. **103:** 3411.
24. FUJITSU, H., E. MATSUMURA, K. TAKESHITA & I. MOCHIDA. 1981. J. Chem. Soc. Perkin Trans. **1:** 2650.
25. FORSTER, D. 1979. Adv. Organomet. Chem. **17:** 225.

CARBONYLATION AND DECARBONYLATION CYCLES
OF ALKYL COMPLEXES IN CATALYTIC REACTIONS*

Jim D. Atwood, Thomas S. Janik, Michael F. Pyszczek, and Patrick S. Sullivan

Department of Chemistry
State University of New York at Buffalo
Buffalo, New York 14214

Homogeneous hydrogenations of olefins have played an important part in the development of organometallic chemistry, especially of phosphine and phosphite complexes. A number of important reaction types have first received attention as part of a catalytic cycle (oxidative addition, reductive elimination, insertions, etc.).[1, 2] The majority of catalytic systems to date have centered on d^8 square planar complexes, or transition metal hydrides, and more recently on metal clusters.[1-3]

The most often studied hydrogenation catalyst is $Rh(PPh_3)_3Cl$.[4-6] Although the mechanism involved in a catalytic cycle is difficult to ascertain, this complex has been studied often and the scheme below is accepted as the mechanism.[7-9]

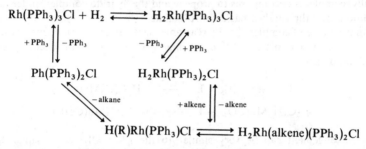

The initial step seems to be oxidative addition of H_2 to the 16 e^- complex $Rh(PPh_3)_3Cl$. A similar mechanism can be written for other square planar 16-electron hydrogenation catalysts. The complex $HRh(CO)(PPh_3)_3$ can serve as the model of hydride complexes in the catalytic cycle, and the accepted mechanism is shown below.[10]

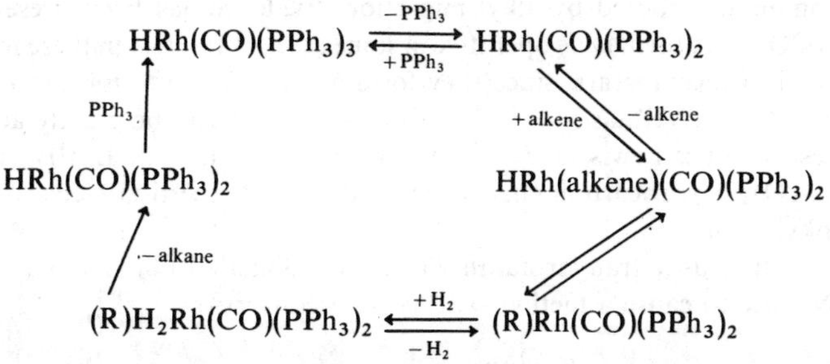

$$HRh(CO)(PPh_3)_3 \underset{+PPh_3}{\overset{-PPh_3}{\rightleftarrows}} HRh(CO)(PPh_3)_2$$

The individual steps in both of these schemes are similar (the primary difference is whether alkene or H_2 is added first), and independent evidence exists for these steps.

* The financial support of the National Science Foundation during the course of this research is gratefully acknowledged.

259

Oxidative addition of hydrogen and dissociation of a phosphine are very commonly observed.[11] This is also true of reductive elimination of an alkane from a hydrido-alkyl complex, although the mechanism is not always straightforward.[12]

Radical mechanisms (17-electron complexes) have been suggested in some catalytic reactions.[13] There is good evidence for hydrogenations by radical mechanisms for the complexes $HCo(CN)_5^{-3}$, $HMn(CO)_5$, and perhaps $HCo(CO)_4$.[13]

Allyl complexes, where the initial step can be η^3-η^1 rearrangement, have been used as catalysts, primarily by Muetterties and co-workers.[14–19] The complex η^3-$C_3H_5Co(P(OMe)_3)_3$ was active for the hydrogenation of benzene at ambient conditions of temperature and pressure.[14, 16] Allyl manganese complexes, η^3-$C_3H_5Mn(CO)_2L_2$ [L = $P(OCH_3)_3$, PEt_3, etc.] and η^3-$C_3H_5Mn(CO)$-$(P(OCH_3)_3)_3$, were active for the hydrogenation of alkenes although with a short catalyst lifetime.[15] Rhodium complexes (η^3-$C_3H_5Rh(P(O-i-Pr)_3)_3)_2$ and η^3-$C_3H_5Rh(P(O-i-Pr)_3)_3$ were active for arene hydrogenation, although rapid decomposition to a hydride limited the studies.[17] Room temperature isomerization of alkenes occurred with $[(\eta^3$-$C_3H_5)Fe(CO)_3]_2$.[18] For this reaction the allyl group probably was cleaved prior to the isomerization.[18] Cleavage of the allyl group may be important in forming the catalyst in the other allyl systems.[19] Certainly, each of the allyl complexes decomposes to propene and the hydrido complex under reaction conditions and the catalyst has a short lifetime.

Another type of complex that by rearrangement can create an open coordination site is a methyl complex.[20, 21] Methyl migration to a CO can occur, opening a coordination site.[20, 21]

$$CH_3Mn(CO)_5 + L \longrightarrow CH_3C(O)Mn(CO)_4L$$

$$\eta^3\text{-}C_3H_5Mn(CO)_4 + L \longrightarrow \eta^1\text{-}C_3H_5Mn(CO)_4L$$

This rearrangement can be very similar to the η^3-η^1 allyl conversions. Methyl migration can occur at markedly different rates depending on the ligand environment and the metal center.[20] The methyl migration reaction has been known for a number of years. Excellent reviews have appeared, and a detailed stereochemical analysis has shown that $MeMn(CO)_5$ reacts by methyl migration, not by CO insertion.[20–22] It has been presumed, based on the manganese results, that reactions of alkyl complexes proceed by alkyl migration. Evidence has been presented that indicates a CO insertion may be operative in some cases.[23] Interest in these migrations has increased in recent years, since they form a possible route for carbon-carbon bond formation. Shriver has shown that alkyl migrations can be greatly accelerated by the presence of a Lewis acid.[24, 25] We have shown that in $LiAlH_4$ reduction of $Cp_2Fe_2(CO)_4$, hydrocarbon chains are built by CO insertion steps,[26] probably prior to alkyl formation.

Solvents such as tetrahydrofuran (THF) or dimethylsulfoxide (DMSO) are sufficiently basic to cause a methyl → acetyl interconversion.[27, 28]

$$CpFe(CO)_2R \xrightarrow{\text{DMSO}} CpFe(CO)(COR)(DMSO)$$

$$CH_3Mn(CO)_5 \xrightarrow{\text{THF}} CH_3C(O)Mn(CO)_4(THF)$$

Each of these complexes was characterized spectroscopically, but could not be isolated.[27, 28] If an alkene were sufficiently basic to cause the methyl → acetyl interconversion,

$$Me-\underset{\underset{CO}{|}}{M} + alkene \longrightarrow Me\overset{\overset{O}{\|}}{C}-M(alkene)$$

an acetyl, alkene transition metal complex would be formed (the acetyl and alkene would be *cis* in a rigid molecule). Loss of a liquid would create an open coordination site for H_2, and a catalytic cycle could be completed. The presence of the methyl group can serve as a mechanistic marker, providing additional information regarding the stereochemistry of key intermediates.

We have found that $CH_3Co(CO)_2(P(OMe)_3)_2$ is active for the catalytic hydrogenation of terminal olefins at the rate of ~500 turnovers/hour at ambient conditions.[29] The catalytic activity is inhibited by 0.25 atmosphere of CO, which leads to $CH_3C(O)Co(CO)_2(P(OMe)_3)_2$, and by the presence of excess $P(OMe)_3$, which leads to $CH_3C(O)Co(CO)(P(OMe)_3)_3$.

$$CH_3Co(CO)_2(P(OMe)_3)_2 + CO \longrightarrow CH_3C(O)Co(CO)_2(P(OMe)_3)_2$$
$$\xrightarrow{\quad P(OMe)_3 \quad} CH_3C(O)Co(CO)(P(OMe)_3)_3$$

The presence of *p*-dinitrobenzene had no effect on the rate of hydrogenation. The hydrides $HCo(CO)_2(P(OMe)_3)_2$ and the dimer $Co_2(CO)_4(P(OMe)_3)_4$ showed no reactivity for hydrogenation or isomerization even under photolysis. The data collected are most consistent with a series of carbonylation/decarbonylation cycles. Such a scheme is shown below.[29]

$$CH_3Co(CO)_2P_2 + alkene \rightleftharpoons CH_3C(O)Co(CO)(alkene)P_2 \qquad (1)$$
$$CH_3C(O)Co(CO)(alkene)P_2 \xrightarrow{\;+H_2\;} CH_3C(O)Co(alkene)(H_2)P_2 + CO \qquad (2)$$
$$CH_3C(O)Co(alkene)(H_2)P_2 \rightleftharpoons CH_3Co(R)(H)(CO)P_2 \qquad (3)$$
$$CH_3Co(R)(H)(CO)P_2 + CO \longrightarrow CH_3Co(CO)_2P_2 + RH \qquad (4)$$
$$P = P(OMe)_3$$

The series of acetyl-methyl interconversions allows each intermediate in the catalytic cycle to have 18 electrons. The geometries, as shown in FIGURE 1, of the six-coordinate intermediates can be assigned with some confidence based on the observations. The addition of H_2 would occur *cis*, and the favored geometry of the two $P(OMe)_3$ ligands would be *trans*. The alkene inserts into the *cis* Co—H bond, and the methyl group migrates to fill the coordination site. This leads to a *cis* disposition of the alkyl group and the hydride with the methyl *trans* to the hydride. This geometry is required since the catalyst has a very long lifetime (>10,000 turnovers) and elimination of CH_4 kills the catalyst.

Two experiments were accomplished to investigate the proposed sequence more thoroughly. Increasing the pressure of hydrogen from 2 atmospheres to 60 atmospheres in the hydrogenation of 1-octene led to an increase of 100 in the rate of formation of octene and isomerized octenes catalyzed by $CH_3Co(CO)_2(P(OMe)_3)_2$. At room temperature and 60 atmospheres of H_2, 30 ml of 1-octene were converted to 23 ml octane, 3 ml *trans*-2-octene, 0.6 ml *cis*-2-octene, and 3 ml of 1-octene in

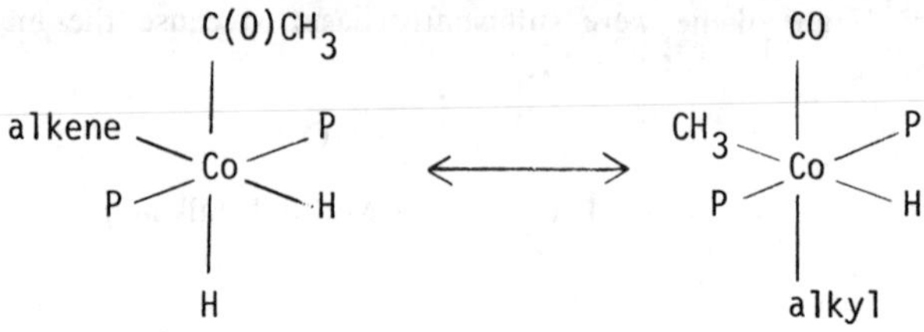

FIGURE 1. Required stereochemistry of Reaction 3 leading to *cis* alkyl and H, but *trans* CH_3 and H [P = P(OMe)$_3$].

one hour. This increase in rate with an increase in hydrogen pressure suggests that Equilibrium 2 is important in determining the rate. That Equilibria 1–4 are established during the hydrogenation is further shown by the deuterium labeling experiment.

$$\text{1-octene} + D_2 \ (2 \text{ atm}) \xrightarrow{\text{CH}_3\text{Co(CO)}_2\text{(P(OMe)}_3\text{)}_2} \text{isomerization and hydrogenation}$$

Up to three deuterium atoms were incorporated into the 1-octene, two were incorporated into *trans*-2-octene, and two were incorporated into *cis*-2-octene. The relative amounts are shown in TABLE 1. The incorporation into 1-octene indicates that the equilibria occur several times without progressing to the hydrogenated product. The incorporation into the 2-octenes is substantial but less than into 1-octene, consistent with deuterium incorporation before isomerization. After isomerization the 2-octene would be lost from the coordination sphere and further deuterium incorporation would not occur. The incorporation into 1-octene also indicates the equilibria exchanging hydrogen and deuterium occur readily since the primary octane formed contained only one deuterium (about three times the amount of $C_8H_{16}D_2$). No octane was formed containing no deuterium. The extensive D_2 incorporation into the octene requires that the equilibria occur readily. Both the pressure dependence and D_2 labeling are consistent with the proposed scheme and suggest that Equilibria 1, 2, and 3 are readily established under reaction conditions. Increased H_2 pressure would shift the equilibria toward hydrogenated product.

To maintain the methyl *trans* to the hydride in the intermediate, $CH_3Co(R)(H)(CO)(P(OMe)_3)_2$, the P(OMe)$_3$ ligands must be *trans* as shown in FIGURE 1. To test this hypothesis we prepared the bidentate analogue of the catalyst, $CH_3Co(CO)_2$(Pom-Pom) [Pom-Pom = $(OMe)_2PCH_2CH_2P(OMe)_2$] and investigated its ability to catalyze the hydrogenation and isomerization of alkenes. This complex functioned very similarly to its monodentate analogue as a catalyst only slower, with a rate of 20 turnovers/hour. A portion of the lowered activity may be due to the decomposition of the catalyst, which is greatly enhanced in comparison

TABLE 1
THE RELATIVE AMOUNTS OF DEUTERIUM INCORPORATION
INTO OCTENES DURING ISOMERIZATION

Alkene	C_8H_{16}	$C_8H_{15}D$	$C_8H_{14}D_2$	$C_8H_{13}D_3$
1-Octene	1	0.9	0.24	0.04
trans-2-Octene	1	0.62	0.12	—
cis-2-Octene	1	0.57	0.03	—

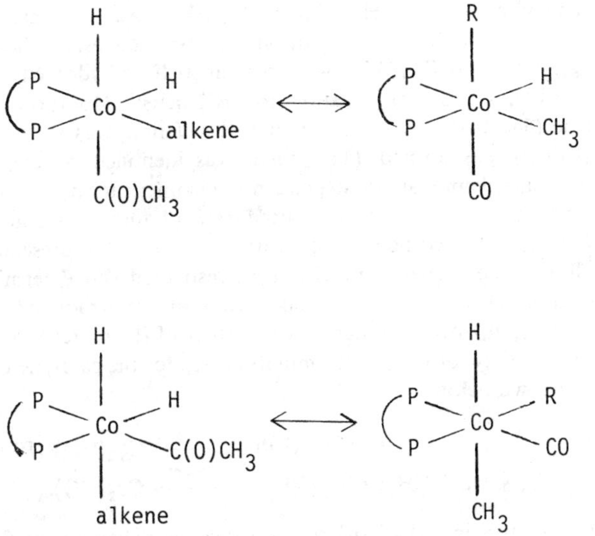

FIGURE 2. Possible stereochemistries of the intermediates in catalytic hydrogenation of alkenes by CH₃Co(CO)₂(Pom-Pom) (R = alkyl, P⌢P = Pom-Pom). The first reaction leads to the isomer with R, H, and CH₃ on the face of the octahedron, which should lose RH and CH₄ at comparable rates.

to the $P(OMe)_3$ complex. The amount of CH_4 and CH_3CHO formed during hydrogenation by $CH_3Co(CO)_2(Pom-Pom)$ was 33% after 2 hours, which can be compared to 5% after 24 hours for $CH_3Co(CO)_2(P(OMe)_3)_2$. The structures that lead to enhanced decomposition are shown in FIGURE 2. Since CH_4/CH_3CHO elimination can occur more readily from the $CH_3Co(CO)_2(Pom-Pom)$ reaction than from $CH_3Co(CO)_2(P(OMe)_3)_2$, the observations are consistent with the proposed scheme. The fact that $CH_3Co(CO)_2(Pom-Pom)$ shows activity confirms that CO dissociation, not $P(OMe)_3$ dissociation, opens the coordination site for H_2.

The monophosphite complex, $CH_3Co(CO)_3P(OMe)_3$, showed no catalytic activity at ambient conditions. In attempting to define the reason for the inactivity we examined the decarbonylation reaction of $CH_3C(O)Co(CO)_3P(OMe)_3$.

$$CH_3C(O)Co(CO)_3P(OMe)_3 \longrightarrow CH_3Co(CO)_3P(OMe)_3 + CO$$

This reaction occurs at a reasonable rate at 60°C in hexene. When $CH_3Co(CO)_3$-$P(OMe)_3$ was placed in the presence of 1-hexene and H_2 at 60°C, hydrogenation and isomerization were observed (100 turnovers/hour) although the catalyst had a short lifetime with CH_4 elimination enhanced. In contrast to $CH_3Co(CO)_2$-$(P(OMe)_3)_2$, the monophosphite species also functioned as a catalyst for cyclohexene with little difference in rate from 1-hexene. This suggests that the selectivity in hydrogenation observed for $CH_3Co(CO)_2(P(OMe)_3)_2$ is steric in nature.

The analogous triphenylphosphine complex, $CH_3Co(CO)_3PPh_3$, shows only slight activity for hydrogenation and isomerization of 1-hexene at 60°C; decarbonylation of $CH_3C(O)Co(CO)_3PPh_3$ occurs at a reasonable rate only above

80°C. The benzyl complex, $C_6H_5CH_2Co(CO)_3PPh_3$, was of considerable interest because the carbonylation-decarbonylation equilibrium is established at 30°C. The complex, $C_6H_5CH_2Co(CO)_3PPh_3$, was active at 30°C for hydrogenation of 1-octene (10 turnovers/hour) and isomerization to cis-2-octene (16 turnovers/hour) and trans-2-octene (56 turnovers/hour) although the lifetime was short and the dimer $Co_2(CO)_6(PPh_3)_2$ was formed. (The dimer was identified by comparison of its properties to a sample prepared independently.) Isomerization of 1-octene occurred in the absence of H_2 at a much reduced rate (2 turnovers/hour). The dimer, $Co_2(CO)_6(PPh_3)_2$, showed no reaction with 1-octene in the presence of H_2 up to 60°C, which was the highest temperature investigated. To determine whether a 17-electron complex could be the active catalyst we irradiated a solution of $Co_2(CO)_6(PPh_3)_2$ in 1-octene under an atmosphere of H_2; no reaction was observed. The dimer is evidently formed in a termination step for the catalytic cycle, probably one of those shown below.

$$2\ HCo(CO)_3PPh_3 \longrightarrow Co_2(CO)_6(PPh_3)_2 + H_2$$
$$HCo(CO)_3PPh_3 + C_6H_5CH_2Co(CO)_3PPh_3 \longrightarrow Co_2(CO)_6(PPh_3)_2 + C_7H_8$$

We have also investigated a series of manganese alkyl complexes. $CH_3Mn(CO)_3$-$(P(OMe)_3)_2$ functions as a hydrogenation catalyst at 70°C and one atmosphere of H_2. 1-Octene was hydrogenated at 10 turnovers/hour and isomerized to trans-2-octene and cis-2-octene at rates of 2 turnovers/hour and 1 turnover/hour, respectively. The complex, $CH_3Mn(CO)_3(P(OMe)_3)_2$, could be recovered in near quantitative yields after 10-hour reaction times. Cyclohexene or cis-2-hexene could not be hydrogenated before decomposition of the catalyst. Methane was slowly evolved during catalytic reactions; eventually a brown oil with no infrared absorbances in the carbonyl stretching region was obtained. The $P(OPh)_3$ analogue, $CH_3Mn(CO)_3$-$(P(OPh)_3)_2$, was inactive at temperatures up to 100°C and could be quantitatively recovered from the reaction mixture. The benzyl complex, $C_6H_5CH_2Mn(CO)_4PPh_3$, showed only very slight activity for hydrogenation or isomerization of terminal alkenes.

These investigations into the potential of alkyl complexes of cobalt and manganese as catalysts for the hydrogenation and isomerization of alkenes have shown that opening two coordination sites by carbonylation/decarbonylation reactions will function in catalytic cycles. All evidence is consistent with the proposed mechanism, while substantial portions of the data are inconsistent with other mechanisms. A radical mechanism can be ruled out by the following data:

1. p-Dinitrobenzene has no effect on the rate of hydrogenation.
2. Dimers are never observed as products, nor is C_2H_6 eliminated from solutions of the methyl complexes.
3. The analogous hydrides are not active in catalytic reactions.
4. The analogous dimers are not catalysts, even under photolysis.

The presence of the alkyl group serves as a probe of the possible geometries of the intermediates, which cannot be directly observed, by comparison of the rates of elimination. The fact that CH_4 is eliminated during the catalysis by $CH_3Co(CO)_2$-$(P(OMe)_3)_2$ at $1/10^6$ of the rate of hydrogenated alkane is quite informative

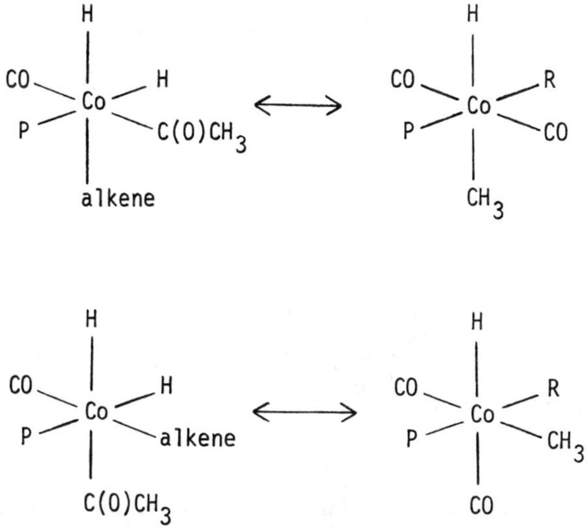

FIGURE 3. Possible stereochemistries in the catalytic hydrogenation of alkenes by $CH_3Co(CO)_3P(OMe)_3$ [R = alkyl, P = $P(OMe)_3$]. The second reaction should lead to loss of CH_4.

regarding the required geometry, and provides strong evidence for the reaction shown in FIGURE 1. That the rate of CH_4 elimination is enhanced relative to hydrogenated product for either $CH_3Co(CO)_2$(Pom-Pom) or $CH_3Co(CO)_3P(OMe)_3$ indicates other geometric possibilities for these complexes. Such possibilities are shown in FIGURES 2 and 3.

This ability to define the stereochemistry of intermediates allows more mechanistic interpretation than is usually possible for a catalytic cycle. Reaction 3 requires that the alkene insert into the *cis* Co—H bond as opposed to a hydride migration and further that this reaction occurs many orders of magnitude more rapidly than elimination of CH_3CHO, which would be a competing reaction. Alkyl migration to the open coordination site occurs. Reductive elimination only occurs from a *cis* geometry, which indicates some rigidity in the six-coordinate complex and rules out bimolecular elimination in this cycle. The data suggest that the alkene adds prior to H_2 since the addition of H_2 would almost certainly lead to loss of CH_4 and $CH_3C(O)H$ as shown in FIGURE 4.

$$CH_3C(O)Co(H_2)(CO)(P(OMe)_3)_2 \longrightarrow CH_3Co(CO)(H_2)(P(OMe)_3)_2$$

Addition of H_2 to a very similar rhodium complex leads to aldehyde elimination in the hydroformylation reaction.[30, 31]

The ligand environment affects the capability of a complex to function as a catalyst both in the rate and in catalyst lifetime. The rate effects appear to be directly related to the effect of the ligands on rates of alkyl migration.[32] Our data suggest that two phosphorus donors are required for reasonable catalyst lifetime.

This research has shown that (1) alkyl complexes can be active in catalytic cycles for hydrogenation and isomerization of alkenes; (2) this activity is seen at

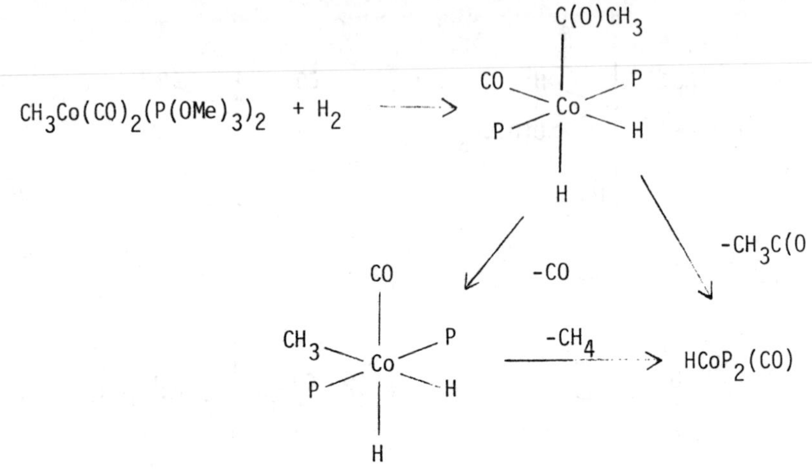

FIGURE 4. Illustration of the addition of H_2 to the cobalt center prior to alkene complexation. This would result in enhanced reductive elimination of CH_4 and $CH_3C(O)H$.

conditions where carbonylation and decarbonylation occur readily; (3) similar hydrido complexes are not active; (4) metal dimers are not active, even under photolysis; (5) p-dinitrobenzene has no effect on the rate of hydrogenation, but CO or $P(OMe)_3$ inhibits catalytic activity; (6) the most reasonable mechanism involves alkyl migration; and (7) the presence of the alkyl group as a mechanistic probe provides *unique* information on the catalytic cycle.

REFERENCES

1. HALPERN, J. 1968. Adv. Chem. Ser. **70**: 1.
2. COLLMAN, J. P. 1968. Acc. Chem. Res. **1**: 136.
3. JAMES, B. R. 1979. Adv. Organomet. Chem. **17**: 376.
4. OSBORN, J. A., F. H. JARDINE, J. F. YOUNG & G. WILKINSON. 1964. J. Chem. Soc. A: 1711.
5. JARDINE, F. M., J. A. OSBORN & G. WILKINSON. 1967. J. Chem. Soc. A: 1574.
6. MONTELATICI, S., A. VAN DER ENT, J. A. OSBORN & G. WILKINSON. 1968. J. Chem. Soc. A: 1054.
7. MEAKIN, P., J. P. JESSEN & C. A. TOLMAN. 1972. J. Am. Chem. Soc. **94**: 3240.
8. HALPERN, J. & C. S. WONG. 1973. J. Chem. Soc. Chem. Commun.: 629.
9. HALPERN, J., F. OKAMOTO & A. ZAKHARIEV. 1976. J. Mol. Catal. **2**: 65.
10. O'CONNOR, C. & G. WILKINSON. 1968. J. Chem. Soc. A: 2665.
11. WERNER, H. & R. FESER. 1979. Angew. Chem. Int. Ed. Engl. **18**: 157.
12. NORTON, J. R. 1979. Acc. Chem. Res. **12**: 139.
13. HALPERN, J. 1979. Pure Appl. Chem. **51**: 2171.
14. STUHL, L. S., M. RAKOWSKI-DUBOIS, F. J. HIRSEKORN, J. R. BLEEKE, A. E. STEVENS & E. L. MUETTERTIES. 1978. J. Am. Chem. Soc. **100**: 2405.
15. STUHL, L. S. & E. L. MUETTERTIES. 1978. Inorg. Chem. **17**: 2148.
16. RAKOWSKI, M. C., F. J. HIRSEKORN, L. S. STUHL & E. L. MUETTERTIES. 1976. Inorg. Chem. **15**: 2379.
17. DAY, V. M., M. F. FREDRICH, G. S. REDDY, A. J. SIVAK, W. R. PRETZER & E. L. MUETTERTIES. 1977. J. Am. Chem. Soc. **99**: 8091.

18. PUTNIK, C. F., J. J. WELTER, G. D. STUCKY, M. J. D'ANIELLO, JR., B. A. SOSINSKY, J. F. KIRNER & E. L. MUETTERTIES. 1978. J. Am. Chem. Soc. **100**: 4107.
19. BLEEKE, J. R. & E. L. MUETTERTIES. 1981. J. Am. Chem. Soc. **103**: 556.
20. CALDERAZZO, F. 1977. Angew. Chem. Int. Ed. Engl. **16**: 299.
21. BERKE, H. & R. HOFFMAN. 1978. J. Am. Chem. Soc. **100**: 7224.
22. NOACK, K. &. F. CALDERAZZO. 1968. Inorg. Chem. **7**: 345.
23. BRUNNER, H. & H. VOGT. 1981. Angew. Chem. Int. Ed. Engl. **20**: 405.
24. BUTTS, S. B., E. M. HOLT, S. H. STRAUSS, N. W. ALCOCK, R. E. STIMSON & D. F. SHRIVER. 1979. J. Am. Chem. Soc. **101**: 5865.
25. BUTTS, S. B., S. H. STRAUSS, E. M. HOLT, R. E. STIMSON, N. W. ALCOCK & D. F. SHRIVER. 1980. J. Am. Chem. Soc. **102**: 5093.
26. WONG, A. & J. D. ATWOOD. 1981. J. Organomet. Chem. **210**: 395.
27. NICHOLAS, K., S. RAYHU & M. ROSENBLUM. 1974. J. Organomet. Chem. **78**: 133.
28. CALDERAZZO, F. & F. A. COTTON. 1962. Inorg. Chem. **1**: 30.
29. JANIK, T. S., M. F. PYZCZEK & J. D. ATWOOD. 1981. J. Mol. Catal **11**: 33.
30. BROWN, C. K. & G. WILKINSON. 1970. J. Chem. Soc. A: 1392.
31. PRUETT, R. L. 1979. Adv. Organomet. Chem. **17**: 1.
32. RUSZCZYK, R. &. J. D. ATWOOD. (Manuscript in preparation.)

THE EFFECT OF NET IONIC CHARGE ON THE CATALYTIC BEHAVIOR OF METAL HYDRIDES[*]

Robert H. Crabtree, Douglas R. Anton, and Mark W. Davis

Yale University
Department of Chemistry
Sterling Chemistry Laboratories
New Haven, Connecticut 06511

One important factor in the activation of a substrate for catalysis is the electronic influence of the metal and associated ligands (ML_n) on the coordinated substrate. The nature of M and L and the oxidation state of M are usually considered to be important factors. The purpose of this paper is to call attention to another factor that can have a profound influence: the net ionic charge.

The net ionic charge on a cation or anion {e.g., $1+$ in the case of $[CoCl_2(NH_3)_4]Cl$} is delocalized over the complex in accordance with Pauling's electroneutrality principle.[1] Hydrogen atoms in the immediate coordination sphere of a metal are particularly effective in this delocalization. It is striking that the vast majority of H_2O and NH_3 complexes are cationic. Even organometallic complexes, which usually do not bind hard ligands very well, can form complexes with H_2O or other hard ligands when a net positive charge is present {e.g., $[(C_6H_6)Os(H_2O)_3]^{2+}$,[2] $[Pddpe(thf)_2]^{2+}$,[3] $[Rh(C_5Me_5)(MeCO)_3]^{2+}$,[4] $[IrH_2(PPh_3)_2(H_2O)_2]^+$}.[5] These lightly stabilized complexes can make useful catalyst precursors.[6]

Hydrogen has an electronegativity (2.2) only very slightly higher than those of the middle and late transition metals (1.5–2.2). Accordingly, one might expect that the $M-H$ bond in the corresponding hydrides would be relatively nonpolar. The generally weak $\nu(M-H)$ stretching frequencies in the IR spectra of these compounds may reflect this state of affairs. The chemical reactivity of $M-H$, sometimes tending to M^+-H^- and sometimes to M^--H^+,[7,8] is understandable on the same basis.

It is among the metal hydrides that one might expect to find the most striking net ionic charge effects. In Grey, Pez, Wallo, and Corsi's most elegant work described in this session (see Reference 7 and this volume), an anionic metal hydride is found to catalytically reduce even carboxylic esters. The complex therefore acts as an H_2-regenerable version of $LiAlH_4$. While it is true that the K^+ counterion plays its part in polarizing the $C=O$ group, a major part of the kinetic facility of this normally very difficult reduction must come from the $\delta-$ character of the hydride ligands, due to delocalization of a part of the net ionic charge.

In our hydrogenation system based on $[Ir(cod)L_2]PF_6$ (cod = 1,5-cyclooctadiene; L = *tert* phosphine) in contrast, the net positive ionic charge influences the chemistry in a different way. In the intermediate in the activation of the catalyst by H_2, $[IrH_2(cod)L_2]PF_6$, the H ligands seem to have a $\delta+$ character as suggested by a ^{13}C NMR study.[9] The catalytic system is a very weak H^- donor

[*] We thank Exxon Educational Foundation and the National Science Foundation for support. RHC thanks the A. P. Sloan and Henry and Camille Dreyfus Foundations, and MWD thanks the F. W. Heyl and Elsie L. Heyl Foundation for fellowships.

as shown by the lack of reaction with chlorinated solvents, ketones, or O_2. In the case of the slightly more electron-withdrawing ligand dct (dct = *sym*-dibenzocyclooctadiene), which like cod is a chelating diolefin, the hydrogen adducts are so acidic that they undergo stereochemical rearrangement by H^+ loss (Equation 1).[10]

$$\text{(dct) Ir L}_2 \xrightarrow{H_2} \text{(dct) Ir} \xrightarrow{-H^+} \text{(dct)IrHL}_2 \xrightarrow{+H^+} \text{(dct) Ir} \qquad [1]$$

	KINETIC PRODUCT		THERMODYNAMIC PRODUCT

Their pKa is ca. 11, compared to a value of at least 19 for the cod analogues.

In addition to the effect on the hydride ligands themselves, a positive charge on a metal complex seems to greatly increase the binding constants for hard ligands, as mentioned above. We have recently discovered an unusual directing effect in catalysis with $[Ir(cod)(PCy_3)py]PF_6$ (Cy = cyclohexyl; py = pyridine) which we ascribe to this effect (Equation 2). Here, the catalyst first binds to the

Pd/C	20	80
Ir	99.9	0.1

OH group on one face of the terpene substrate. The catalyst then reduces the substrate essentially only from that side. In contrast, Pd/C leads to preferential addition from the opposite face in ethanol, and an approximately 50:50 mixture of products is formed in cyclohexane. The homogeneous catalyst used is unusual, not only in showing a directing effect, but also in being able to reduce tri- and tetrasubstituted olefins so that the effect can be expressed in the stereochemistry of the products. The effect is probably due to chelation[12] (see also Reference 13).†This type of chelation is also important in asymmetric hydrogenation, in which cationic complexes are widely used.[14]

Interestingly, the catalyst reduces the homoallylic terpene mentioned above, but not the unsaturated alcohol, cholesterol,[15] in which the C=C bond is in a different ring and much more remote from the directing −OH group. This strongly suggests that the catalyst will quite generally direct H_2 addition to vicinal, rather than distal, C=C groups.

A further advantage of ionic character in a catalyst is also illustrated by the same iridium catalyst mentioned above. Catalyst separation, easy in heterogeneous cases, can become difficult in homegeneous ones. The soluble catalyst or dissociated

---

† We have recently learned that G. Stork *et al.* have observed very similar effects with the iridium catalyst.[11]

ligands may contaminate the products. In the iridium case, provided the organic product is soluble in hexane or ether, it can be extracted entirely free of catalyst from the evaporated residues of the catalytic reaction with either of these solvents.[15]

CONCLUSION

In this short review, we hope to have illustrated some of the special effects of net ionic charge, especially positive sharge, on the catalytic and stoichiometric behavior of metal hydrides. The extension of these ideas to complexes other than metal hydrides and to reactions other than hydrogenation may also have interesting consequences.

ACKNOWLEDGMENTS

We would like to thank Sarah Danishefsky for communicating some results that led us to reexamine the possibility of directing effects in the iridium system after initially unpromising results.

REFERENCES

1. PAULING, L. 1948. J. Chem. Soc. 150: 1461.
2. HUNG, Y., W.-J. KUNG & H. TAUBE. 1981. Inorg. Chem. 20: 157.
3. DAVIES, J. A., F. R. HARTLEY & S. G. MURRAY. 1980. J. Chem. Soc. Dalton Trans.: 2246.
4. WHITE, C., S. J. THOMPSON & P. M. MAITLIS. 1977. J. Chem. Soc. Dalton Trans.: 1654.
5. CRABTREE, R. H., P. C. DEMOU, D. EDEN, J. M. MIHELCIC, C. A. PARNELL, J. M. QUIRK & G. E. MORRIS. 1983. J. Am. Chem. Soc. 104: 6994.
6. DAVIES, J. A. & F. R. HARTLEY. 1981. Chem. Rev. 81: 79.
7. GREY, R. A., G. P. PEZ, A. WALLO & J. CORSI. 1980. Chem. Commun.: 783.
8. McNEILL, E. A. & F. R. SCHOLER. 1977. J. Am. Chem. Soc. 99: 6243.
9. CRABTREE, R. H. & J. M. QUIRK. 1980. J. Organomet. Chem. 199: 99.
10. ANTON, D. R. & R. H. CRABTREE. 1983. Organometallics 2: 621.
11. STORK, G & D. KAHNE. 1983. J. Am. Chem. Soc. 105: 1072.
12. CRABTREE, R. H. & M. W. DAVIS. 1983. Organometallics 2: 681.
13. BROWN, J. M. & R. G. NAIK. 1982. Chem. Commun.: 348.
14. JAMES, B. R. 1979. Adv. Organomet. Chem. 17: 319.
15. SUGGS, J. W., S. D. COX, R. H. CRABTREE & J. M. QUIRK. 1981. Tetrahedron Lett. 22: 303.

MODELING HETEROGENEOUS CATALYSTS WITH HOMOGENEOUS CATALYSTS: MODELING THE HYDRODENITROGENATION REACTION*

Richard M. Laine

Physical Organic Chemistry Department
SRI International
Menlo Park, California 94025

INTRODUCTION

Within the past decade, organometallic chemists have developed a growing fascination for metal cluster chemistry, especially metal cluster catalysis chemistry. The sources of this fascination are the potential benefits cluster catalysis can provide (1) to further the development and understanding of heterogeneous catalysis chemistry through modeling studies, (2) to advance the development of new, industrially important, homogeneous catalysis chemistry, and (3) to provide a sound basis for the development of heterogenized metal cluster catalysis chemistry.

In the past decade, beginning with Muetterties,[1] Ugo,[2] and Lewis and Johnson,[3] numerous reviews, papers, and seminars have presented the idea that catalytic reactions in which the active catalyst species were polynuclear metal complexes or "metal clusters" could be useful models of analogous reactions catalyzed by heterogeneous catalysts.[4-10] These proposals were based on the rationale that many heterogeneously catalyzed reactions require multiple metal sites for the reaction to occur. Because metal clusters contain several metal atoms, it should be possible to use metal clusters to obtain homogeneous multiple-site catalysis. Kinetic and mechanistic studies of homogeneously catalyzed reactions as well as spectral analyses of these reactions are relatively facile by comparison with similar studies performed on heterogeneous catalytic reactions. Consequently, the ability to model heterogeneous catalytic reactions using homogeneous catalysts could provide valuable kinetic and mechanistic insights into the functioning of heterogeneous catalysts that would be difficult if not impossible to obtain using the techniques available to the surface chemist or physicist.

Unfortunately, the experimental evidence to date that supports the validity of these proposals is limited despite the extensive efforts of many researchers. Moreover, the evidence that does exist has not led to any startling or useful new insights into heterogeneous catalysis chemistry. Part of the problem undoubtedly results from the basic difficulty in identifying a homogeneously catalyzed reaction as being a cluster-catalyzed reaction.[11] A second and perhaps greater difficulty may lie with the initial choice of heterogeneous catalytic reactions that researchers have sought to evaluate through development of homogeneous cluster catalysis models.

The majority of researchers in the field of metal cluster catalysis chemistry have

* We would like to thank the Army Research Office and the National Science Foundation for the generous support of these studies.

271

concentrated their initial efforts on developing homogeneous cluster catalysts that activate CO for hydrogenation, homologation, hydroformylation, and hydrocarboxylation.[12–16] The key problem with this choice is that CO is an excellent ligand. It forms some of the most stable organometallic complexes found in nature. By its very nature, CO will promote cluster fragmentation to mononuclear species because it has the propensity to stabilize reactive fragments. Therefore, studies of cluster-catalyzed reactions in which CO must be used under pressure are quite likely to result in cluster fragmentation to mononuclear species. These effects will at best make identification and characterization of cluster-catalyzed reactions difficult.[17]

Considering these problems, it seems logical to approach the challenge of validating the modeling hypothesis through studies of catalytic reactions that do not require CO as a substrate. In this vein, Muetterties has begun a serious effort to study the catalytic hydrogenation of alkenes and alkynes using rhodium phosphite clusters and has obtained novel and useful results.[18,19] Our own efforts in this area, which are the subject of this presentation, have been to develop homogeneous cluster-catalyzed reactions in which clusters activate the C—H and C—N bonds of saturated tertiary amines and to show that the reactivity patterns of the homogeneous catalysts with tertiary amines correlate extremely well with the reactivity patterns of the same amines with heterogeneous catalysts. In this presentation, we explore the development of these reactivity patterns for several catalyst systems and demonstrate how the results allow us to draw important mechanistic conclusions about C—H and C—N bond activation of considerable use in explaining the mechanisms of the industrially important hydrodenitrogenation (HDN) reaction.

HDN is the process by which nitrogen is removed (as NH_3) from crude oil, coal, oil shale, and tar sands when these materials are upgraded (refined) to give hydrocarbon fuels or petrochemical feedstocks.[20,21]

BACKGROUND

Our initial research in catalytic activation of amines began as an outgrowth of our studies on the homogeneous catalysis of the water-gas shift reaction (WGSR).[22] We observed that when we attempted to catalyze the WGSR in the presence of tertiary amines using rhodium or ruthenium catalysts and D_2O as in Equation 1, deuterium was incorporated in the alkyl groups of the amines.[23,24]

$$D_2O + CO + Et_3N \xrightarrow{\text{Rh}_6\text{(CO)}_{16} \text{ or Ru}_3\text{(CO)}_{12}} D_2 + CO_2 + Et_2NCHDCD_3 \quad (1)$$

Concurrent with these studies, Murahashi was exploring the catalytic reactivity of tertiary amines with palladium black. Murahashi observed the following two reactions:[25,26]

$$Me_2NCH_2Pr + Me_2NCD_2Pr \xrightarrow[200°C]{\text{Pd black}} Me_2NCH_{2-x}D_xPr \quad x = 0, 1, 2 \quad (2)$$

$$R_3N + R'_3N \xrightarrow[200°C]{\text{Pd black}} R_2NR' + R'_2NR \quad (3)$$

In our initial studies of rhodium catalysis of Reaction 1, we proposed a cluster-catalyzed mechanism for deuterium for hydrogen exchange as shown in

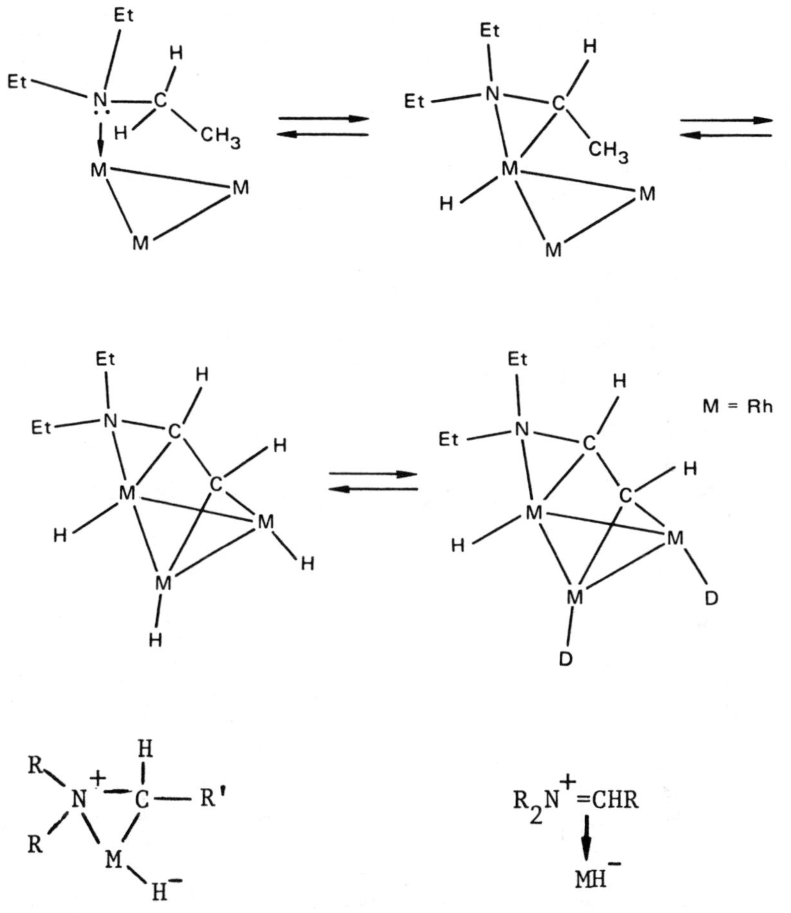

SCHEME 1

SCHEME 1. Murahashi's proposed mechanism for the palladium black–catalyzed deuterium for hydrogen exchange in Reaction 2 requires the intermediacy of an iminium ion/palladium complex, **2**, similar to the metalloazocyclopropane intermediate, **1**, proposed in SCHEME 1. Because of the apparent mechanistic similarities between Murahashi's heterogeneously catalyzed reactions and the homogeneous rhodium-catalyzed reaction, we suggested that it might be possible to model palladium black catalysis of Reaction 2 using the homogeneous rhodium system used in Reaction 1. We have since determined that the ruthenium rather than the rhodium homogeneous catalyst system is a more suitable model for both Reactions 2 and 3. For example, the mechanism for deuterium (D) for hydrogen (H) exchange as catalyzed by ruthenium (SCHEME 2) differs

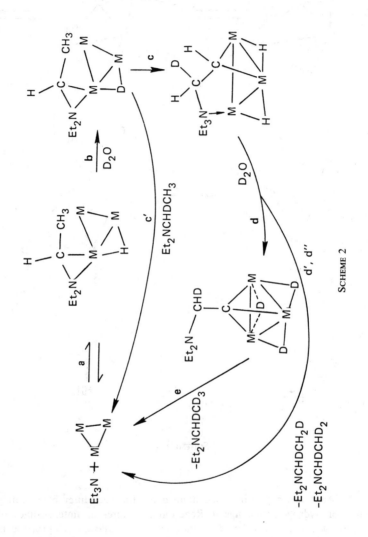

SCHEME 2

from that of rhodium in that the initial hydrogen exchange is always an α hydrogen ($\alpha\beta\beta\beta$ exchange mechanism) in contrast to the rhodium system where the α hydrogen that is exchanged is always the last hydrogen to exchange ($\beta\beta\beta\alpha$ exchange mechanism).

From our modeling studies of the catalytic reactions of palladium black with tertiary amines, we developed a set of reaction parameters that allow us to define the reactivity patterns of tertiary amines for D for H exchange as catalyzed by either heterogeneous or homogeneous catalysts.[24] These reaction parameters can be simply represented as follows:

1. *Site selectivity.* Defined as the preference of the catalyst for the α C—H versus the β C—H positions in D for H exchange catalysis. Analyzed by ^{2}D NMR.

$$Et_2NCH_2CH_3 + D_2O \xrightarrow{\text{catalyst}} Et_2NCHDCD_3 + HDO$$

2. *Product selectivity.* Defined as the preference of the catalyst for a particular deuterium exchange product. Thus, two major D for H exchange products are observed for Et_3N, Et_3N-d_4, and Et_3N-d_3. The products from Pr_3N or Bu_3N are either $d_3 > d_2 > d_1$ or $d_1 > d_2 > d_3$ depending on the catalyst. Analyzed by electron impact (EI) fragmentation patterns.

3. *Catalyst D for H exchange activity for Et_3N, Pr_3N, or Bu_3N.* Defined as the turnover frequency (TF) or number of moles of amine that exchange D for H per mole of catalyst-hour.

4. *Transalkylation catalyst activity.* A fourth reaction parameter that can be used to define a reactivity pattern for a given catalyst does not involve D for H exchange, but is reliant upon C—H activation. This parameter is defined as the ability of the catalyst to catalyze Reaction 3. Measured as the mole percent of mixed-alkyl amine per total moles of amine. The maximum yield obtainable from a statistical redistribution of alkyl groups is 73.3 %.[25]

As a prelude to our discussions of our HDN modeling studies, it is useful to briefly review the application of these reaction parameters to modeling the catalytic reactions of palladium black using a homogeneous ruthenium catalyst.[24] TABLE 1 shows the site selectivities of D for H exchange for the three alkylamines Et_3N, Pr_3N, and Bu_3N for the palladium black catalyst and the homogeneous catalyst derived from $Ru_3(CO)_{12}$. The product selectivities and the catalyst exchange activities for these same three amines are listed in TABLE 2. Note that a comparison of the homogeneous and heterogeneous catalyst activities

TABLE 1

CATALYTIC DEUTERIUM FOR HYDROGEN EXCHANGE ON TERTIARY AMINES: SITE SELECTIVITIES ($\alpha:\beta$ RATIOS) FOR VARIOUS CATALYSTS AT 150°C*

	Et_3N	Pr_3N	Bu_3N
$Ru_3(CO)_{12}$	0.29:1	1.30:1	2.1:1
Pd black†	0.38:1	0.80:1	1.6:1 (4:1)‡
Rh black	—	—	4:1‡

* The values shown represent $\alpha:\beta$ site selectivity ratios determined by ^{2}H NMR.

† Pd black is recovered quantitatively.

‡ Data from Reference 25. Reactions were run at 200°C.

TABLE 2

PATTERNS OF DEUTERIUM FOR HYDROGEN EXCHANGE FOR VARIOUS GROUP 8 METALS AND TERTIARY AMINES AND RELATIVE CATALYTIC ACTIVITIES AT 150°C*

Catalyst	Et_3N-d_1	Et_3N-d_2	Et_3N-d_3	Et_3N-d_4	Turnover Frequency for Amines†
$Rh_6(CO)_{16}$	β(0.5)	β(1.4)	β(5.6)	α(9.5)	3.6
$Ir_4(CO)_{12}$‡	?(3.5)	?(2.5)	?(1.1)	?(019)	1.7
$Ru_3(CO)_{12}$	α(0.6)	β(1.9)	β(4.9)	β(6.05)	5.8
	β(0.6)	β(1.9)	β(4.9)	α(6.05)	—
$Os_3(CO)_{12}$§	α(0.7)	β(0.7)	β(1.2)	β(2.6)	0.5
Pd black	α(0.6)	β(1.2)	β(3.3)	β(4.6)	0.12

	Pr_3N-d_1		Pr_3N-d_2	Pr_3N-d_3	
$Ru_3(CO)_{12}$	α(15.0)		β(8.0)	β(5.0)	5.0
Pd black	α(5.5)		β(2.6)	β(2.1)	0.08

	Bu_3N-d_1		Bu_3N-d_2	Bu_3N-d_3	
$Ru_3(CO)_{12}$¶	α(20.9)		β(8.3)	β(3.1)	8.4
Pd black	α(30.3)		β(13.0)	β(5.1)	0.3

* The substitution patterns were determined by a combination of mass spectral analysis of the amine fragmentation patterns by electron impact mass spectroscopy (11 eV and 70 eV) and ^2H NMR. Numbers in parentheses represent product as a percent of total tertiary amines recovered, usually 98%.

† These values are (mmol amine reacted)/[mmol catalyst) (hour)] and are the average of at least two different runs.

‡ Mass spectral analysis did not give a clear-cut exchange pattern.

§ Reaction time was 40 hours.

¶ Reaction time was 16 hours.

can only be qualitative because the catalytic activity of the palladium black undoubtedly depends on the initial catalyst surface area.

The data contained in the two tables show a very high correlation between the two catalysts. The notable features are (1) the parallel changes that favor α exchange over β exchange that occur as the amines change from Et_3N to Pr_3N to Bu_3N, (2) the parallel changes in product selectivity on going from Et_3N to Bu_3N, and (3) the parallel changes in catalyst activity obtained as the amine is varied between Et_3N and Bu_3N.

At this point in our studies it seemed likely that we had a valid homogeneous model of the catalytic reactions of amines with palladium black. As a final proof, we rationalized that a true homogeneous model should not only exhibit similar reactivity patterns for catalysis of D for H exchange on tertiary amines, but also should model other reactions of amines catalyzed by palladium black such as Reaction 3. As shown in TABLE 3, not only does ruthenium catalyze transalkylation, Reaction 4, but at rates and temperatures superior to those obtained with palladium black.

$$Et_3N + Pr_3N \xrightarrow[\text{M = Ru or Os}]{M_3(CO)_{12}/125°C} Et_2NPr + Pr_2NEt \qquad (4)$$

Of importance to this study was the fact that catalytically activated ruthenium metal did not catalyze D for H exchange. Moreover, addition of soluble palladium

TABLE 3

CATALYTIC DEUTERIUM FOR HYDROGEN EXCHANGE ON TERTIARY AMINES:
SITE SELECTIVITIES ($\alpha:\beta:\gamma$ RATIOS) FOR VARIOUS CATALYSTS*†

Catalyst	Temperature (°C)	Et_3N	Pr_3N	Bu_3N
Pd black	150	0.38:1	0.80:1	1.50:1
$Ru_3(CO)_{12}$	150	0.29:1	1.30:1	2.10:1
$Rh_6(CO)_{16}$	150	0.25:1	0.33:1	0.46:1:0.20
$Rh_6(CO)_{16}$	200	0.38:1	0.40:1:0.06	0.44:1:0.7
Rh metal‡	232	0.22:1		
CoMo†	237	0.24:1	0.25:1	0.36:1
CoMo	260	0.20:1	0.27:1:0.10	0.29:1:0.10
Mo metal§	230	0.26:1		

* The values shown represent $\alpha:\beta:\gamma$ site selectivity ratios determined by ^2H NMR spectrometry using a Nicolet 360-MHz instrument. $CDCl_3$ was used as an internal standard. Solutions were heated at listed temperatures for times listed in TABLE 4.

† Blanks run with alumina showed negligible deuterium incorporation.

‡ At this temperature the $Rh_6(CO)_{16}$ catalyst visibly decomposed to fine metal particles that catalyze D for H exchange.

§ Mo metal was formed by decomposition of $Mo(CO)_6$.

species [such as $PdCl_2(PhCN)_2$] did not promote any catalysis of D for H exchange as observed with palladium black.

The evidence in favor of homogeneous cluster catalysis is as follows. First, as the CO pressure in Reactions 1 and 4 is reduced, the rate of reaction increases significantly† for all active catalysts. Second, a bridging carbene complex, **3**, analogous to that proposed for step c of SCHEME 2, has been isolated and structurally

characterized from the reaction of $Os_3(CO)_{12}$ with Et_3N.[28] Third, Yin and Deeming, in their studies on the reactions of $PhN(CH_3)_2$ with $Os_3(CO)_{12}$, observed only the formation of amine-cluster complexes even through their reactions were run at higher temperatures than those used in our studies.[29] Finally, we have isolated the cluster complex **4** from reactions similar to Reactions 1 and 4.[30]

† We recently described criteria for identifying cluster-catalyzed reactions.[11] One criterion is that the occurrence of increased catalytic activity under conditions that favor cluster formation [low CO pressures and third-row carbonyl complexes, e.g., $Os_3(CO)_{12}$] is suggestive of cluster catalysis. The CO pressures used in the present studies are used solely to inhibit Reaction 4.

From this initial study, we drew the following conclusions:

The similarities in the reactivity patterns indicate that both the homogeneous ruthenium catalyst and the heterogeneous catalyst, palladium black, function by the same or similar reaction mechanisms.

Thus, homogeneous metal cluster catalysts can be used to model the catalytic actions of heterogeneous catalysts.

Therefore, the catalytic and stoichiometric reactions of metal clusters can be used to describe the mechanistic features of heterogeneous catalysts at least for the catalytic reactions of tertiary amines.

With these conclusions in mind, we sought to extend our modeling studies to include other heterogeneously catalyzed reactions of amines. A key observation in the above modeling studies was the efficient catalytic activation and cleavage of $C-N$ single bonds in tertiary amines at low temperatures. This observation suggested that it might be possible to model the industrially important HDN reaction

TABLE 4

PRODUCT SELECTIVITIES OBTAINED FROM CATALYSIS OF D FOR H EXCHANGE
BY CoMo AND RHODIUM FOR SEVERAL TERTIARY AMINES*

Catalyst	Temperature (°C)	Run Time (hours)	Et_3N-d_1	Et_3N-d_2	Et_3N-d_3	Et_3N-d_4
$Os_3(CO)_{12}$	150	20	$\alpha(0.5)$	$\beta(0.7)$	$\beta(1.2)$	$\beta(2.6)$
$Rh_6(CO)_{16}$	150	20	$\beta(0.5)$	$\beta(1.4)$	$\beta(5.6)$	$\alpha(9.5)$
$Rh_6(CO)_{16}$	200	3.5	$\beta(0.1)$	$\beta(0.1)$	$\beta(0.4)$	$\alpha(1.0)$
Rh†	232	2.5	?(16.2)	?(14.4)	?(9.5)	?(5.3)
CoMo‡	237	20	$\beta(0.5)$	$\beta(3.4)$	$\beta(12.1)$	$\alpha(6.3)$
CoMo‡	260	6	$\beta(0.1)$	$\beta(1.3)$	$\beta(4.78)$	$\alpha(3.6)$
Mo metal§	230	20	$\beta(0.0)$	(0.2)	$\beta(0.6)$	$\alpha(0.7)$

Catalyst	Temperature (°C)	Run Time (hours)	Pr_3N-d_1	Pr_3N-d_2	Pr_3N-d_3	Pr_3N-d_4
$Ru_3(CO)_{12}$	150	20	$\alpha(15.0)$	$\beta(8.0)$	$\beta(5.0)$	
$Rh_6(CO)_{16}$	150	20	$\beta(0.3)$	$\beta(0.7)$	$\alpha(1.6)$	
$Rh_6(CO)_{16}$	200	3.5	$\beta(0.0)$	$\beta(0.4)$	$\alpha(2.6)$	?(0.3)
CoMo	237	20	$\beta(0.2)$	$\beta(3.0)$	$\alpha(1.7)$	?(0.5)
CoMo	260	20	$\beta(2.7)$	$\beta(6.1)$	$\alpha(3.3)$	(1.2)

Catalyst	Temperature (°C)	Run Time (hours)	Bu_3N-d_1	Bu_3N-d_2	Bu_3N-d_3	Bu_3N-d_4
$Ru_3(CO)_{12}$	150	20	$\alpha(20.9)$	$\beta(8.3)$	$\beta(3.1)$	
$Rh_6(CO)_{16}$	150	20	$\beta(0.2)$	$\beta(0.4)$	$\alpha(0.5)$	
$Rh_6(CO)_{16}$	200	3.5	$\beta(0.5)$	$\beta(1.0)$	$\alpha(2.4)$	?(0.3)
CoMo	237	40	$\beta(3.7)$	$\beta(7.4)$	$\alpha(6.1)$	?(1.8)
CoMo	260	20	$\beta(0.8)$	$\beta(1.3)$	$\alpha(1.0)$	?(0.5)

* Values in parentheses represent percent of total amines recovered (usually 98%).

† At this temperature, the $Rh_6(CO)_{16}$ decomposes giving visible metal particles that actively catalyze D for H exchange.

‡ Blank reactions with Condea alumina given negligible D for H exchange.

§ Mo metal was formed by decomposition of $Mo(CO)_6$. Note that Mo metal has the same product selectivity as the rhodium catalysis.

using homogeneous cluster catalysts since the major objective in HDN is to remove nitrogen from crude oils as NH_3 via catalytic cleavage of $C-N$ single bonds. Such a modeling study would be valuable in developing a mechanistic understanding of how heterogeneous HDN catalysts function, at present an unknown process, and would perhaps provide the information necessary to develop new, more efficient catalysts.

Our approach to modeling heterogeneously catalyzed HDN[31,32] was to assume that if the catalytic reactivity patterns of tertiary amines with an industrial HDN catalyst could be correlated with those of one of the homogeneous systems we had previously explored, then a valid model system might be obtained.

For our HDN modeling studies, we chose a Ketjenfine cobalt-molybdenum oxide catalyst (CoMo) supported on gamma alumina as the industrial catalyst. Reaction of the activated catalyst with Et_3N, Pr_3N, and Bu_3N under suitable conditions gave the site selectivities shown in TABLE 3 and the product selectivities listed in TABLE 4. We also examined the catalytic activity of CoMo for Reaction 4 as shown in TABLE 5.

From the data in TABLES 4 and 5, it is obvious that the best model for CoMo is the $Rh_6(CO)_{16}$-based catalyst system. Unlike the results from the palladium black modeling studies, the site selectivities for the CoMo- and $Rh_6(CO)_{16}$-catalyzed reactions do not change dramatically on changing the amine from Et_3N to Pr_3N to Bu_3N. There appears to be a slight preference for α exchange relative to β exchange as the steric bulk of the amine increases. Similarly, the product selectivities do not change significantly as the amines increase in size, again unlike those in palladium black modeling studies. Both the CoMo- and $Rh_6(CO)_{16}$-catalyzed reactions give the same two major products. The only difference is that the second major product in the rhodium-catalyzed reaction is always the major product in the CoMo-catalyzed reaction and vice versa.

TABLE 5

EXTENT OF FORMATION OF MIXED ALKYL OR DIALKYL-AMINES FROM CATALYTIC ALKYL EXCHANGE BETWEEN Et_3N AND Pr_3N*

Catalyst	Temperature (°C)	Et_2NH	Et_2NPr	$EtNPr_2$	$HNPr_2$	Et_2NBu
$Os_3(CO)_{12}$	150	—	26.7	28.4	—	—
Pd black†	150	—	0.2	0.6	—	—
$Rh_6(CO)_{16}$	150	—	1.6	1.6	—	—
$Rh_6(CO)_{16}$	200	0.2	5.2	4.6	0.1	0.6
CoMo	260	3.9	3.1	4.2	5.2	—
$Rh_6(CO)_{16}$‡	200	0.4	—	—	—	0.2‡
$Os_3(CO)_{12}$‡	200	4.6	—	—	—	0.4‡
CoMo‡	260	2.0	—	—	—	1.8‡

* The numerical values correspond to (mol mixed amine/total mol amine) × 100 and are reproducible to within 10% of the indicated value.

† 1.0 mmol catalyst.

‡ Reaction run with Et_3N only. Note that the ethyl lost in the Et_3N $C-N$ bond cleavage results in the formation of small quantities of Et_2NBu and ethanol in both the homogeneously and heterogeneously catalyzed reactions. We observe no formation of ethane in the gases above the reaction solutions.

It is not clear if these differences warrant an explanation because most model systems are not expected to be perfect.

Unfortunately, despite repeated efforts, it was impossible to generate activated CoMo catalysts that gave reproducible catalytic activities. Consequently, we were not able to apply the catalyst activity reaction parameter described above to the CoMo model study.

The transalkylation modeling studies (TABLE 5) indicate that neither the rhodium nor the CoMo catalysts are particularly active as transalkylation catalysts. However, one important observation from these studies is that CoMo catalysis of the first step in the HDN of Et_3N and Pr_3N occurs as evidenced by the formation of Et_2NH and Pr_2NH, which are indicative of C—N bond cleavage. Of further importance is that these same products are also observed in the rhodium-catalyzed reactions. Moreover, the ethyl groups are not lost as ethane, as would normally be expected for CoMo-catalyzed HDN of Et_3N. The ethyl groups appear in the reaction solution as EtOH and Et_2NBu. The identical products are observed in the rhodium-catalyzed reaction.

These results strongly support reliable modeling of the catalytic reactivity patterns of CoMo with tertiary amines by homogeneous rhodium cluster catalysts.

Attempts to catalyze the D for H exchange reactions with rhodium metal gave totally different site and product selectivities from those observed with the homogeneous rhodium catalysts. Attempts to catalyze D for H exchange reactions using a soluble molybdenum complex $[Mo(CO)_6]$ gave no exchange until the complex had completely decomposed to heterogeneous species.

Evidence in favor of rhodium-cluster-catalyzed reactions includes the spectroscopic observation of $Rh_5(CO)_{15}^-$, $Rh_6(CO)_{15}^{2-}$, and $Rh_{12}(CO)_{30}^{2-}$ as the only species in the reaction solution and the identification of rhodium clusters as the active catalysts for the hydrogenation of benzaldehyde under quite similar conditions to those used in the present studies.[32]

COMMENTS ON THE MECHANISMS OF CATALYTIC HDN

Because it is difficult to study heterogeneous catalytic HDN of crude oil, model studies are often performed using model compounds such as pyridine[21] or quinoline.[20] It is thought that these compounds resemble the nitrogen-containing compounds found in crude oil.

To date, modeling studies of heterogeneously catalyzed HDN of quinoline have resulted in the determination of rate constants and thermodynamic equilibria for the formation and disappearance of all the species shown in SCHEME 3.[20,21,33-36] Most of the quinoline undergoes HDN by the heavily lined but more costly (in terms of H_2 requirements) pathway. No one has attempted to describe mechanisms for C—N bond cleavage as it occurs in HDN. If it were possible to understand the mechanisms for C—N bond cleavage in CoMo-catalyzed HDN, rationally it should be possible to improve the CoMo catalysis of HDN or to develop better catalysts.

Our observation of the catalytic cleavage of saturated C—N bonds in the transalkylation reaction (4),[27] coupled with our success with modeling the catalytic reactions of amines with heterogeneous catalysts,[24] including the HDN catalyst,

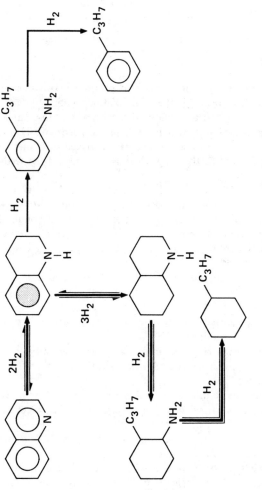

SCHEME 3. Quinoline HDN reaction network.

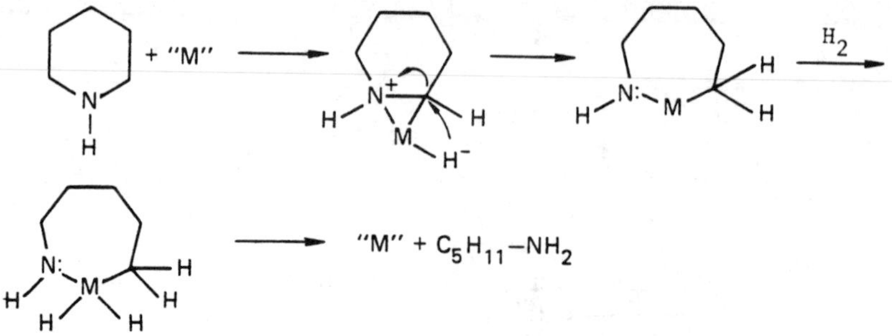

SCHEME 4

has provided us with enough information to propose reasonable mechanisms for C—N bond cleavage as it occurs in the HDN of model compounds or crude oil. It is even possible to explain the anomalous behavior of CoMo-catalyzed HDN that occurs in the presence of H_2S and H_2O. That is, generally, H_2S or even water in the feedstock poisons most heterogeneously catalyzed reactions. In contrast, the rate of CoMo catalysis of HDN is significantly enhanced if the feedstock contains H_2S or H_2O.[21,33-38]

Arguments have been presented that suggest that, at least in the case of sulfur, the CoMo molybdenum oxides are slowly transformed into MoS_2, which, because of its different physical properties, is a better catalyst. Although this is at least partially true (see Reference 39), there are several other alternative mechanisms in which H_2S and H_2O act as nucleophiles to facilitate ring opening, as we will illustrate shortly.

To discuss CoMo-catalyzed C—N bond scission, we assume that piperidine is representative of the saturated heteroaromatic rings that undergo C—N bond cleavage as part of the HDN process. We can then propose a set of mechanisms for catalytic cleavage based on the mechanisms we and others have proposed for the transalkylation reaction (4).[25,26]

For simple, unassisted ring opening, at least two pathways are possible. One requires a metal alkyl intermediate (SCHEME 4) and the other a metal alkylidene intermediate (SCHEME 5).

There are undoubtedly other mechanistic choices. Our intent here, as well as with the mechanisms described below, is to provide logical mechanistic pathways to explain literature observations based on our results. As with any proposed mechanism, a true test of that mechanism's viability is that it must explain the characteristic anomalies of the reaction.

As noted above, CoMo catalysis of HDN is greatly enhanced by the presence of H_2S and H_2O, which are normally catalyst poisons. Of key importance is the observation by Satterfield that the rate enhancement occurs in the C—N bond-breaking/ring-opening step, not in the hydrogenation step.[20,37] In the following schemes we propose reaction mechanisms for HDN that account for the anomalous increases in catalytic activity that occur when HDN is performed in the presence of H_2S or H_2O.

The first mechanism (SCHEME 6) considers HDN catalysis as passing through an iminium complex intermediate where, in the presence of H_2S, ring opening proceeds via nucleophilic attack to form the thiohemiaminal. If the reaction

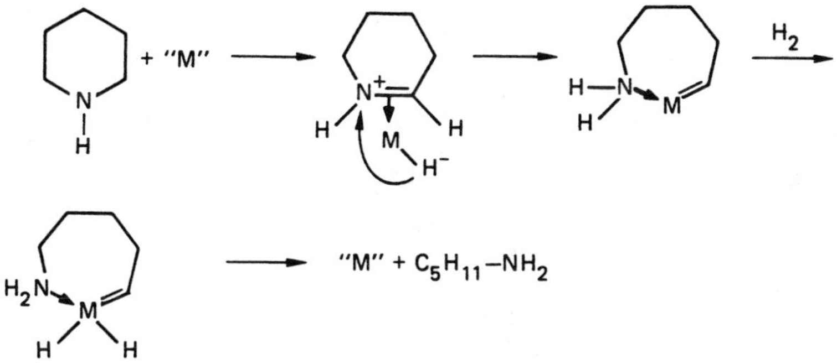

SCHEME 5

proceeds via a metalloazocyclopropane in the presence of H_2O, then a simpler mechanism ensues (SCHEME 7). An analogous reaction pathway can be written for H_2S. Alternatively, secondary amines are known to form Schiff's base complexes with metals[40] as shown in SCHEME 8. Similar mechanisms can be written for nucleophilic attack by H_2O.

Thermodynamic calculations predict that the formation of ethylamine is favored at low temperatures:[41]

$$NH_3 + CH_2{=}CH_2 \longrightarrow EtNH_2 \qquad \Delta G_{298°} = -3.5 \, kcal/mole \qquad (5)$$

At slightly higher reaction temperatures the reaction becomes thermoneutral, and finally, at temperatures of 450°C, primary amines are readily cracked to alkenes and NH_3 in the presence of alumina. Consequently, the difficult step in HDN is $C{-}N$ bond scission to ring-opened products. Any primary amines formed will rapidly give NH_3 and alkene. The alkene will then be hydrogenated to alkane. Thus, our mechanisms have concentrated on an explanation of how the ring opens.

A key observation that supports nucleophilic-assisted ring opening is the observation by Klemm that CoMo catalysts can catalyze the formation of sulfur-carbon bonds under HDN conditions:[42]

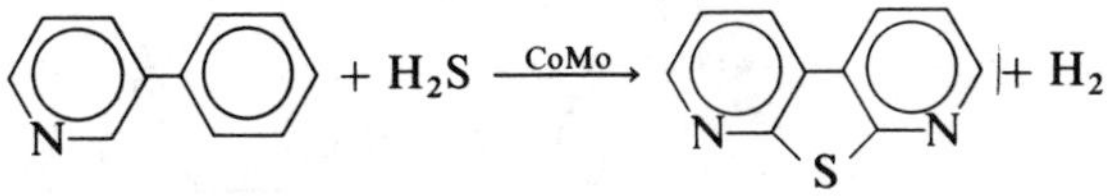

The formation of the aminal intermediates and their ring opening to give imines, thialdehydes, and aldehydes is analogous to chemistry that has been known for decades and most commonly found in sugar chemistry.

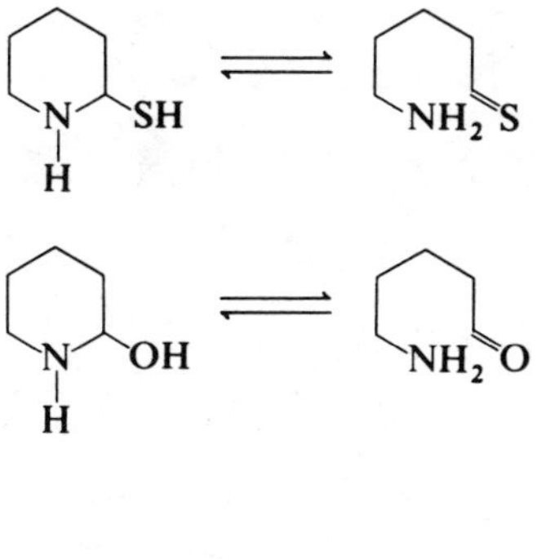

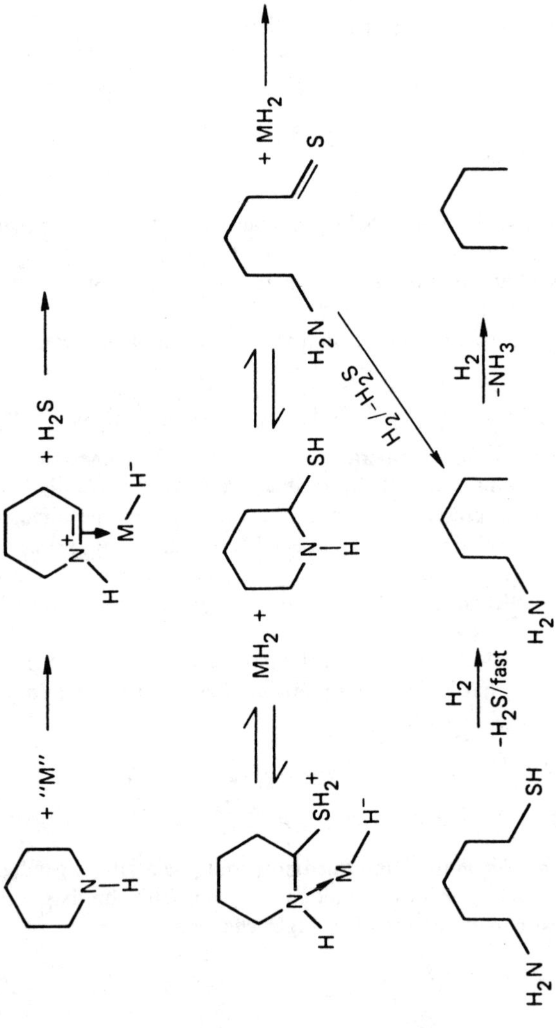

SCHEME 6

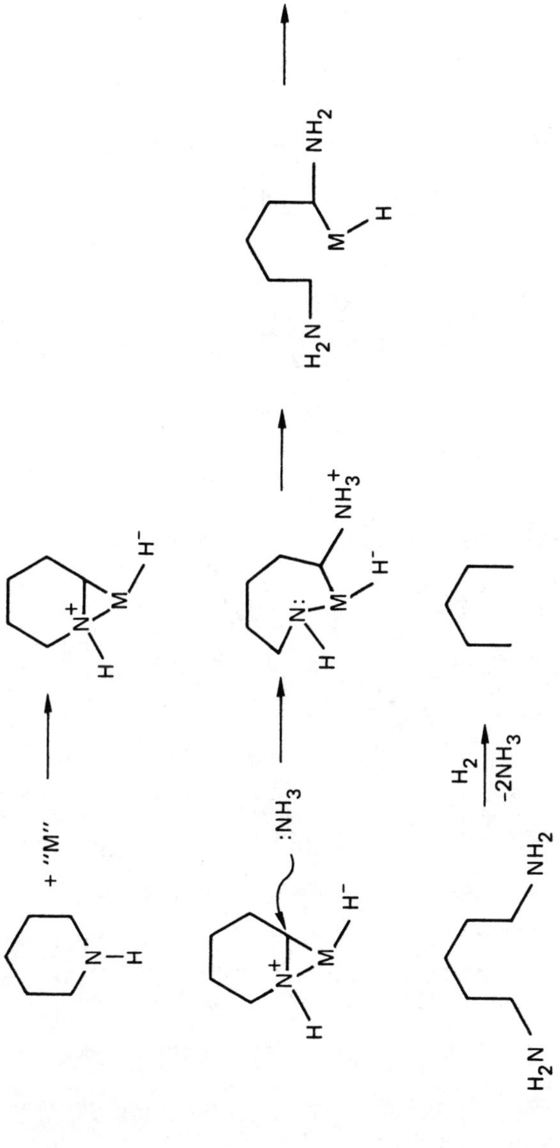

SCHEME 7

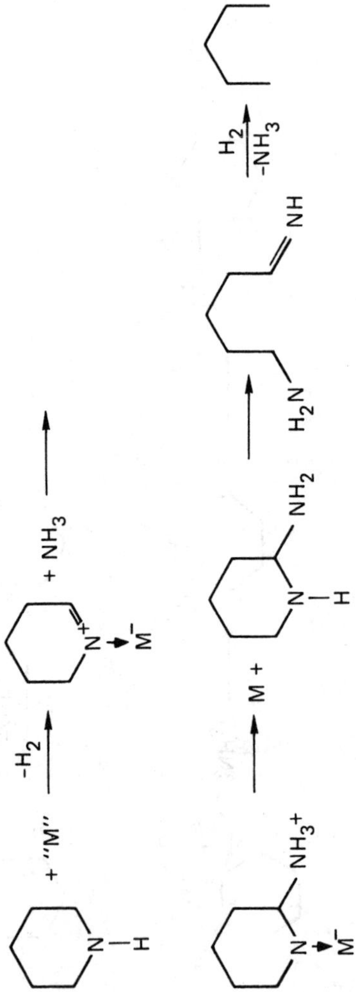

SCHEME 8

Thialdehyde desulfurization can occur at 0°C on supported molybdenum catalysts.[15] Only hydrodeoxygenation of the aldehyde or the related primary alcohol hydrogenation product has not been demonstrated for heterogeneous catalysts, although we have shown that our homogeneous rhodium cluster catalyst can hydrodeoxygenate amino alcohols to amines.[43]

If in SCHEME 8 piperidine were substituted for H_2S, in either of the mechanisms, the expected end product would not be pentane but rather n-pentylpiperidine (see SCHEME 9) because the tertiary amine formed following ring opening would not be as susceptible to HDN as the primary amines. In fact, Sonnemans et al. see n-pentylpiperidine as the major intermediate in the CoMo-catalyzed HDN of either pyridine or *piperidine*.[21]

Finally, we find that rhodium cluster catalysts will homogeneously hydrogenate pyridine, and effect HDN, Reaction 5, to give 1,5-(bispiperidino)pentane formed in better than 60% yield,[43] a product analogous to n-pentylpiperidine.

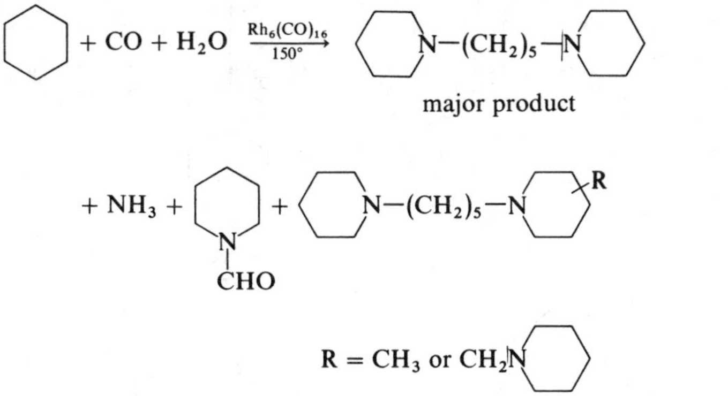

major product

(6)

$$R = CH_3 \text{ or } CH_2N$$

Experimental Details

General Methods

Solvents were purified via distillation from suitable drying agents under N_2. The tertiary amines were purchased from Aldrich Chemical Co. and distilled from CaH_2 under N_2 before use. D_2O was purchased from either Bio-Rad Laboratories or Aldrich and used as received. Catalyst precursors, with the exception of the CoMo, were purchased from Strem Chemicals and used as received. The CoMo catalyst was purchased from Ketjenfine and consisted of 4% CoO and 12% MoO_3 supported on gamma-alumina.

Analyses

Analyses for the results shown in TABLE 1 were performed on a Hewlett-Packard 5711 chromatograph equipped with flame ionization detection (FID) using a 1.5 m by 0.325 cm column packed with 5% Carbowax 20M on acid-washed Chromosorb W. Infrared spectra were obtained using a Perkin-Elmer 281 IR

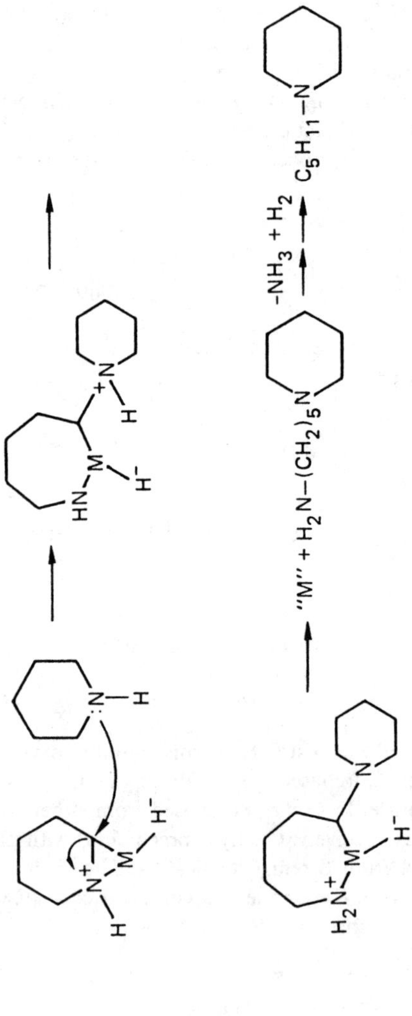

SCHEME 9

spectrophotometer. Deuterium NMR spectra were taken on either a Nicolet 200- or 360-MHz instrument. Mass spectral (MS) and gas chromatograph/mass spectral (GC/MS) analyses were performed using an LKB-9000 mass spectrometer or a Ribermag R 10–10C. Electron impact (EI) mass spectral fragmentation patterns for the deuterated amines were recorded at both 12 eV and 70 eV to thoroughly identify both the parent and fragment species. All mass spectral fragmentation data were corrected for natural abundance isotopes, as described previously.

Homogeneous Catalytic Exchange of Deuterium for Hydrogen

In all reactions listed in TABLE 1 through 4, the reaction conditions are identical unless indicated. In a magnetically stirred, quartz-lined bomb reactor of 34 mL volume (Parr General Purpose 45 mL volume bomb reactor) were mixed 43 mmol of amine (Et_3N or Pr_3N or Bu_3N), 100 mmol of D_2O, and 0.1 mmol of catalyst precursor. The reactor is sealed and degassed via three pressurization/depressurization cycles with 100 psi CO and finally charged to 400 psi CO and heated at 150°C or 200°C for 20 hours. The reactor was cooled, and the deuterium incorporation was determined by ^2H NMR and by analysis of the mass spectral fragmentation patterns at 12 eV and 70 eV.

Heterogeneous Catalytic Exchange of Deuterium for Hydrogen

Reactions for the heterogeneous studies were run the same as those for the homogeneous studies except that the palladium reactions were run with 2.0 mmol Pd under 100 psi N_2 and the CoMo reactions were run, after catalyst activation (see below), with 0.3 to 0.7 g of CoMo under 250 psi of D_2 or H_2 (no observable difference) at temperatures of 230–260°C. The same conditions were used in the $Mo(CO)_6$ studies.

Catalyst Activation

The Ketjenfine CoMo catalyst comes as pellets. The pellets were crushed and sieved before activation. CoMo activation was accomplished by heating the crushed material at 475°C for 20 hours in flowing H_2. Despite attempts to maintain consistent conditions, the catalyst activity varied from batch to batch; thus reliable reaction rate data could not be obtained.

Homogeneously Catalyzed Transalkylation

In a magnetically stirred, quartz-lined bomb reactor were mixed 14 mmol of Et_3N, 14 mmol of Pr_3N, 0.1 mL of H_2O, and 0.05 mmol of catalyst precursor. After degassing with N_2, the reactions were heated to 125°C or more commonly 150°C for 20 hours, then analyzed by GC. Note that to prevent decomposition we ran the rhodium-catalyzed reactions under 400 psi CO at temperatures of 150°C or 200°C.

Heterogeneously Catalyzed Transalkylation

Reactions were run as for the homogeneous studies except that 0.075 mmol of Pd black was used. The CoMo-catalyzed reactions were run with 0.5 g of catalyst under 250 psi H_2.

$H(\eta^2\text{-}CH_3C{=}N\text{-}Et)\ Ru_3(CO)_{10}$

A mixture of 400 mg of $Ru_3(CO)_{12}$, 2.0 mL of Et_3N, and 6.0 mL of benzene are refluxed under N_2 for 4 hours. The reaction solution is cooled, and the yellow solution is separated from an intractable dark red gum. Rotary evaporation of the solvent followed by chromatography on silica gave 15 mg of a crystalline yellow solid.

C, H, N analysis for $Ru_3(CO)_{10}C_5H_9N$ ($M_r = 654.3$): Calculated C, 25.67; H, 1.37; N, 2.14. Found C, 25.20; H, 1.37; N, 2.15. NMR (CDCl$_3$) δ = q, 2.70(2H); S, 2.11(3H); t, 1.13(3H); S, −16(1H); νCO(CH$_2$Cl$_2$) = 2028(w), 2060(s), 2012(ms), 1995(m, br), 1966(mw, br) cm^{-1}.

SUMMARY

We recently demonstrated that it is possible to model the catalytic reactions of tertiary amines with heterogeneous catalysts by studying the reactivity patterns of tertiary amines with homogeneous group 8 transition metal catalysts. In one study we modeled the catalytic reactions of the industrial hydrodenitrogenation (HDN) catalyst cobalt-molybdenum (CoMo) using a homogeneous rhodium catalyst. This included the catalytic cleavage of saturated carbon-nitrogen bonds.

On the basis of our modeling studies and evidence in the literature, we can, for the first time, propose mechanisms that describe how HDN catalysts remove nitrogen as NH$_3$ from the nitrogenous materials found in crude oil or model compounds. Moreover, our proposed mechanisms can also account for the anomalous behavior exhibited by HDN catalysts in the presence of normal catalyst poisons, such as H$_2$S or H$_2$O, that enhance HDN catalysis. We propose that H$_2$S, H$_2$O, or related compounds enhance catalysis by promoting heterocyclic ring opening via nucleophilic attack on the metal-complexed heterocycle.

REFERENCES

1. MUETTERTIES, E. L. 1977. Science **196**: 839.
2. UGO, R. 1975. Catal. Rev. Sci. Eng. **11**: 225.
3. LEWIS, J. & B. F. G. JOHNSON. 1975. Pure Appl. Chem. **44**: 43.
4. SMITH, A. K. & J. M. BASSET. 1977. J. Mol. Catal. **2**: 279.
5. JOHNSON, B. F. G. & J. LEWIS. 1977. Colloq. Int. CNRS **281**: 101.
6. MINGOS, D. M. P. 1974. J. Chem. Soc. A: 133.
7. MUETTERTIES, E. L. 1981. Catal. Rev. Sci. Eng. **23**: 69.
8. MOSKOVITS, M. 1979. Acc. Chem. Res. **12**: 229.
9. ICHIKAWA, M. 1979. J. Catal. **56**: 127.
10. LAINE, R. M. 1980. Am. Chem. Soc. Div. Pet. Chem. Prepr. **25**: 704.
11. LAINE, R. M. 1982. J. Mol. Catal. **14**: 137–169.

12. DEMITRAS, G. C. & E. L. MUETTERTIES. 1977. J. Am. Chem. Soc. **99:** 2796–2797.
13. DOYLE, G. 1981. J. Mol. Catal. **13:** 237–247.
14. SLOCUM, D. W. 1980. Catal. Org. Syn.: 245–275. (Academic Press. New York, N.Y.)
15. ALPER, H. & K. E. HASHEM. 1981. J. Am. Chem. Soc. **103:** 6514–6515.
16. SUSS-FINK, G. 1980. J. Organomet. Chem. **193:** C20–C22.
17. BRADLEY, J. S. 1979. J. Am. Chem. Soc. **101:** 7419–7421.
18. BURCH, R. R., E. L. MUETTERTIES, R. G. TELLER & J. M. WILLIAMS. 1982. J. Am. Chem. Soc. **104:** 4257–4258.
19. SIVAK, A. & E. L. MUETTERTIES. 1979. J. Am. Chem. Soc. **101:** 4878–4887.
20. SATTERFIELD, C. N. & S. GULTEKIN. 1981. Ind. Eng. Chem. Proc. Des. Dev. **20:** 62–68.
21. SONNEMAS, J., G. H. VAN DEN BERG & P. MARS. 1973. J. Catal. **31:** 220–230.
22. UNGERMANN, C., R. G. RINKER, P. C. FORD, V. LANDIS, S. A. MOYA & R. M. LAINE. 1979. Adv. Chem. Ser. **173:** 81.
23. LAINE, R. M., D. W. THOMAS, L. W. CARY & S. E. BUTTRILL. 1978. J. Am. Chem. Soc. **100:** 6527–6528.
24. SHVO, Y., D. W. THOMAS & R. M. LAINE. 1981. J. Am. Chem. Soc. **103:** 2461–2463.
25. MURAHASHI, S.-I, T. HIRANO & T. YANO. 1978. J. Am. Chem. Soc. **100:** 348–350.
26. MURAHASHI, S.-I. & T. WATANABE. 1979. J. Am. Chem. Soc. **101:** 7429–7430.
27. SHVO, Y. & R. M. LAINE. 1989. J. Chem. Soc. Chem. Commun.: 753–754.
28. SHAPELY, J. R., M. TACHIKAWA, M. R. CHURCHILL & A. R. LASHEWYCK. 1978. J. Organomet. Chem. **162:** C39–C40.
29. YIN, C. C. & A. J. DEEMING. 1977. J. Organomet. Chem. **133:** 123.
30. SHVO, J. & R. M. LAINE. Unpublished work. (See Experimental Details.)
31. LAINE, R. M., D. W. THOMAS & L. W. CARY. 1982. J. Am. Chem. Soc. **104:** 1763–1765.
32. CHO, B. R. & R. M. LAINE. 1982. J. Mol. Catal. **15:** 383–387.
33. SONNEMANS, J., W. J. NEYENS & P. MAX. 1974. J. Catal. **34:** 230–234.
34. SATTERFIELD, C. N. & J. F. COCCHETTO. 1981. Ind. Eng. Chem. Proc. Des. Dev. **20:** 53–61.
35. COCCHETTO, J. F. & C. N. SATTERFIELD. 1981. Ind. Eng. Chem. Proc. Des. Dev. **20:** 49–52.
36. SHIH, S. S., J. R. KATZER, H. KWART & A. B. STILES. 1977. Am. Chem. Soc. Dev. Pet. Chem. Prepr. **22:** 919.
37. SATTERFIELD, C. N. & D. L. CARTER. 1981. Ind. Eng. Chem. Proc. Des. Dev. **20:** 538–540.
38. BALLANTINE, J. A., H. PURNELL, M. RAYANAKORN, J. M. THOMAS & K. J. WILLIAMS. 1981. J. Chem. Soc. Chem. Commun.: 9.
39. TSIGDINOS, G. A. 1977. Molybdenum Chemistry Bulletin Cdb-17. Climax Molybdenum. Climax, Colo.
40. HIRAI, H., H. SAWAI & S. MAKISHIMA. 1973. Bull. Chem. Soc. Jpn. **43:** 1148–1153.
41. Thermodynamics of Organic Compounds. Wiley and Sons. New York, N.Y.
42. KLEMM, L. H., J. J. KARCHESY & D. R. McCOY. 1979. Phosphorus Sulfur **7:** 933.
43. LAINE, R. M., D. W. THOMAS & L. W. CARY. 1979. J. Org. Chem. **44:** 4964–4968.

HOMOGENEOUS CATALYTIC HYDROGENATION. 3. SELECTIVE REDUCTIONS OF POLYNUCLEAR AROMATIC AND HETEROAROMATIC NITROGEN COMPOUNDS CATALYZED BY TRANSITION-METAL CARBONYL HYDRIDES*

Richard H. Fish

Lawrence Berkeley Laboratory
University of California
Berkeley, California 94720

INTRODUCTION

Many of the synthetic fuel processes, such as coal liquefaction and oil shale retorting, generate products that require additional hydroprocessing to reduce their nitrogen and sulfur content. Since these synthetic fuel products contain a wide variety of polynuclear aromatic and polynuclear heteroaromatic constituents, it becomes extremely important to understand the reactivity of model synthetic fuel compounds under various homogeneous catalytic hydrogenation conditions.

The recent reports by Pettit *et al.* on the utilization of transition-metal carbonyl compounds as catalysts in the hydroformylation of olefins and the reduction of nitroarenes[1-3] prompted us to study these catalysts with polynuclear aromatic and heteroaromatic nitrogen compounds that are known to be present in various synthetic fuels.

In addition, the intermediacy of transition-metal carbonyl hydrides from the corresponding transition metal carbonyls was also postulated under water-gas shift (CO, H_2O), synthesis gas (CO, H_2), and strictly hydrogenation conditions (H_2 alone),[4-7] which further increased our interest in generating these hydrides for the purpose of testing their reactivities and selectivities with model synthetic fuel compounds.

Previous studies on the reductions of polynuclear aromatic compounds, with transition-metal carbonyl hydrides as catalysts, date back to the 1950s with the observations by Friedman *et al.*[8] They reacted a variety of polynuclear aromatic compounds under synthesis gas conditions (CO/H_2, 1:1, 3,000 psi, 135–200°C) with $Co_2(CO)_8$ as the catalyst. For example, anthracene is reduced to 9,10-dihydro-anthracene (100%), while other polynuclear aromatics such as pyrene and phenanthrene were more difficult to hydrogenate. Other workers, for example, Taylor and Orchin and Sweany *et al.*, reacted $HCo(CO)_4$ and $HMn(CO)_5$, respectively, with 9,10-dimethylanthracene to show both *cis* and *trans* 9,10-dihydro-9,10-dimethylanthracene ($\approx$1:1), thus implicating a free radical process for these reductions.

* This study was jointly supported by the Director, Office of Energy Research, Office of Basic Energy Science, Chemical Sciences Division, and the Assistant Secretary of Fossil Energy, Office of Coal Research, Liquefaction Division of the U.S. Department of Energy through the Pittsburgh Energy Technology Center under Contract No. DE–ACO3–76SF00098.

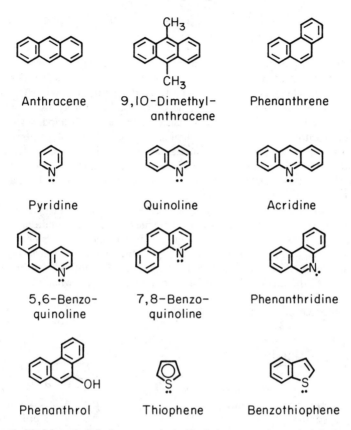

FIGURE 1. Model synthetic fuel compounds used in the homogeneous catalytic hydrogenation reactions.

It was surprising to find that polynuclear heteroaromatic nitrogen compounds had not been studied, under homogeneous hydrogenation conditions, to any significant extent. These studies were limited to the hydrogenation of quinoline, and under either synthesis gas (CO/H_2)[11,12] with $Mn(CO)_8(Bu_3P)_2$ or H_2 conditions with $RhCl_2Py_2(dmf)BH_4$ as catalyst,[13] 1,2,3,4-tetrahydroquinoline was formed.

Recently, we communicated on the homogeneous hydrogenation of a wide variety of polynuclear aromatic and heteroaromatic nitrogen compounds under water-gas shift (wgs), synthesis gas (sg), and H_2 (alone) conditions using transition-metal carbonyl hydrides as catalysts.[14] In addition, we discovered that Wilkinson's catalyst, $(\phi_3P)_3RhCl$, was also an excellent catalyst for the hydrogenation of polynuclear heteroaromatic compounds[15] and Lynch et al. reported similar results with $Fe(CO)_5$ as the catalyst under wgs conditions.[16]

In this paper, we will amplify on our initial studies with transition-metal carbonyls and their reactivity with both polynuclear aromatic and heteroaromatic nitrogen compounds[14] (FIGURE 1).

Results

A wide variety of transition-metal carbonyl compounds were used to generate, *in situ*, the corresponding transition-metal carbonyl hydrides under wgs (CO, H_2O, base) conditions with anthracene, phenanthrene, and pyrene as substrates.

FIGURE 2 details the mechanism of formation of transition-metal carbonyl hydrides under wgs conditions from the corresponding carbonyls.[4-6] Our hope was that the carbonyl hydrides would transfer hydrogen to the model synthetic fuel compounds rather than reductively eliminate H_2 gas. TABLE 1 demonstrates the results with the above-named polynuclear aromatic substrates and clearly shows that anthracene, a linear polynuclear aromatic, is reduced to 9,10-dihydroanthracene but in poor yields. The bent polynuclear aromatics pyrene and phenanthrene were totally unreactive under wgs conditions.

In contrast, we found that the polynuclear heteroaromatic nitrogen compounds were more reactive than their carbon analogues under wgs conditions. TABLE 2 provides the data to show this reactivity trend and also demonstrates the regioselectivity of the reductions, where only the nitrogen heterocyclic ring is reduced. Under these wgs conditions, only Fe, Mn, and Co carbonyls gave products with the model synthetic fuel compounds studied, while those carbonyls of W, Cr, Ru, Rh, Re, Mo, and Os gave CO_2 and H_2 (wgs reaction) but gave no reduced products.

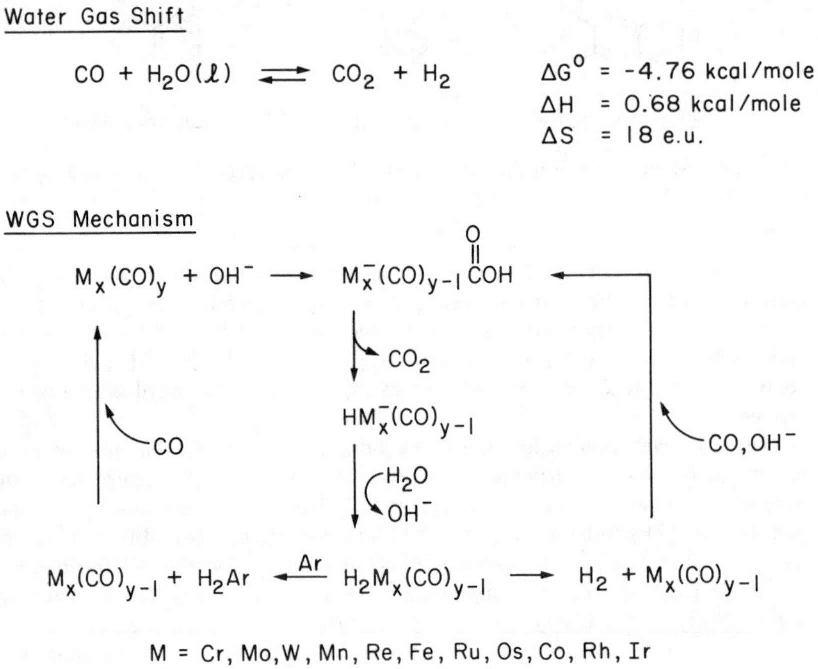
Water Gas Shift
CO + H2O(ℓ) ⇌ CO2 + H2
ΔG° = -4.76 kcal/mole
ΔH = 0.68 kcal/mole
ΔS = 18 e.u.
WGS Mechanism
Mx(CO)y + OH⁻ → Mx⁻(CO)y-1 COH
CO2
HMx⁻(CO)y-1
H2O
OH⁻
CO
CO,OH⁻
Mx(CO)y-1 + H2Ar ← Ar H2Mx(CO)y-1 → H2 + Mx(CO)y-1
M = Cr, Mo, W, Mn, Re, Fe, Ru, Os, Co, Rh, Ir

FIGURE 2. Mechanism of the wgs reaction for the formation of transition-metal carbonyl hydrides.

TABLE 1

TABLE 1

HYDROGENATION OF POLYNUCLEAR AROMATICS UNDER WATER-GAS SHIFT CONDITIONS*

Catalyst	Temperature (°C)	Substrate	Product
$Rh_6(CO)_{16}$	180	anthracene	none
$Ru_3(CO)_{12}$	180	anthracene	none
$Os_3(CO)_{12}$, $Cr(CO)_6$ $Mo(CO)_6$, $W(CO)_6$, $Re_2(CO)_{10}$	200	anthracene	none
$Co_2(CO)_6(PPh_3)_2$	160	anthracene	3% 9,10-dihydroanthracene
$Fe(CO)_4(PBu_3)$	200	anthracene	8% 9,10-dihydroanthracene
$Mn_2(CO)_8(PBu_3)_2$	160	anthracene	4% 9,10-dihydroanthracene
$Mn_2(CO)_8(PBu_3)_2$	180	anthracene	8% 9,10-dihydroanthracene
$Mn_2(CO)_8(PBu_3)_2$	200	anthracene	13% 9,10-dihydroanthracene
$Mn_2(CO)_8(PBu_3)_2$	200	phenanthrene	none
$Mn_2(CO)_8(PBu_3)_2$	200	pyrene	none
$Mn_2(CO)_{10}$	180	anthracene	6% 9,10-dihydroanthracene
$Mn_2(CO)_6(PBu_3)_4$	180	anthracene	7% 9,10-dihydroanthracene
$Mn_2(CO)_4(PBu_3)_6$	180	anthracene	9% 9,10-dihydroanthracene
$Mn_2(CO)_2(PBu_3)_8$	180	anthracene	7% 9,10-dihydroanthracene

* P_{CO} = 350 psi; 12 ml tetrahydrofuran; 3 ml 0.2 M KOH; substrate/catalyst ratio, 20:1.

We then turned our attention to sg conditions ($CO/H_2 = 1$). We found that both Mn and Co carbonyls exhibited greater reactivity when water was removed and replaced with hydrogen gas. Again, anthracene was far more reactive than pyrene or phenanthrene, which were unreactive under sg conditions. In turn, the polynuclear heterocyclic nitrogen compounds demonstrated far greater reactivity than their carbon analogues. FIGURE 3 illustrates the reactivity of both polynuclear aromatics and heteroaromatic nitrogen compounds with $Co_2(CO)_6(\phi_3P)_2$ as the catalyst under sg conditions (200°C, 1 hour, 350 psi H_2, 350 psi CO, $CO/H_2 = 1$, substrate/catalyst = 20:1). Again, as under wgs conditions, the reductions under sg conditions provide high regioselectivity for the nitrogen heterocyclic ring.

Mechanistically, we found that the rate of hydrogenation of polynuclear aromatic and heteroaromatic nitrogen compounds with Fe, Co, and Mn carbonyls as catalysts was independent of the partial pressure of CO. This suggests that substrate does not need to coordinate to these metal centers for hydrogenation to proceed.

In experiments to define the role of carbon monoxide, base, and H_2 alone, we found that removal of carbon monoxide had a profound effect on the reductions of the model synthetic fuel compounds in the presence of ruthenium carbonyls. We can illustrate this with quinoline using $H_4Ru_4(CO)_{12}$ as the catalyst and varying the CO partial pressure from 0 psi to 350 psi with the H_2 partial pressure remaining constant at 350 psi (FIGURE 4). A similar effect of carbon monoxide inhibition was also found in the hydrogenation of ethylene with $H_4Ru_4(CO)_{12}$ as the catalyst.[17]

It is interesting to note that anthracene is reduced to 1,2,3,4-tetrahydroanthracene; however, this occurs only in the presence of base and hydrogen gas with ruthenium carbonyl catalyst, i.e., $RuCl_2(CO)_2(\phi_3P)_3$ and $Ru_3(CO)_6(\phi_3P)_6$.

TABLE 2

REDUCTIONS OF POLYNUCLEAR HETEROAROMATIC NITROGEN COMPOUNDS UNDER WATER-GAS SHIFT CONDITIONS*

Substrate	Catalyst	Substrate/Catalyst	Temperature (°C)	Time (hours)	Product (%)†
5,6-Benzoquinoline	$Mn_2(CO)_8(Bu_3P)_2$	20	200	2	1,2,3,4-tetrahydro-5,6-benzoquinoline (4)
5,6-Benzoquinoline	$Fe(CO)_4Bu_3P$	10	180	5	1,2,3,4-tetrahydro-5,6-benzoquinoline (1)
7,8-Benzoquinoline	$Mn_2(CO)_8(Bu_3P)_2$	20	200	2	no product
Phenanthridine	$Mn_2(CO)_8(Bu_3P)_2$	20	200	2	9,10-dihydrophenanthridine (1)
Acridine	$Fe(CO)_5$‡	10	180	2	9,10-dihydroacridine (100)
Acridine	$Mn_2(CO)_8(Bu_3P)_2$	10	200	2	9,10-dihydroacridine (38)
Quinoline	$Mn_2(CO)_8(Bu_3P)_2$	20	200	5	1,2,3,4-tetrahydroquinoline (4)

* Reactions run in tetrahydrofuran (12 ml) with 0.2 M KOH (3 ml), 350 psi CO.
† Determined by capillary gas chromatography using a digital integration (HP5880A). Isolated by column chromatography (Florisil) and identified by gas chromatography/mass spectroscopy and ^{1}H; 250 MHz NMR spectroscopy.
‡ 800 psi CO.

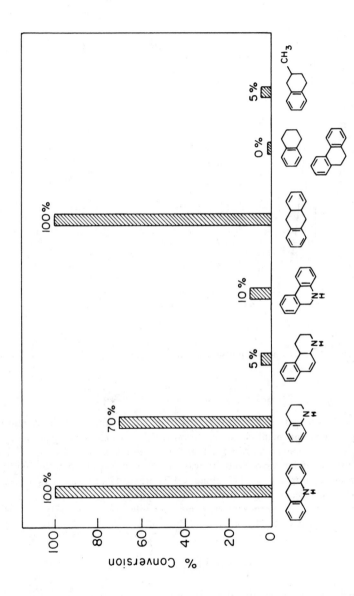

FIGURE 3. Percent conversion in the homogeneous hydrogenation of model synthetic fuel compounds by $Co_2(CO)_6(Ph_3P)_2$ at 200°C for one hour, substrate/catalyst = 20:1, under sg conditions.

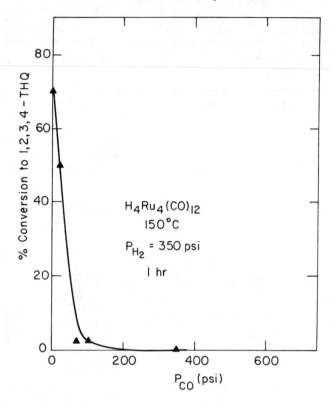

FIGURE 4. Effect of CO partial pressure on $H_4Ru_4(CO)_{12}$-catalyzed hydrogenation of quinoline.

The speculation is that base (KOH) is required to open up coordination sites on the ruthenium metal center by forming CO_2. Consequently, anthracene must bind via four hapto coordination and this results in selective outer ring hydrogenation. A similar regioselectivity was shown by a ruthenium anion hydride in the reduction of anthracene with H_2 alone (see Reference 18 and references therein).

With the polynuclear nitrogen heterocyclic compounds, selective reduction also occurred at the nitrogen heterocyclic ring and can be illustrated with $H_4Ru_4(CO)_{12}$ as the catalyst (H_2, alone) (FIGURE 5).

DISCUSSION

Our results clearly indicate that polynuclear heteroaromatic nitrogen compounds are more reactive under wgs, sg, or H_2 (alone) conditions than their polynuclear aromatic counterparts. More importantly, the high regioselectivity in the reduction of the nitrogen heterocyclic ring in polynuclear heteroaromatic nitrogen compounds has important implications in the use of the transition-metal carbonyls for

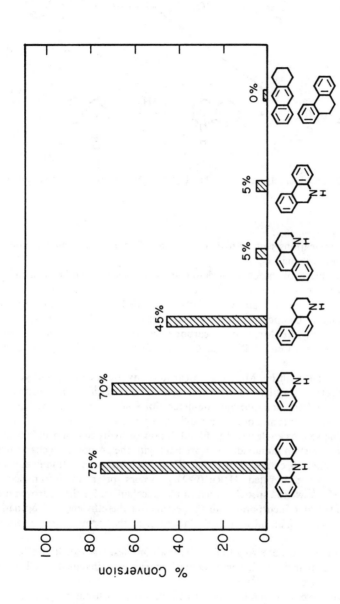

FIGURE 5. Percent conversion in the homogeneous hydrogenation of model synthetic fuel compounds by $H_4Ru_4(CO)_{12}$ at 150°C for one hour, substrate/catalyst = 10:1, H_2 (alone).

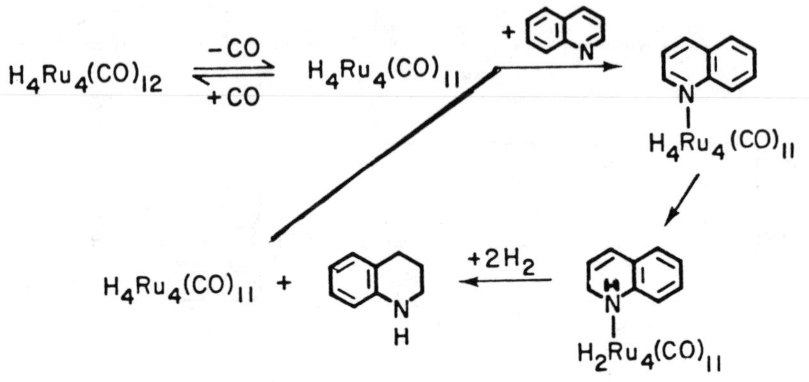

FIGURE 6. Proposed mechanism for the $H_4Ru_4(CO)_{12}$-catalyzed hydrogenation of quinoline.

synthetic fuel applications such as hydroprocessing of coal liquids and shale oils as well as hydrodenitrogenation.

It is of interest to emphasize the mechanistic differences between Fe, Mn, and Co carbonyls and those of Ru mononuclear and cluster carbonyls. Under either wgs or sg conditions, the Fe, Mn, and Co carbonyls reduce both polynuclear aromatic and heteroaromatic compounds, while other transition-metal carbonyls are unreactive. Additionally, Fe, Mn, and Co carbonyls were found to be poor water-gas shift catalysts, while those of Rh, Ru, Cr, Mo, and W were more active shift catalysts.[14]

This implies that with Fe, Mn, and Co carbonyl hydrides transfer of hydrogen is either by electron transfer[3] or hydrogen radical transfer,[19] with coordination of substrate to metal not an important rate-determining step. Alternatively, substrate coordination to ruthenium mononuclear[20] and cluster carbonyls must be an important rate-determining step as evidenced by the inhibition of hydrogenation of substrate in the presence of carbon monoxide. In agreement with the proposed mechanism for ethylene hydrogenation, catalyzed by $H_4Ru_4(CO)_{12}$, where coordination of ethylene to coordinatively unsaturated $H_4Ru_4(CO)_{11}$ occurs prior to intramolecular hydrogen transfer,[17] we can speculate on a similar mechanism for the hydrogenation of quinoline. FIGURE 6 incorporates all the features of the ethylene hydrogenation mechanism,[17] including an intramolecular addition to the carbon-nitrogen double bond.[15]

We are in the process of trying to isolate these complexes as well as clarify many of the other mechanistic details with deuterium gas experiments as well as the evaluation of other catalysts.[15]

Finally, we have recently heterogenized, on divinylbenzene/polystyrene, the homogeneous catalysts $H_4Ru_4(CO)_{12}$, $(\phi_3P)_3RhCl$, and $RuCl_2(\phi_3P)_3$ and have found, under strictly H_2 conditions, excellent activity for these heterogenized homogeneous catalysts in the selective reductions of polynuclear heteroaromatic nitrogen compounds.[21] These latter results provide further definitive evidence for the potential utilization of these catalysts in future synthetic fuel processes.

EXPERIMENTAL

The reactions were conducted in Parr minireactors (45 ml) constructed of stainless steel and equipped with a 2,000 psi pressure gauge, 2,000 psi rupture disc with a magnetic stirring bar and either a septum attachment for sampling under pressure and temperature or a setup without this assembly, and a needle valve for controlling pressurization and flushing of gases. The reactors are placed in a constant temperature bath ($\pm 1°C$). After reaction, the reactor body was cooled in water and the reaction mixture worked up in a standard manner.

The reaction mixtures were analyzed by capillary gas chromatography (12 m × 0.1 mm OV 101) using the HP 5880A gas chromatograph. We identified products by gas chromatography/mass spectroscopic analysis and isolation by Florisil chromatography followed by 250 MHz ^1H NMR spectroscopy. TABLES 1 and 2 define concentrations of catalyst and substrates as well as reaction conditions including CO and H_2 partial pressures.

Gases were analyzed by transferring to a gas cylinder and using a CEC model 21103 mass spectrometer to ascertain CO, CO_2, and H_2 concentrations.

ACKNOWLEDGMENTS

The experimental work reported in this paper was performed by Dr. Gregg A. Cremer and Arne D. Thormosen. We wish to thank Dr. Heinz Heinemann of Lawrence Berkeley Laboratory for encouragement and support of this research project.

REFERENCES

1. KANG, H. G., C. H. MAULDIN, T. COLE, W. SLEGEIR, K. CANN & R. PETTIT. 1977. J. Am. Chem. Soc. 99: 8323.
2. CANN, K., T. COLE, W. SLEGEIR & R. PETTIT. 1978. J. Am. Chem. Soc. 100: 3969.
3. COLE, T., R. RAMAGE, K. CANN & R. PETTIT. 1980. J. Am. Chem. Soc. 102: 6182.
4. GRICE, N., S. C. KAO & R. PETTIT. 1979. J. Am. Chem. Soc. 101: 1627.
5. DARENSBOURG, D. J., B. J. BALDWIN & J. A. FROELICH. 1980. J. Am. Chem. Soc. 102: 4688.
6. FORD, P. C. 1981. Acc. Chem. Res. 14: 31.
7. KAESZ, H. D. & R. B. SAILLANT. 1972. Chem. Rev. 72: 231 (and references therein).
8. FRIEDMAN, S., S. METLIN, A. SVEDI & I. WENDER. 1959. J. Org. Chem. 24: 1287.
9. TAYLOR, P. D. & M. ORCHIN. 1972. J. Org. Chem. 37: 3913.
10. SWEANY, R., S. C. BUTLER & J. HALPERN. 1981. J. Organomet. Chem. 213: 487.
11. DERENCSENYI, T. T. & T. VERMUELEN. 1980. Chem. Abst. 93: 188929.
12. DERENCSENYI, T. T. 1979. Ph.D. Thesis, University of California. Berkeley, Calif. (Lawrence Berkeley Laboratory Report 9777.)
13. JARDINE, I. & F. J. McGUILLIN. 1970. Chem. Commun.: 626.
14. FISH, R. H., A. D. THORMODSEN & G. A. CREMER. 1982. J. Am. Chem. Soc. 104: 5234.
15. FISH, R. H. & A. D. THORMODSEN. 1983. J. Am. Chem. Soc. (Submitted for publication.)
16. LYNCH, T. J., M. BANAH, M. McDOUGALL, H. D. KAESZ & C. R. PORTER. 1982. J. Mol. Catal. 17: 109.
17. DOI, Y., K. KOSHIZUKA & T. KEII. 1982. Inorg. Chem. 21: 2732.
18. GREY, R. H., G. P. PEZ & A. H. WALL. 1980. J. Am. Chem. Soc. 102: 5948.
19. FEDER, H. M. & J. HALPERN. 1975. J. Am. Chem. Soc. 97: 7186.
20. WILCZYNSKI, R., W. A. FORDYCE & J. HALPERN. 1983. J. Am. Chem. Soc. 105: 2066.
21. FISH, R. H., A. D. THORMODSEN & H. HEINEMANN. 1983. Organometallics. (Submitted for publication.)

TRANSITION METAL HYDRIDE COMPLEXES VIA
DISTAL C—H ACTIVATION

J. C. Calabrese, M. C. Colton, T. Herskovitz, U. Klabunde,
G. W. Parshall, D. L. Thorn, and T. H. Tulip*

*Central Research and Development Department
E. I. du Pont de Nemours and Company
Experimental Station
Wilmington, Delaware 19898*

Distal carbon-hydrogen bond activation (DCHA) is a generic term for intra-molecular metal activation of a remote (γ, δ, etc.) C—H site in a hydrocarbyl ligand.[1] Although DCHA is a subset of the well-known cyclometallation reaction, it has only recently attracted attention, in sharp contrast to the long-standing, widespread interest in its more proximal analogues, α- and β-hydrogen activations (eliminations).[2,3] This paper documents our findings on the isolation, properties, and reactions of the products of DCHA in group VIII metals, hydridometallacyclic complexes. To place these results in perspective, we also include a brief description of our prior studies pertaining to catalytically relevant hydride complexes. Our investigations of catalysis by transition metal compounds have often focused on hydride complexes, both as catalysts themselves, or precursors thereto, and as isolable models for suggested catalytic intermediates (see, e.g., Reference 4). We have enjoyed particular success in using iridium hydrides as mimics for the products of various C—H activation processes, and DCHA is presented here as a recent chapter in the "iridium saga," a logical extension of our earlier work.[5,6]

Our introduction to hydridoiridium complexes came during an examination of H/D exchange between arenes and D_2. A number of polyhydride complexes, including $IrH_5(PMe_3)_2$, were found to catalyze the reaction.[7]† FIGURE 1 depicts the general mechanism we envision for this exchange. The cycle is initiated by generation of a coordinatively unsaturated metal hydride, schematically represented by **IH**, which in turn undergoes sequential D_2 oxidative addition and HD elimination to its deuteride counterpart, **ID**. In the critical step of the cycle an arene is bound, possibly in an η^2 fashion, **IIH**, and a C—H bond is thereby brought into proximity to the metal. Insertion of the metal into this adjacent C—H bond yields an aryl metal complex which can decompose to produce products as shown. Arene precoordination is apparently required as alkane C—H bonds, weaker than those of arenes, do not undergo H/D exchange with these catalysts, presumably because they lack a site of unsaturation for ligation.

After a number of years, we returned to the use of iridium complexes containing strongly donating ligands in our investigation of CO_2 fixation by transition metals.[8-13] A series of Ir(I) and Rh(I) complexes of PMe_3, dmpe, depe, or diars, coordinate carbon dioxide as η^1, C-bound CO_2 or as the coupled ligand C_2O_4.[8]

---

* To whom correspondence should be addressed.
† Abbreviations: Me = CH_3, Et = CH_2CH_3, Ph = C_6H_5, dmpe = $Me_2PCH_2CH_2PMe_2$, depe = $Et_2PCH_2CH_2PEt_2$, diars = 1,2-$(Me_2As)_2C_6H_4$, THF = tetrahydrofuran, Ac = $C(O)CH_3$.

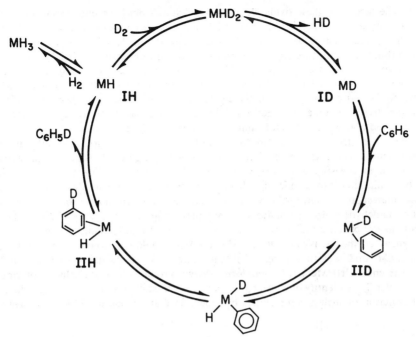

FIGURE 1. Proposed scheme for H/D exchange between benzene and D_2 catalyzed by metal polyhydrides.

Both forms have been crystallographically verified.[8, 12] Bound CO_2 can be fixed by external reagents, such as in the addition of FSO_3Me to $IrCl(dmpe)_2CO_2$, which yields the methoxycarbonyl complex $[IrCl(CO_2Me)(dmpe)_2]FSO_3$, by methyl cation attack at oxygen.[11] However, it was an unexpected side reaction that focused our attention on the insertions of electron-rich iridium centers into C—H bonds. Treatment of $[Ir(PMe_3)_4]Cl$, **1**, with CO_2 in acetonitrile yields not the expected carbon dioxide adduct but rather a *trans* hydrido cyanoacetate complex (Equation 1), which results from apparent CO_2 insertion into the Ir—C bond of a cyanomethyl-hydride precursor.[9] Similar reactions were observed with other activated hydrocarbons. Complex **1** does in fact react with acetonitrile to form *cis*-$[IrH(CH_2CN)(PMe_3)_4]Cl$, but this hydride complex does not undergo carboxylation. Carbon

$$[IrP_4]Cl + CO_2 + NCCH_3 \longrightarrow \textit{trans-}[IrH(O_2CCH_2CN)P_4]Cl \qquad (1)$$

$$\textbf{1} \qquad \qquad P = PMe_3$$

dioxide therefore plays an as-yet-undefined role in the C—H insertion process. Similar results were observed in CO_2-promoted intramolecular C—H activation, an example of which is shown in Equation 2. In the absence of CO_2, the cyclo-metallated form of Complex **2** is not observed.

$$IrCl(PEt_3)_3 + CO_2 \longrightarrow \overline{IrCl(H)(O_2CCH_2CH_2\dot{P}Et_2)}(PEt_3)_2 \qquad (2)$$

$$\textbf{2}$$

The facility of these oxidative additions results from a unique combination of reactivity and stability in the Ir(I) and Ir(III) complexes, respectively. Complexes 1 and 2 are electron rich owing to the strong donor ligands, while remaining coordinatively unsaturated, 16-electron species.[14] In contrast, the Ir(III) oxidative addition products are octahedral d^6, 18-electron complexes with strong Ir—H, and often Ir—C, bonds. The stability of these latter complexes is enhanced by the small, tightly bound ancillary ligands which do not readily dissociate, thereby limiting associative or elimination reactions.[15, 16] For these reasons the iridium complexes described above are excellently suited for studies of C—H activation, especially in cases where the C—M—H products would normally undergo rapid subsequent reactions, as is required in catalytic processes. A striking example of this application came in our investigation of the coordination chemistry of formaldehyde.[6, 17, 18] This study, like that into carbon dioxide chemistry mentioned above, was prompted by an interest in evaluating this C_1 molecule as a source of multicarbon chemicals. Formaldehyde is also attractive as it may play an important role in CO reduction and its use promised to allow us a glimpse at possible intermediates in synthesis gas (syngas) chemistry. We initially hoped that CH_2O would bind to our Ir(I) complexes in a C-bound, η^1 form analogous to that found for CO_2. As shown in Equation 3, treatment of $[Ir(PMe_3)_4]PF_6$ with formaldehyde produces instead the hydridoformyl complex 3, the identity of which has been verified crystallographically (see FIGURE 2). Formation of such a complex has precedent in that similar C—H bond cleavage

$$[IrP_4]^+ + H_2CO \longrightarrow cis\text{-}[IrH(CHO)P_4]^+ \qquad (3)$$

$$P = PMe_3 \qquad\qquad 3$$

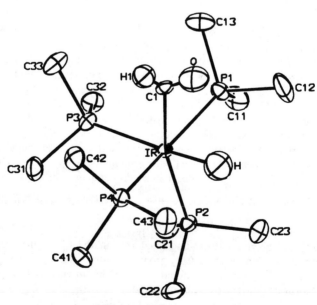

FIGURE 2. A perspective view of the cation of cis-$[IrH(CHO)(PMe_3)_4][PF_6]$. Methyl hydrogen atoms have been omitted for clarity. Here and in FIGURES 4–6 the vibrational ellipsoids are drawn at the 50% level.

reactions are known[19] and must form the basis for the catalytic decarbonylation of aldehydes by rhodium and iridium complexes.[20] However, the stability of Complex 3, which does not decompose after hours at 60 °C, is remarkable. Many formyl complexes have recently been synthesized to test suggestions that formyl groups are present in Fischer-Tropsch and related CO reduction reactions.[21] Generally these complexes decompose rapidly by hydrogen transfer to form hydrido carbonyl products. The combination of thermodynamic stability and kinetic inertia in Complex 3 apparently prevents both hydrogen transfer and reductive elimination, which would liberate formaldehyde. While conversion of our hydridoformyl complexes to hydrido-carbonyl products can be realized by thermal or chemical generation of a vacant coordination site, more interesting reactions relevant to CO reduction have also been observed. Equation 4 illustrates the reduction of the hydridoformyl complex to a *cis*-hydridomethyl compound, one of only a few isolable hydridoalkyl complexes

$$
\begin{array}{ccc}
\overset{\displaystyle H}{\underset{\displaystyle |}{Ir}}-CHO & \xrightarrow{BH_3 \cdot THF} & \overset{\displaystyle H}{\underset{\displaystyle |}{Ir}}-CH_3 \\[2em]
\Big\downarrow {\scriptstyle NaBH_4} & & \Big\uparrow {\scriptstyle [H]} \\[2em]
\overset{\displaystyle H}{\underset{\displaystyle |}{Ir}}-CH_2OH & \xrightarrow[-H_2O]{[H]} & \overset{\displaystyle H}{\underset{\displaystyle |}{Ir}} \cdots CH_2 \\
\mathbf{4} & & \mathbf{5}
\end{array}
\qquad (4)
$$

reported (see, e.g., Reference 22). We postulate that this hydrogenation proceeds via the hydroxymethyl complex **4**, which has also been isolated and crystallo-graphically characterized, and the hydridocarbenoid species **5**, which collapses by hydrogen migration to produce a methyl fragment.[17, 18, 23-26] The transient carbene species has been trapped and similar carbenoid complexes have been generated which exhibit C—C bond-forming reactions reminiscent of steps postulated in Fischer-Tropsch synthesis (see, e.g., References 27-30). In each of these examples, an unusual ligand array is stabilized by coordination to the robust iridium hydride fragment, thereby enabling us to either isolate the novel complex or to observe directly some aspect of its chemistry. As mentioned above, this outline is very cursory and the interested reader is urged to consult References 6, 17, 18, and 23-26 for the full details of the Ir-formaldehyde story.

With these results as background we were primed to apply iridium hydride chemistry to DCHA, an interest that, like the examples above, had its beginnings in a model for a catalytic process. Today, hydrocarbon cracking and isomerization over platinum catalysts, which convert raw petroleum fractions to usable fuel, are among our most important chemical processes. With the advent of coal-based liquid fuel and chemical technologies our reliance on these transformations promises to increase. For these reasons these reactions have been the subject of numerous mechanistic investigations. As organometallic chemists we were intrigued by the mechanism depicted in FIGURE 3 for a model reaction, the skeletal isomerization of neopentane.[31, 32] Each step of the sequence has precedent in reactions known

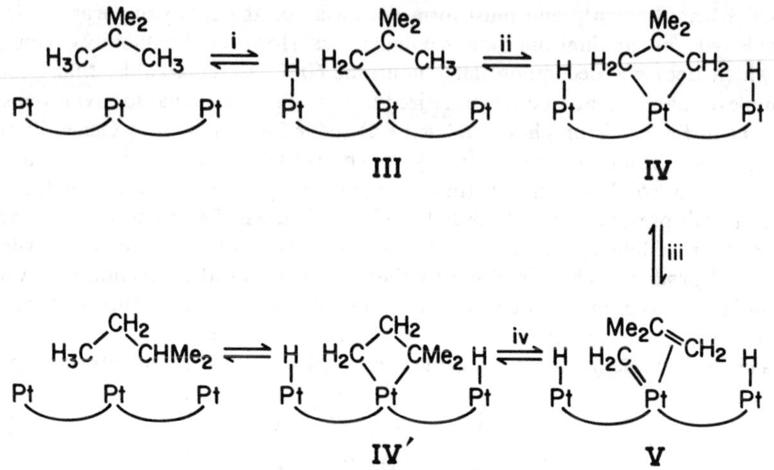

FIGURE 3. Organometallically attractive mechanism for neopentane isomerization to 2-methylbutane on Pt surface.

for discrete organometallic complexes. It is widely accepted that platforming reactions are initiated by a C—H bond activation, Step i, which produces surface alkyl and hydride fragments, such as in **III**. Recently such reactions have been well documented using organotransition metal and organolanthanide complexes.[33-37] The conversion of **IV** to **IV′**, which is known for isolated platinacyclobutane complexes, resembles the key transformation of the olefin metathesis reaction.[38-40] In the hypothetical Step iii, metallacycle **IV** is opened to a carbene, olefin intermediate **V**, which can undergo alkene rotation and ring closure to yield **IV′**. Double C—H reductive elimination would then generate the isomerization product, 2-methylbutane. But how is metallacycle **IV** produced? Distal C—H activation at a γ-site, as shown in Step ii, is an attractive possibility. Just such cyclometallations have recently been observed, as shown in Equations 5 and 6.[41, 42] In the first reaction,

$$P_2Pt(CH_2CMe_3)_2 \xrightarrow{155\,°C} P_2\overline{Pt(CH_2CMe_2CH_2)} + CMe_4 \qquad (5)$$

$$P = PEt_3$$

$$Ru_2Cl(OAc)_4 + P + Mg(CH_2EMe_3)_2 \longrightarrow P_4\overline{Ru(CH_2EMe_2CH_2)} \qquad (6)$$

$$P = PMe_3; E = C, Si$$

one of an elegant series of studies by Whitesides and co-workers described elsewhere in this volume, thermolysis of a bis(phosphine)dineopentylplatinum complex yields a platinacyclobutane via a hydridoalkylmetallacycle intermediate which is not observed because it readily eliminates alkane.[41] A similar mechanism is thought to pertain in Equation 6 after initial generation of an Ru(II) dialkyl species.[42] In each of these cases the presence of a sacrificial alkyl ligand, which acts as a leaving group, prevents direct observation of the C—H activation step. Our intent was to define a system containing no sacrificial ligand, and to thereby isolate the distal C—H activation.

Based on our previous experience with C—H activation by iridium complexes, Complex 7, Equation 7, seemed ideally suited for the purpose. By variation in

$$1 + \text{LiCH}_2\text{CMe}_3 \longrightarrow \left[\text{P}_3\text{Ir} \diagup\!\!\!\diagdown \right] \longrightarrow \textit{fac-}\text{P}_3(\text{H})\text{Ir} \diamondsuit \quad (7)$$

$$\text{P} = \text{PMe}_3 \qquad\qquad 7 \qquad\qquad\qquad\qquad 8$$

conditions we hoped to discern which factors influenced the projected equilibrium with the hydridocyclobutane **8**. Our approach to **7** was simply treatment of [Ir(PMe₃)₄]Cl, **1**, with neopentyl lithium. The reaction proceeds smoothly and quantitatively but not to the expected Ir(I) alkyl. Our initial disappointment gave way to elation when we realized that the product was in fact Complex 8. This material was uniquely characterized by its IR and ¹H, ¹³C, and ³¹P NMR spectra as the *fac* isomer, and this stereochemistry was verified crystallographically for its tris(trimethylarsine) counterpart (FIGURE 4). Like the hydridoformyl complex described above, the metallacyclobutane hydride is remarkably stable. In any all-*cis* dialkyl hydride rapid reductive elimination, either C—H or C—C, would be expected, but we find that a solution of the compound is unaffected by heating at 150 °C for days. As the complex does not react with CO at moderate pressures, which should trap any Ir(I) species present, we do not believe that reductive elimination proceeds even under relatively forcing conditions. Consistent with this, we have been unable to observe **7** in the preparation of **8** even at low temperatures. However we are confident that the mechanism depicted in Equation 7 is accurate because we have directly observed a silyl analogue of **7**, Ir(CH₂SiMe₃)(PMe₃)₃, and find that it smoothly undergoes a parallel γ—C—H activation over 24 hours at room temperature.

Given the unexpected facility of these DCHA reactions, we then surveyed other iridium hydrocarbyl combinations for similar processes. Benzylmagnesium reagents react with Complexes 1 and 2 as shown in Equation 8. In neither case is the presumed

$$1, 2 + \text{BrMgCH}_2\text{Ph} \longrightarrow \longrightarrow \textit{fac-}\text{P}_3(\text{H})\text{Ir} \diagdown\!\!\!\!\!\bigcirc$$

$$9$$

$$\text{P} = \text{PMe}_3$$

$$\longrightarrow \text{P}_3\text{Ir}\!-\!\!\overset{\displaystyle \text{Me}}{\bigcirc} \quad (8)$$

$$\text{P} = \text{PEt}_3$$

Ir(I) benzyl intermediate observed. Metal attack at a γ, *ortho* C—H site, as in orthometallation of arylphosphines, yields a benzoiridacyclobutene complex, which is stable when the supporting ligands are trimethylphosphines. Use of triethylphosphine yields the product of effective ring to methylene hydrogen transfer, the ring-opened *o*-tolyl complex, which we believe is formed by reductive elimination from **9** (P = PEt₃). As discussed by Milstein, generation of a five-coordinate intermediate may be the determining step in reductive elimination from octahedral complexes.[15] The increased steric demand of triethyl- vs. trimethylphosphine should

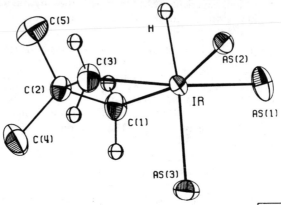

FIGURE 4. A perspective view of the inner coordination sphere of *fac*-IrH(CH$_2$CMe$_2$CH$_2$)-(AsMe$_3$)$_3$. The methyl hydrogen atoms have been omitted.

result in enhanced dissociation to such an intermediate. However, angular strain in the unsaturated four-membered ring must also be important as other tris(triethylphosphine)iridacycle hydrides, such as **8** and **11** (P = PEt$_3$), are relatively robust and can easily be isolated. This ring strain is apparent in the distorted angular parameters [e.g., Ru—C(2)—C(7), 144.2(2)°; C(1)—Ru—C(2), 65.0(1)°] of the structure of a related ruthenacyclobutene complex, shown in FIGURE 5. The preparation of this complex, Equation 9, supports the suggested mechanism for Equation 6.

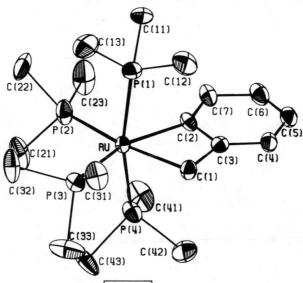

FIGURE 5. A perspective view of Ru(2-CH$_2$C$_6$H$_4$)(PMe$_3$)$_4$. The hydrogen atoms have been omitted.

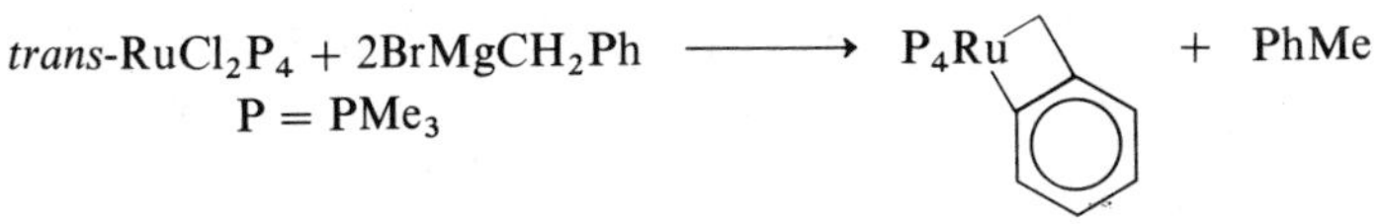
trans-$RuCl_2P_4$ + $2BrMgCH_2Ph$
P = PMe_3
P_4Ru
+ PhMe
(9)

We have also prepared the previously reported ruthenacyclobutane complexes in a similar fashion. Finally, in an attempt to prepare the rhodium analogue of Complex 9, we isolate a dynamic η^3-benzyl complex, **10**. A similar coordination mode in the Ir complexes of Equation 8 may bring the *ortho* C—H bond into proximity with the metal and initiate cyclometallation.

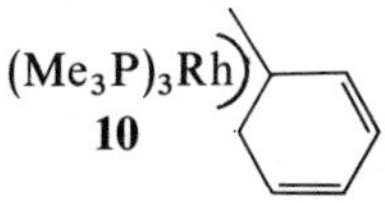
$(Me_3P)_3Rh$
10

DCHA is not limited to the γ-site; rather, δ-activations are markedly favored. As shown in Equation 10, we observe exclusive (>95%) formation of a five-membered

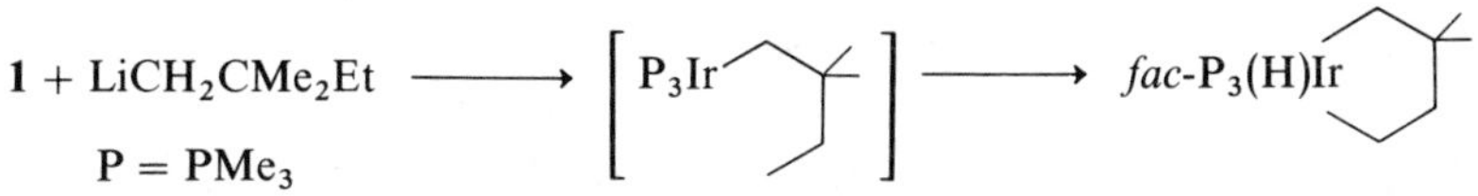
1 + $LiCH_2CMe_2Et$
P = PMe_3
P_3Ir
fac-$P_3(H)Ir$
(10)

metallacycle from a 2,2-dimethylbutyliridium intermediate. No metallacyclobutane products are produced despite the 8 to 3 statistical advantage of γ- to δ-hydrogen atoms. Similar selectivity has been reported for Pt cyclometallations by Whitesides and co-workers,[41] and we find the same result using the Ru system of Equation 9. Note that β-elimination does not proceed in the iridacyclopentane product, presumably due to the lack of a vacant coordination site. In contrast, when $[Rh(PMe_3)_4]Cl$ is substituted for **1** in Equation 10, the products are $RhH(PMe_3)_4$ and 3,3-dimethylbutene, which we believe result from sequential β- and reductive eliminations (of unknown order) from an unobserved metallacycle intermediate. We find a similar divergence in the DCHA products of reactions of Ir(I) and Rh(I) halides with 2-methyl-2-phenyl(neophyl)lithium (Equation 11). In both cases,

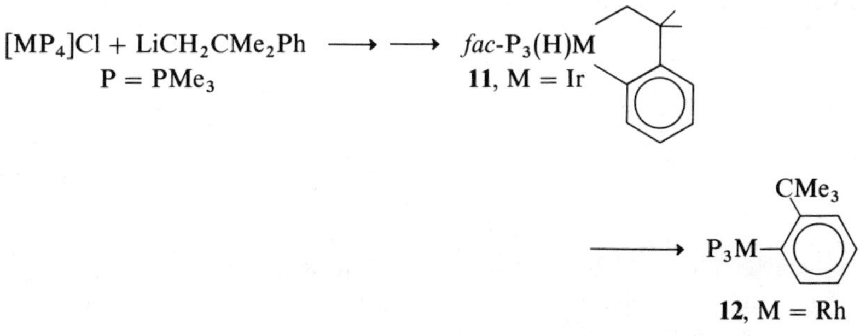
$[MP_4]Cl$ + $LiCH_2CMe_2Ph$
P = PMe_3
fac-$P_3(H)M$
11, M = Ir
CMe_3
P_3M
12, M = Rh
(11)

δ-activation at an *ortho* aryl site leads to the benzannelated metallacyclopentene hydride, **11**, which is extremely stable for M = Ir. For M = Rh, **11** is unobserved even in the presence of excess trimethylphosphine and apparently undergoes facile reductive elimination to the isomeric product **12**. Here, as in the examples above, it appears that whereas Ir(III) hydrido hydrocarbyl complexes are relatively stable, their Rh(III) analogues, while kinetically accessible, are unstable with respect to Rh(I) isomers.

The hydrocarbyl ligands described above were chosen because they lack β-hydrogens and thus cannot undergo β-elimination. A class of β-elimination-stabilized ligands that have received little attention are the enolates of methyl ketones.[43, 44] The lithium or potassium salts of acetone enolate react cleanly with Complex 1 to produce the hydrido iridacyclobutanone complex shown in Equation 12, presumably via DCHA in an intermediate Ir(I)-enolate complex. The enolates

$$1 + MOC(CH_2)CH_3 \longrightarrow \left[P_3Ir \overset{\triangle}{\diamond} =O \right] \longrightarrow \mathit{fac}\text{-}P_3(H)Ir \overset{\square}{\diamond} =O \quad (12)$$
$$M = Li, K$$

13

P = PMe₃

14

15

of pinacolone and acetophenone undergo similar cyclometallation to generate the iridacyclopentanone hydrides **13** and **14**, respectively. Use of [Rh(PMe₃)₄]Cl in place of Complex 1 in Equation 12 again yields only an Rh(I) product, **15**. Complex 15 is O-bound in the solid state and at −80 °C in solution, but like the benzyl complex described above, its NMR spectra at higher temperature indicate a dynamic process, involving both η³-oxoallyl and η¹ C-bound coordination modes.

In DCHA, as in intermolecular C—H activation, aromatic carbon-hydrogen bonds are most susceptible to metal insertion.[5, 7] While the benzyl precursors in Equation 8 are readily cyclometallated, neither the isomeric o-tolyl nor its bulkier 2-mesityl counterpart forms a hydridometallacycle. At 25 °C, bis(triethylphosphine)-dineophyl platinum smoothly δ-cyclometallates at an *ortho* site as shown in Equation 13. In contrast, δ-DCHA of the analogous saturated bis(2,2-dimethylbutyl) platinum

$$P_2Pt(CH_2CMe_2Ph)_2 \longrightarrow P_2Pt \quad + PhCMe_3 \quad (13)$$
$$P = PEt_3$$

complex requires 126 °C.[45] With these results in mind and the knowledge that Complexes 1 and 2 are active for intermolecular arene C—H activation,[46] we wondered whether aromatic DCHA could compete with β-elimination. Treatment of Complex 1 with the β-hydrogen-containing 2-phenethyl or 2-phenylpropyl magnesium halides leads solely to δ-DCHA products (Equation 14). FIGURE 6

1 + BrMg(CH₂CHRPh) $\longrightarrow$ $\left[\begin{array}{c} \text{P}_3\text{Ir} \end{array} \right]$

R = H, Me; P = PMe₃

16

$\longrightarrow$ *fac*-P₃(H)Ir   (14)

17

contains a perspective drawing of the molecular structure of one of the isomers of **17** (R = Me). The reactions are rapid and quantitative, and we have observed no products of β-elimination. This DCHA requires the presence of the aryl group; the analogous reaction of Complex 1 with 2-ethylbutyl reagents, in which the ratio of δ:β hydrogen atoms is 6:1, yields only the products of β-elimination, IrH(PMe₃)₄ and 2-ethylbutene. The presumed iridium(I) intermediate **16** does not eliminate reversibly prior to ring closure. We have independently prepared the product of such a β-elimination for R = H, IrH(η²-styrene)(PMe₃)₃, and find that it is not converted to **17** (R = H). The presence of a γ, rather than δ, *ortho* aryl C—H is not sufficient to preclude β-elimination. Complex 1 yields IrH(η²-styrene)(PMe₃)₃ when treated with α-methylbenzyl reagents, and no cyclometallated complex is observed.

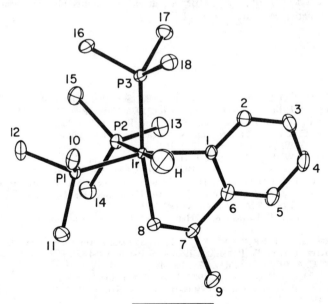

FIGURE 6. A perspective view of *fac*-IrH(CH₂CHMe-2-C₆H₄)(PMe₃)₃. The unaccompanied numerals represent carbon atoms. Hydrogen atoms other than the hydride have been omitted.

Just as in the intermolecular arene activation described above, this preference for intramolecular aryl C—H activation probably results from η^2-arene precoordination. Recent studies, including isolation of a rhenium-η^2-arene complex, strongly support this argument.[47, 48] We envision an intermediate such as 18, in which the coordinated arene would block the vacant coordination site required for β-elimination and position an *ortho* C—H bond in proximity to the metal. We have prepared Complex

18

19   P = PMe₃

19, as a model for this η^2-coordinated intermediate and find that it adopts the pyramidal geometry shown. In such a geometry an *ortho* C—H bond is adjacent to the vacant axial site and thereby primed for activation. Aliphatic ligands lacking unsaturation could not coordinate so as to preclude β-elimination, and η^3-complexation, which would be expected for substituted benzyl ligands, must be less favored than the σ, η^2 chelate coordination of 18.

In this paper, we have shown that distal C—H activation can provide a route to novel hydridometallacyclic complexes, which themselves often undergo interesting subsequent reactions. The reactions are facile, accommodate a variety of ligand types, and may proceed even in the presence of β-hydrogens. This work has been the natural successor to our earlier studies in catalytically relevant hydride complexes and, while not descriptive of catalysis per se, gives us insights into the fundamental workings of active catalysts.

REFERENCES

1. TULIP, T. H. & D. L. THORN. 1981. J. Am. Chem. Soc. 103: 2448–2450.
2. DAVIDSON, P. J., M. F. LAPPERT & R. PEARCE. 1976. Chem. Rev. 76: 219–242.
3. SCHROCK, R. R. & G. W. PARSHALL. 1977. Chem. Rev. 76: 243–268.
4. SEIDEL, W. C. & C. A. TOLMAN. Ann. N.Y. Acad. Sci. (This volume.)
5. PARSHALL, G. W. 1977. Catalysis 1: 335–368.
6. PARSHALL, G. W., D. L. THORN & T. H. TULIP. 1982. Chem. Tech.: 571–576.
7. KLABUNDE, U. & G. W. PARSHALL. 1972. J. Am. Chem. Soc. 94: 9081–9087.
8. HERSKOVITZ, T. & L. J. GUGGENBERGER. 1976. J. Am. Chem. Soc. 98: 1615–1616.
9. ENGLISH, A. D. & T. HERSKOVITZ. 1977. J. Am. Chem. Soc. 99: 1648–1649.
10. HERSKOVITZ, T. 1977. J. Am. Chem. Soc. 99: 2391.
11. HARLOW, R. L., J. B. KINNEY & T. HERSKOVITZ. 1980. J. Chem. Soc. Chem. Commun.: 813–814.
12. CALABRESE, J. C., T. HERSKOVITZ & J. B. KINNEY. 1983. J. Am. Chem. Soc. 105: 5914–5915.
13. HERSKOVITZ, T. & G. W. PARSHALL. 1976. U.S. Patent 3,954,821.
14. TOLMAN, C. A. 1977. Chem. Rev. 77: 313–348.
15. MILSTEIN, D. & J. C. CALABRESE. 1982. J. Am. Chem. Soc. 104: 3773–3774.
16. TATSUMI, K., R. HOFFMANN, A. YAMAMOTO & J. K. STILLE. 1981. Bull. Chem. Soc. Jpn. 54: 1857–1867.
17. THORN, D. L. 1980. J. Am. Chem. Soc. 102: 7109–7110.
18. THORN, D. L. 1982. Organometallics 1: 197–204.

19. CLARK, G. R., C. E. L. HEADFORD, K. MARSDEN & W. R. ROPER. 1982. J. Organomet. Chem. **231**: 335–360.
20. TSUJI, J. 1977. *In* Organic Synthesis via Metal Carbonyls. I. Wender & P. Pino, Eds. **2**: 595–654. Wiley-Interscience. New York, N. Y.
21. GLADYSZ, J. A. 1982. Adv. Organomet. Chem. **23**: 1–38.
22. CHAN, A. S. C. & J. HALPERN. 1980. J. Am. Chem. Soc. **102**: 838–840.
23. THORN, D. L. & T. H. TULIP. 1981. J. Am. Chem. Soc. **103**: 5984–5986.
24. THORN, D. L. 1982. J. Mol. Catal. **17**: 279–288.
25. THORN, D. L. 1982. Organometallics **1**: 879–881.
26. THORN, D. L. & T. H. TULIP. 1982. Organometallics **1**: 1580–1586.
27. ROFER-DEPOORTER, C. K. 1981. Chem. Rev. **81**: 447–474.
28. MASTERS, C. 1979. Adv. Organomet. Chem. **17**: 61–103.
29. MUETTERTIES, E. L. & J. STEIN. 1979. Chem. Rev. **79**: 479–490.
30. HENRICI-OLIVE, G. & S. OLIVE. 1976. Angew. Chem. **88**: 144–150.
31. AMIR-EBRAHIMI, V., F. GARIN, F. WEISGANG & F. G. GAULT. 1979. Nouv. J. Chim. **3**: 529–532.
32. PARSHALL, G. W. 1980. Homogeneous Catalysis. Wiley-Interscience. New York, N.Y.
33. WATSON, P. L. 1983. J. Am. Chem. Soc. **105**: 6491–6493.
34. JANOWICZ, A. H. & R. G. BERGMAN. 1983. J. Am. Chem. Soc. **105**: 3929–3939.
35. HOYANO, J. K. & W. A. G. GRAHAM. 1982. J. Am. Chem. Soc. **104**: 3723–3725.
36. CRABTREE, R. H. 1982. Chemtech.: 506–512.
37. BAUDRY, D., M. EPHRITIKHINE, H. FELKIN & J. ZAKREWSKI. 1982. J. Chem. Soc. Chem. Commun.: 1235–1236.
38. PUDDEPHATT, R. J. 1982. Comments Inorg. Chem. **2**: 69–95.
39. GRUBBS, R. H. 1978. Prog. Inorg. Chem. **24**: 1–50.
40. CALDERON, N., J. P. LAWRENCE & E. A. OFSTEAD. 1979. Adv. Organomet. Chem. **17**: 449–492.
41. WHITESIDES, G. M., R. H. REAMEY, R. L. BRAINARD, A. N. IZUMI & T. J. MCCARTHY. Ann. N.Y. Acad. Sci. (This volume.) (And references therein.)
42. ANDERSEN, R. A., R. A. JONES & G. WILKINSON. 1978. J. Chem. Soc. Dalton Trans.: 446–453.
43. ITO, Y., M. NAKATSUKA, N. KISE & T. SAEGUSA. 1980. Tetrahedron Lett. **21**: 2873–2876.
44. AMSTUTZ, R., W. B. SCHWEIZER, D. SEEBACH & J. D. DUNITZ. 1981. Helv. Chim. Acta **64**: 2617–2621.
45. DICOSIMO, R., S. S. MOORE, A. F. SOWINSKI & G. M. WHITESIDES. 1982. J. Am. Chem. Soc. **104**: 124–133.
46. TULIP, T. H. Unpublished data.
47. JONES, W. D. & F. J. FEHER. 1982. J. Am. Chem. Soc. **104**: 4240–4242.
48. SWEET, J. R. & W. A. G. GRAHAM. 1983. Organometallics **2**: 135–140.

EQUILIBRIA INVOLVING SYNTHESIS GAS AND
TRANSITION METAL DERIVATIVES

D. W. Slocum

Department of Chemistry and Biochemistry
Southern Illinois University
Carbondale, Illinois 62901

INTRODUCTION

Homogeneous catalyst systems are more easily studied than are heterogeneous ones from the relative ease of implementing an experiment to the greater likelihood of gleaning relevant mechanistic information. Particularly appropriate for reactions of high exothermicity, heat dispersal can be handled with greater facility in homogeneous systems, which offers the potential of higher conversions at lower temperatures. In the simplest case, the homogeneous reduction of carbon monoxide to methanol,[1-3] a desirable goal is to devise a low-temperature, moderate-pressure system, thereby taking advantage of the fact that ΔG for such a conversion at atmospheric pressure is negative below 140 °C.[4]

Since hydrides have been implicated or identified as the catalytic species in many homogeneous reduction processes, it seems appropriate that we examine the generation and concentration of hydrides in such systems. In a sense, generation of the catalytic hydride is the preequilibrium step necessary to enter into the catalytic cycle. Of specific concerns are the nature, activity, and concentration of the hydride species relative to variation in temperature and pressure of synthesis gas (syngas).

Direct homogeneous reduction of carbon monoxide to methanol occurs under high temperatures and high pressures of syngas using soluble complexes of cobalt,[5-10] rhodium,[9, 13] and ruthenium[8, 10, 14-19] as catalyst precursors. Other transition metal carbonyls have been examined, but none offer the potential of the three just mentioned. Turnover frequencies reported for two state-of-the-art ruthenium systems are 8.3×10^{-3} s^{-1} and 1.05×10^{-2} s^{-1}.[20-22] However, the former system suffers from use of extraordinarily high pressure (1,300 atm) and the other from use of a solvent (acetic acid) that reacts with the product; both also employ temperatures >250 °C. The rates of several of the existing heterogeneous processes for the conversion of syngas to methanol lie in the range of 0.1–1.0 g methanol/(g catalyst) (h),* which is also roughly the converted rates of the turnover frequencies cited above.

COBALT SYSTEMS

Disproportionation of Dicobalt octacarbonyl

In order to form the reduction catalyst, $HCo(CO)_4$, in soluble cobalt systems, $Co_2(CO)_8$ must be present. For this to happen the known disproportionation reaction of $Co_2(CO)_8$[23] must be reversible under conditions of concern. In methanol the reaction is

$$12CH_3OH + 3Co_2(CO)_8 \rightleftharpoons 2[Co(HOCH_3)_6]^{+2} + 4Co(CO)_4^- + 8CO \quad (1)$$

* For a summary, see Table I in Reference 2.

An equilibrium constant can be written for this equation:

$$K_{eq} = \frac{[Co(HOCH_3)_6^{+2}]^2[Co(CO)_4^-]^4[CO]^8}{[Co_2(CO)_8]^3[CH_3OH]^{12}}$$

Since methanol is the solvent and is present in large excess, the term involving its concentration can be set at unity. Thus the equilibrium is primarily responsive to the concentration of carbon monoxide in the solvent, which in turn is related to the partial pressure of carbon monoxide. (We shall neglect the question of the solubility of carbon monoxide in different solvents at various pressures, although this is a significant point which needs to be addressed.) At 4,000 psi syngas (2,000 psi partial pressure of carbon monoxide), a common pressure for the examination of syngas reactions in such solvents as methanol, it is known that this equilibrium is far to the left, but there are few quantitative data to support this contention. To test the reversibility of the $Co_2(CO)_8$/CO system, an experiment was performed in which a 4:1 molar ratio of methanol to $Co_2(CO)_8$ in toluene was allowed to undergo disproportionation at STP.[24] The autoclave was quickly closed and brought to a temperature of 180°C, and the pressure was brought to 2,000 psi of carbon monoxide. A gradual decrease in pressure was observed. This final pressure reading could be converted to a calculated uptake of 2.5 mol CO/mol of $Co_2(CO)_8$; the theoretical uptake, according to Equation 1, was 2.67 mol CO/mol of $Co_2(CO)_8$. Thus the disproportionated $Co_2(CO)_8$ system was demonstrably reversible to a nearly quantitative degree at high pressures of carbon monoxide. It would be most desirable to examine this equilibrium at lower pressures of carbon monoxide and at other temperatures.

Further evidence for reversibility was obtained by Orchin and co-workers.[25] In this study $Co_2(CO)_8$ was isolated in > 50% yield from separately disproportionated systems in methanol and benzyl alcohol after each had been subjected to 1,400 psi of carbon monoxide at 175 °C for one hour. The conclusion was reached that the disproportionated system had reverted to form once again $Co_2(CO)_8$. An otherwise identical study in ethylene glycol did not permit the same conclusion, since only 1.3% of the charged $Co_2(CO)_8$ was isolated. Feder and Rathke have also demonstrated the reversibility of Equation 1 in the solvents acetonitrile and dimethylformamide.[26]

One last concern. The disproportionation of $Co_2(CO)_8$ is accelerated by a variety of substances including certain tertiary amines,[23] oxygen bases,[23] and halide ions.[27] The effect these substances have on the equilibrium under high pressure of carbon monoxide may be to render the system less reversible. Reaction with strong base such as piperidine yields a disproportionated salt but without elimination of carbon monoxide. Thus there is no obvious reason to see why this form of the reaction should be reversible under high pressure of carbon monoxide. In the instance of the halide ions, this reverse reaction has not been investigated. The effect of halide ions is of salient concern, since many soluble cobalt salt–catalyzed reactions of synthesis gas involve a halide ion promoter, chiefly iodide ion, whose function is often disputed. Interpretations include iodide ion reacting with an organic precursor as in the methanol homologation[28, 29] and methanol carbonylation (see for example Reference 30) reactions as well as an interpretation where the iodide ion is tied up as a ligand on a series of cobalt(II) carbonyl anions.[31, 32] The effect of tertiary phosphines such as PBu$_3$[33] and DIPHOS[34] might also be questioned in this regard, since they are examples of a common class of promoters which are utilized in soluble cobalt salt–catalyzed carbonylations. Such phosphines are less basic and better transition metal ligands than are amines; they *may* exert a decided influence on the disproportionation equilibrium.

Our concern here has been simply that to generate the catalytic species hydrido-cobalt tetracarbonyl, the known disproportionation of $Co_2(CO)_8$ under a variety of conditions must be reversible under applied pressure of carbon monoxide. Relatively little is known about this reverse reaction under appropriate conditions.

Equilibria Involving Dicobalt octacarbonyl and Hydrogen

Apparently the equilibrium between $Co_2(CO)_8$ and hydrogen is more complex than the simple one transcribed here:

$$Co_2(CO)_8 + H_2 \rightleftharpoons 2HCo(CO)_4 \tag{2}$$

Aside from the several studies dealing with the modes of dissociation of $Co_2(CO)_8$ itself,[35, 36] a few important studies have elaborated upon the complexities of this equilibrium. Although the following equations represent current thinking derived chiefly from kinetic data,[37, 38]† the recent observation of the autocatalytic effect of $Co_2(CO)_8$ on the decomposition of $HCo(CO)_4$ has suggested that the reaction sequence may involve radical intermediates such as $Co(CO)_4\cdot$.[39] (The likely radical nature of reactions involving $HCo(CO)_4$ and the lack of hydride functionality of the

$$Co_2(CO)_8 \rightleftharpoons Co_2(CO)_7 + CO \tag{3}$$

$$H_2 + Co_2(CO)_7 \rightleftharpoons H_2Co_2(CO)_7 \tag{4}$$

$$H_2Co_2(CO)_7 \rightleftharpoons HCo(CO)_4 + HCo(CO)_3 \tag{5}$$

$$HCo(CO)_3 + CO \rightleftharpoons HCo(CO)_4 \tag{6}$$

molecules are questions that have preoccupied Orchin and co-workers for a number of years.)[40, 41] Note that the sum of reactions shown in the sequence (3) through (6) is Equation 2. Recently, several theoretical calculations have been performed on the hydridocobalt tetracarbonyl system.[42–45] Estimates of the bond energy of the H—Co bond have been deduced. Unfortunately, none of the authors attempted to relate their findings to the principal preoccupation of the present author, namely, that significant concentrations of the *in situ*-formed catalyst, $HCo(CO)_4$, must be generated under any and all conditions utilized before any real catalytic activity can occur.

To obtain some semiquantitative data regarding the extent of formation of $HCo(CO)_4$ under appropriate conditions, the following experiment was performed. A system was rapidly brought up to 180 °C and 4,000 psi 1:1 syngas after having first dissolved 1 equivalent of $Co(CO)_8$ and 4 equivalents of methanol in toluene such that disproportionation took place (see above). Assuming complete solution of all cobalt species in the solvent, certainly an oversimplification, a calculated decrease in pressure amounting to 2.67 mol CO + 1.0 mol H_2 = 3.67 mol gas based on

---

† For a discussion of the stability of the system $HCo(CO)_4 + Co_2(CO)_8$ in relation to temperature and carbon monoxide partial pressure see Reference 3, page 17.

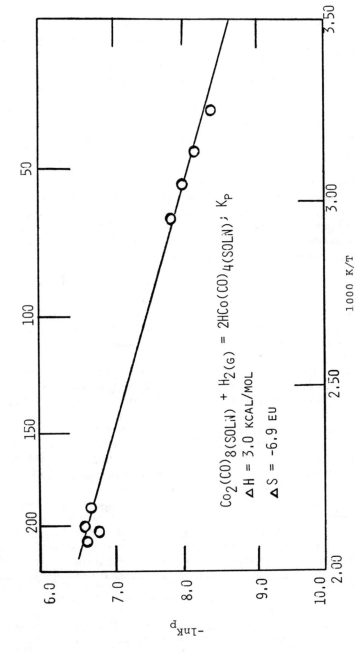

FIGURE 1. Equilibrium constant vs. temperature plot for $Co_2(CO)_8 + H_2$ system. (Courtesy of H. Feder and J. Rathke, Argonne National Laboratories.)

Equations 1 and 2 can be made. The observed gradual decrease in pressure to the final leveling off value led to a computed decrease of 3.1 mol.[24] This in conjunction with the observed 2.5 mol decrease when only carbon monoxide was used as the gas strongly suggests that both the reverse of the disproportionation reaction of $Co_2(CO)_8$ and the hydrogen-fixing reaction are occurring in conjunction, each to a significant extent. Data such as these at other temperatures and pressures and including systems containing appropriate promoters and cocatalysts would be most desirable.

Overall equilibrium data for Equation 2 for one set of conditions can be gained from the graph illustrated in FIGURE 1. This plot was made from a combination of the low-temperature data obtained by Ungvary[37] and points at 200 °C determined by Feder and Rathke[26] by titration of rapidly cooled equilibrated systems using methylene blue.[46] A published determination of similar equilibrium data is known to contain an error,[47] the contribution of the vapor pressure of hydridocobalt tetracarbonyl having been inadvertently omitted.[48] The same is true of a second publication by these same authors involving the determination of thermodynamic data for this equilibrium in the presence of tri-n-butylphosphine.[49] However, these errors may not be as significant as first thought since the boiling point of $HCo(CO)_4$ has been recently determined to be some 35°C higher than the previously accepted value.[50]

Our own interest in this system was to attempt to reduce carbon monoxide under modest pressures of 1:1 syngas. Runs using $Co(acac)_2$ as a source of soluble cobalt at 1,500 psi and 200 °C for one or two hours gave no indication of providing any products of carbon monoxide reduction. However, some recent work on methanol homologation brings these results into sharper focus.[51] Methanol was reacted at 77 atm of 1:1 syngas and 125°C for 26 hours (methyl iodide and phosphine promoters). The mixture of products obtained was reminiscent of the product mixture obtained normally under more intensive conditions. The salient feature was that a turnover number of 17.6 could be calculated for this experiment indicating that the reaction was still catalytic, even under these relatively mild conditions. Interpretation of these data must mean that a concentration of hydridocobalt tetracarbonyl was generated sufficient to catalyze the homologation reaction. Thus, although there are no quantitative data available on the $Co_2(CO)_8$, hydrogen equilibrium in the 1200–1500 psi region, there would appear to be sufficient hydridocobalt tetracarbonyl generated to provide a moderate rate of catalysis.

In conclusion it can be suggested that the reason that reduction rates and homologation rates fall off at low pressures and temperatures is simply that lower concentrations of the active catalyst [presumably $HCo(CO)_4$] are generated. If a modified method of generating this catalyst could be found, one that would operate efficiently at low temperatures and pressures, more efficient, less energy intensive cobalt-catalyzed carbon monoxide reductions and homologations could be developed.

RUTHENIUM SYSTEMS: EQUILIBRIA INVOLVING SOLUBLE RUTHENIUM
SPECIES AND HYDROGEN

Studies of hydride formation for the ruthenium carbonyl system are less extant than those for the cobalt carbonyl system. King has detected $H_2Ru_4(CO)_{13}$ and

$H_4Ru_4(CO)_{12}$ at 180°C and 100 atm using a high pressure IR cell.[19] Knifton has deduced the presence of $[HRu_3(CO)_{11}]^-$, $[H_3Ru_4(CO)_{12}]^-$, $[HRu_4(CO)_{13}]^-$, and $[HRu_6(CO)_{18}]^-$ in his fluid melt system.[52] $H_2Ru(CO)_4$, first described in such systems a number of years ago,[53, 54] has been rejected by King[19] as a contributing species, but has been utilized by Dombeck.[21, 22] In perhaps the most definitive study to date, Dombeck has demonstrated an accelerating effect of iodide ion on the reduction of carbon monoxide to methanol and has inferred the following equilibrium:[55]

$$7/3Ru_3(CO)_{12} + 3I^- + H_2 \rightleftharpoons 2HRu_3(CO)_{11}^- + Ru(CO)_3I_3^- + 3CO \quad (7)$$

It appears that a mixture of the two species, the hydride and the iodide-containing ions, provides the best system for catalysis of carbon monoxide reduction. These species are each known compounds;[56, 57] also in a footnote to this communication, reference is made to work that has demonstrated the hydridic character of the $[HRu_3(CO)_{11}]^-$ ion.‡ All this is a step in the right direction and in some respects renders this a better-characterized reduction system than the corresponding cobalt system.

In many of the citations made in the Introduction, systems involving catalysis of carbon monoxide reduction by other metals have been reported. In these instances even less speculation as to the nature and mechanisms of operation of the catalytic intermediates has been included.

Within the context of developing a less energy intensive route to methanol, examination of these soluble ruthenium catalysts at relatively low pressures provided some interesting results.[58] Previous descriptions of such systems (see above) involved experiments performed at much higher pressures. The idea that hydride species are involved similar or identical to the ones described in these papers can be inferred from the fact that methanol is produced. Unfortunately, the rate of carbon monoxide reduction fell off so drastically at temperatures below ca. 200°C that experiments below this temperature were discontinued. Our results (TABLE 1) show that both $Ru(acac)_3$ and $Ru_3(CO)_{12}$ in tetrahydrofuran (THF) under 1,500 psi 1:1 syngas at 200–215°C afforded methanol with the best rates approximating 0.01 (g methanol)/ (g Ru catalyst) (h), i.e., at about 1/100th the rate of commercial processes (for a summary see Table 1 in Reference 4).

TABLE 1
RUTHENIUM CARBONYL-CATALYZED REDUCTION OF CO AT LOW PRESSURE*

Catalyst (mmol)	Tempera- ture (°C)	Pressure (psi)	Selectivities (wt%)†			Weight CH_3OH/ (Weight Ru) × (Hours)
			CH_3OH	CH_3CHO	CH_3CH_2OH	
$Ru(acac)_3$ (2.0)	215	1500	78.1	12.5	9.4	0.016
$Ru(acac)_3$ (2.0)‡	215	1500	75.9	10.3	13.8	0.008
$Ru(acac)_3$ (2.0)	215	2500	100	—	—	0.004
$Ru(acac)_3$ (2.0)	215	1000	100	—	—	0.002
$Ru_3(CO)_{12}$ (1.0)	215	1500	100	—	—	0.002

* All runs in 80 ml THF under 1,500 psi 1:1 syngas for 2.0 h.
† GC analysis; components identified by retention times.
‡ 3.5 h.

‡ This work has now been communicated.[65]

TABLE 2

RUTHENIUM CARBONYL–CATALYZED REDUCTION OF CO: EFFECT OF NaI PROMOTION*

Catalyst (mmol)	Promoter (mol)	Selectivities (wt %)†			Rate‡
		CH_3OH	CH_3CHO	CH_3CH_2OH	
$Ru(acac)_3$ (2.0)	NaI (4.0)	64.9	2.5	31.7	0.06
$Ru(acac)_3$ (2.0)	NaI (6.0)	75.0	4.8	20.2	0.04
$Ru(acac)_3$ (2.0)	NaI (2.0)	74.6	7.5	17.9	0.03
$Ru_3(CO)_{12}$ (0.7)	NaI (1.4)	88.2	—	11.8	0.02
$Ru_3(CO)_{12}$ (0.7)§	NaI (1.4)	92.3	—	7.7	0.03
$Ru_3(CO)_{12}$ (1.0)	NaI (6.0)	100	—	—	0.02

* All runs in 80 ml THF under 1,500 psi 1:1 CO/H_2 at 215°C for 2.0 h.
† GC analysis; component identities assigned by retention times only.
‡ Weight CH_3OH/(weight Ru catalyst) (h).
§ 4,000 psi.

Detailed examination of the $Ru(acac)_3$ system (TABLE 1) revealed the following modest increase in pressure brought no increase in rate, neither did a decrease in pressure. When $Ru(acac)_3$ was run at 215°C, small amount of materials identified as acetaldehyde and ethanol were detected. These two runs afforded the highest rates recorded on this table, rates that would be even higher if the rates of formation of the C_2 products were included in their calculation.

A roughly fourfold increase in rate with little change in selectivity was brought about by use of sodium iodide as promoter (TABLE 2). Although unpromoted $Ru_3(CO)_{12}$ was a poorer catalyst than was $Ru(acac)_3$ (TABLE 1), use of an iodide promoter accelerated both systems to the same overall rate. There did not seem to be much of an effect in variation of the ruthenium to iodide ratio from 1:1 to 1:3. A surprising result was the observation that at 4,000 psi, no significant acceleration in rate was noted. This promoted system tended to produce in addition to methanol significant quantities of ethanol along with small quantities of acetaldehyde; in these runs, which represent a variety of conditions, a 2:1 molar ratio of methanol to ethanol was found. When the fact that as much carbon monoxide was going to produce C_2 products as there was to produce the C_1 product (methanol) was taken into account, these low-pressure, promoted ruthenium systems were found to manufacture this mixture at rates between 1/10th and 1/20th that of commercial methanol processes.

Although no evidence for the structure of the catalytic intermediates in these low-pressure reactions was obtained (nor was there any attempt to attain such information), the steady production of methanol in these systems strongly supports the hypothesis of the formation of an active hydride species even under these conditions. Speculation as to the structure(s) of these hydride intermediates is possible. The recent observations of King[19] and the earlier data of Whyman[54] suggest that in the unpromoted ruthenium system (TABLE 1) where the conditions were 200–215°C at 1,500 psi, the known tetranuclear hydrides $H_2Ru_4(CO)_{13}$ and $H_4Ru_4(CO)_{12}$ coexist. At higher pressures of carbon monoxide it seems likely that $Ru(CO)_5$ fragments would be split out of these species, thereby creating hydrides of lower nuclearity. Such hydrides may well prove to be part of the catalytic system in those studies where ruthenium carbonyl precursors were utilized in conjunction with pressures of from 300 to 1,500 atm. On the other hand, available data suggest that for anionic clusters significant stability exists with respect to variation of temperature and pressure of carbon monoxide. Thus for the iodide-promoted system under these mild conditions (TABLE 2), the same catalyst equilibrium as illustrated in Equation 7[55] could well be formed and operating, even though the original observations were made at much higher pressures.

THE TANDEM CATALYST APPROACH

Several examples of carbon monoxide reduction have been reported in which a cocatalyst has been utilized. In most instances the function of the cocatalyst is relatively empirical, but in at least one case involving methanol homologation to ethanol, details as to the function of the synergistic system components have been unraveled.[59] Using a related approach—the concept of the tandem reduction catalyst system, wherein one catalyst serves to fix carbon monoxide while the other serves to reduce the fixed species—two related systems have been developed that reduce carbon monoxide to methanol and in several instances produce a higher homologue along with the methanol.[60] Once more, these systems can be made to operate even under ambient temperature and pressure.

It was natural in this investigation to begin with the bis-cyclopentadienyl zirconium system, since stoichiometric reduction of carbon monoxide under ambient conditions had been reported for such a system.[61] Moreover, many of the features involved in the mechanistic pathway of this system have been worked out.[62, 63] Our intention was to evolve this two-step stoichiometric process into a catalytic process by the expedient of coupling the carbon monoxide–fixing zirconium reagent with a known homogeneous reduction catalyst. Criteria for the reduction catalyst were that it operate at ambient or near ambient conditions and that its mechanism of reduction be known. Wilkinson's catalyst was deemed the catalyst of choice.

Two differently modified bis-cyclopentadienyl zirconium complexes have been found to afford methanol directly when used in conjunction with the soluble reduction reagent Wilkinson's catalyst. The simple scheme depicted in Equation 8 was envisioned:

$$CO/H_2 + (C_5H_5)_2Zr \quad + HRh- \quad \underset{\text{Rh component}}{\overset{\text{Zr component}}{\rightleftharpoons}} \quad CH_3OH \qquad (8)$$

It was thought likely that the reduction would take place on the η^2-coordinated carbon monoxide complex of the zirconium moiety. Recycle of both the zirconium component and the rhodium component such that the process could be repeated would render the reaction catalytic. Unfortunately, the reaction was so slow that no run could be allowed to proceed for a period sufficiently long that a turnover number ≥ 1 could be achieved. Nevertheless, methanol was directly formed using this rationale.

Salient data for the better of the two zirconium systems devised, bis-cyclopentadienyl zirconium bis-borohydride, are recorded in TABLE 3. Our best result is that for the run where, with a 2:1 ratio of zirconium to rhodium component, a pressure of 275 psi and a temperature of 100°C, a rate for the production of methanol of $1.05 \times 10^{-5} s^{-1}$ was observed. Since the concentrations of methanol obtained in these runs were so small, confirmation of the identity of the peak observed by gas

TABLE 3

Cp$_2$Zr(BH$_4$)$_2$/WILKINSON'S CATALYST COUPLED REDUCTION OF CO TO METHANOL*

Cp$_2$Zr(BH$_4$)$_2$ (mmol)	(PPh$_3$)$_3$RhCl (mmol)	Period (hours)	Conditions (psi, °C)	Moles MeOH/ (Moles Zr × Seconds)
3.4	1.7	4	850, 100	9.2 × 10^{-6}
6.8	0.85	4	275, 100	—
3.4	1.7	16	200, 25	—
3.4‡	1.7	4	275, 100	†
3.4	1.7	4	275, 100	1.05 × 10^{-5}
3.4	3.4	4	275, 100	§
		24	14.7, 25	§
		4	270, 100	2 × 10^{-6}

* All runs in 80 ml THF under 1:1 CO/H$_2$.
† GC/mass spectroscopy confirmation of the presence of methanol.
‡ Isolated and resublimed Cp$_2$Zr(BH$_4$)$_2$; in all other runs a THF solution of Cp$_2$ZrCl$_2$ and NaBH$_4$ was utilized.
§ CH$_3$OH detected.

chromatography (GC) to have the same retention time as methanol was sought. This was accomplished by use of GC-coupled mass spectroscopy; the resulting spectrum left no doubt that the material produced by this tandem catalyst system was indeed methanol.

Attempts to accelerate the reaction proved frustrating. Increasing the pressure to 850 psi did not increase the rate whatsoever, while even a modest increase in temperature to 125°C brought decomposition. Most remarkable, however, was the observation that the reduction proceeded under even ambient conditions. This last rate was extremely low; it certainly was less than 10^{-6} s^{-1}.

For most runs (C$_5$H$_5$)$_2$Zr(BH$_4$)$_2$ was prepared in situ by treating a THF solution of zirconocene dichloride with two equivalents of sodium borohydride. In an attempt to better characterize the catalyst system, (C$_5$H$_5$)$_2$Zr(BH$_4$)$_2$ was isolated and purified by sublimation.[64] When this isolated material was dissolved in THF and run through the reaction cycle, the system functioned poorly or not at all. Analytical data for the sublimed material, however, were not in accord with the calculated percentages of the elements.

A second bis-cyclopentadienyl zirconium system also afforded some success in the homogeneous reduction of carbon monoxide. "Zirconocene," generated in situ by the sodium amalgam–induced reduction of zirconocene dichloride, was coupled with Wilkinson's catalyst; the entire system in a solution of THF in a Schlenk tube under one atmosphere of 1:1 syngas at ambient conditions produced directly detectable amounts of methanol. The rate of appearance of methanol for this system was lower by an order of magnitude when compared to the rate of the borohydride system. Here also no reaction was run for a period sufficient to accomplish at least one turnover.

CONCLUSIONS

It has been our intention herein to raise the question of the variation in the amount of hydride formation with variation in temperature and pressure in a number of transition metal–catalyzed syngas systems. For $Co_2(CO)_8$, not only is there a lack of data for lower pressures ($< 4,000$ psi) for the equilibrium depicted in Equation 2, but also there are many questions regarding the reversibility of the disproportionation reaction under such modest conditions, particularly in the presence of certain promoters. Some recent data suggest that there is a utilizable concentration of $HCo(CO)_4$ at pressures as low as 77 atm. With ruthenium carbonyl the situation is much the same, although there is some question as to the actual ruthenium species that is causing reduction catalysis. Dombeck has found that $[HRu_3(CO)_{11}]^-$ is not much of a catalyst in its own right, but that it is significantly promoted by $[Ru(CO)_3I_3]^-$.[55] This same system has been found to provide reasonable rates of methanol production at a low pressure, 1,500 psi; arguments are advanced that support the likelihood that the synergistic mixture of the two cluster anions still exist under such modest pressure and provide the reduction catalysis. Lastly, a system is described that presents another approach to the problem, namely, the coupling of a carbon monoxide–fixing catalyst with a known low temperature and pressure reduction catalyst. This system was found to provide reduction of carbon monoxide to methanol even under ambient conditions.

This brief persual of this question leads us to the conclusion that for these reduction/homologation systems, a principal reason for low rates at modest pressures is that only low concentrations of the appropriate catalytic hydride are generated. If an alternate strategy for the generation of these catalytic species could be devised, one that would provide higher concentrations of the hydride at low pressure, it seems quite possible that less energy intensive routes to C_1 and C_2 molecules can be developed.

ACKNOWLEDGMENTS

The author is grateful to the Pittsburgh Energy Technology Center (Department of Energy) and to the Argonne National Laboratories for his various tenures and to Gulf Research and Development for release of data cited. Particular gratitude is expressed to Ms. M. Woehrel for typing the manuscript and to Dr. D. L. Beach, Dr. R. Laine, and Prof. M. Baird for many helpful discussions.

REFERENCES

1. MUETTERTIES, E. L. & J. STEIN. 1979. Chem. Rev. 76: 479.
2. EISENBERG, R. & D. E. HENDRIKSEN. 1979. Adv. Catal. 28: 95.
3. FALBE, J. 1980. New Syntheses with Carbon Monoxide. Springer-Verlag. New York, Heidelberg & Berlin.
4. KUNG, H. H. 1980. Catal. Rev. Sci. Eng. 22: 235.
5. RATHKE, J. W. & H. M. FEDER. 1978. J. Am. Chem. Soc. 100: 3623.
6. FEDER, H. M. & J. W. RATHKE. 1980. Ann. N.Y. Acad. Sci. 333: 45.
7. RATHKE, J. W. & H. M. FEDER. 1981. Catalysis in Organic Reactions. Marcel Dekker, Inc. New York & Basel.
8. KEIM, W., M. BERGER, A. EISENBEIS, J. KADELKA & J. SCHLUPP. 1981. J. Mol. Catal. 13: 95.
9. FAHEY, D. R. 1981. J. Am. Chem. Soc. 103: 136.
10. KEIM, W., M. BERGER & J. SCHLUPP. 1980. J. Catal. 61: 359.
11. KAPLAN, L. & W. E. WALKER. 1977. German Offen. 2643971.
12. PARKER, D. G., R. PEARCE & D. W. PREST. 1982. Chem. Commun.: 1193.
13. DELUZARCHE, A., R. FONSECA, G. JENNER & A. KIENNEMANN. 1979. Erdoel Kohle Erdgas Petrochem. 32: 313.

14. JENNER, G., A. KIENNEMANN, E. BAGHERZADAH & A. DELUZARCHE. 1980. React. Kinet. Catal. Lett. **15**: 103.
15. DOMBECK, B. D. 1979. European Patent Application 0013008.
16. DOMBECK, B. D. 1982. U.S. Patent 4323513.
17. KNIFTON, J. F. 1982. U.S. Patent 4332914.
18. WHYMAN, R. 1980. European Patent Application 0033425.
19. KING, R. B., A. D. KING, JR. & K. TANAKA. 1981. J. Mol. Catal. **10**: 75.
20. BRADLEY, J. S. 1979. J. Am. Chem. Soc. **101**: 7419.
21. DOMBECK, B. D. 1980. J. Am. Chem. Soc. **102**: 6855.
22. DOMBECK, B. D. 1981. ACS Symp. Ser. **152**: 213.
23. WENDER, I., H. W. STERNBERG & M. ORCHIN. 1952. J. Am. Chem. Soc. **74**: 1216.
24. SLOCUM, D. W. & S. METLIN. Unpublished results.
25. BORTINGER, A., R. J. RUSSE & M. ORCHIN. 1978. J. Catal. **52**: 385.
26. FEDER, H. M. & J. W. RATHKE. Personal communication.
27. BRATERMAN, P. S., B. S. WALKER & T. H. ROBERTSON. 1977. Chem. Commun.: 651.
28. WENDER, I. 1976. Catal. Rev. Sci. Eng. **14**: 97.
29. SLOCUM D. W. 1980. *In* Catalysis in Organic Synthesis 1979. W. Jones, Ed.: 245. Academic Press. New York, N.Y.
30. FRIEDERICH, H. 1959. German Patent 921938.
31. MIZOROKI, T. & M. NAKAYAMA. 1965. Bull. Chem. Soc. Jpn. **38**: 1876.
32. MIZOROKI, T. & M. NAKAYAMA. 1968. Bull. Chem. Soc. Jpn. **41**: 1628.
33. SLAUGH, L. H. 1979. British Patent 1546428.
34. SUGI, Y., K. BANDO & Y. TAKAMI. 1981. Chem. Lett.: 63.
35. FORBUS, N. P., R. OTEIZA, S. G. SMITH & T. L. BROWN. 1980. J. Organomet. Chem. **193**: C71.
36. ABSI-HALABI, M., J. D. ATWOOD, N. P. FORBUS & T. L. BROWN. 1980. J. Am. Chem. Soc. **102**: 6248.
37. UNGVARY, F. 1972. J. Organomet. Chem. **36**: 363.
38. BOR, G. & U. K. DIETLER. 1980. J. Organomet. Chem. **191**: 295.
39. UNGVARY, F. & L. MARKO. 1980. J. Organomet. Chem. **193**: 383.
40. ORCHIN, M. 1981. Acc. Chem. Res. **14**: 259.
41. MATSUI, Y. & M. ORCHIN. 1982. J. Organomet. Chem. **236**: 381.
42. BELLAGAMBA, V., R. ERCOLI, A. GAMBA & G. B. SUFFRITTI. 1980. J. Organomet. Chem. **190**: 381.
43. DEDIEU, A. 1980. Inorg. Chem. **19**: 375.
44. GRIMA, J. P., F. CHOPLIN & G. KAUFMANN. 1977. J. Organomet. Chem. **129**: 221.
45. FEDER, H. M., J. W. RATHKE, M. J. CHEN & L. A. CURTISS. 1981. ACS Symp. Ser. **152**: 19.
46. IWANAGA, R. 1962. Bull. Chem. Soc. Jpn. **35**: 247.
47. ALEMDAROGLU, N. H. & J. M. L. PENNINGER. 1976. Monatsh. Chem. **107**: 1043.
48. PENNINGER, J. M. L. Personal communication.
49. VAN BAVEN, M., N. ALEMDAROGLU & J. M. L. PENNINGER. 1975. J. Organomet. Chem. **84**: 65.
50. ROTH, J. A. & M. ORCHIN. 1980. J. Organomet. Chem. **187**: 103.
51. MARTIN, J. T. & M. C. BAIRD. 1983. Organometallics **2**: 1073.
52. KNIFTON, J. F. 1981. J. Am. Chem. Soc. **103**: 3959.
53. COTTON, J. D., M. I. BRUCE & F. G. A. STONE. 1970. J. Chem. Soc. A: 901.
54. WHYMAN, R. 1973. J. Organomet. Chem. **56**: 339.
55. DOMBECK, B. D. 1981. J. Am. Chem. Soc. **103**: 6508.
56. JOHNSON, B. F. G., J. LEWIS, P. R. RAITHBY & G. SUSS. 1979. J. Chem. Soc. Dalton Trans.: 1356.
57. CLEARE, M. J. & W. P. GRIFFITH. 1969. J. Chem. Soc. A: 372.
58. SLOCUM, D. W., D. HRNCIR & R. HAETER. Unpublished results.
59. CHEN, M. I., H. M. FEDER & J. W. RATHKE. 1982. J. Am. Chem. Soc. **104**: 7346.
60. SLOCUM, D. W. & D. HRNCIR. Unpublished results.
61. SHOER, L. I. & J. SCHWARTZ. 1977. J. Am. Chem. Soc. **99**: 5831.
62. WOLCZANSKI, P. T. & J. E. BERCAW. 1980. Acc. Chem. Res. **13**: 121.
63. GAMBAROTTA, S., C. FLORIANI, A. CHIESI-VILLA & C. GUASTINI. 1983. J. Am. Chem. Soc. **105**: 1691.
64. NANDA, R. K. & M. G. H. WALLBRIDGE. 1964. Inorg. Chem. **3**: 1798.
65. BRICKER, J. C., C. C. NAGEL & S. G. SHORE. 1982. J. Am. Chem. Soc. **104**: 1444.

Index of Contributors

325